GOAT PRODUCTION

• Prof (Dr) P.N. Bhat • Dr B.U. Khan

Centre for Integrated Animal Husbandry Dairy Development,
Flat No. 205, Block No. F – 64/C9,
Sector – 40, Noida – 201 301

2009

Studium Press (India) Pvt. Ltd.

GOAT PRODUCTION

ISBN: 978-93-80012-01-8

Published by:

Studium Press (India) Pvt. Ltd.
4735 / 22, 2nd Floor, Prakash Deep Building
(Near Delhi Medical Association),
Ansari Road, Darya Ganj, New Delhi-110 002
Tel.: 23240257, 65150447; Fax: 91-11-23240273;
jngovil@studiumpress.com; shray@studiumpress.com

Printed at:

Salasar Imaging Systems
Delhi-110035 (India)

ABOUT THE SERIES

According to the 2003 Census data, the country had 485 million livestock population and 489 million poultry population, having the second highest number of cattle (185 million), the highest number of buffaloes (97 million), the third highest number of sheep (61 million), the second highest number of goats (124 million), the sixth highest number of camels (632 million) the fifth highest numbers of chickens (489 million) and the fourth highest number of ducks (33 million) in the world.

Livestock sector has been playing an important role in Indian economy and is an important sub-sector of Indian agriculture. The contribution of livestock to GDP was 4.36% in 2004–05 at current prices. According to CSO estimates, gross domestic product from livestock sector at current prices was about Rs. 935 billion during 1990–2000, (about 22.51% of agriculture and allied GDP). This rose to Rs.1239 billion during 2004–05 with 24.72% share in agriculture and allied GDP. But the share of livestock sector in the plan allocation hovered at around 7% of the agricultural out lay.

This sector plays an important and vital role in providing nutritive food, rich in animal protein to the general public and in supplementing family incomes and generating gainful employment in the rural India, particularly among the landless, small, marginal farmers and women. Distribution of livestock wealth in India is more egalitarian, compared to land. Hence, from the equity and livelihood perspectives, it is an important component in poverty alleviation programmes. This fact however has not been appreciated by Policy Planners and implementers.

The development of animal husbandry has been envisaged as an integral part of system of diversified agriculture. With its large livestock population, India has vast potential for meeting the growing need of millions, in respect of livestock products such as milk, eggs, meat and wool. This sector has the greatest potential in creating new self sustaining jobs in villages, if the knowledge base in veterinary and animal husbandry technology is based on real time data and is used in transforming rural India and other similarly placed developing countries.

Livestock production systems are based on low cost agro by-products as nutritional inputs, using traditional technologies. The spectacular growth of livestock products especially in milk, meat, eggs and poultry meat is attributable to the initiatives taken by Government and the organized private sector and has primarily been driven by horizontal increase in numbers. It has been observed that with increasing income, demand for cereals is decreasing, which is causing a livestock revolution, which is demand driven dissemination.

With the livestock sector assuming an important role in the national economy, there is a requirement to improve the present state of knowledge gathering and information dissemination. Although considerable resources have been directed towards collecting and disseminating information on basic crops, little attention has been given to collecting, analyzing and disseminating information on livestock.

It is necessary that livestock units are made financially viable through generating a service provider industry which becomes a technology catalyser in a low educated farming community. This requires several initiatives, one of the major initiative is to upgrade the knowledge base and make the information available to student, teachers, planners and farmers.

The **Studium Press (India) Pvt. Ltd.** has decided in association with **Centre for Integrated Animal Husbandry and Dairy Development (CIAH&DD)** to bring out a series of books for under graduate and post graduate scholars in Animal and Veterinary Sciences in several volumes under the chief editorship of Prof (Dr) P.N. Bhat, Former Vice Chancellor and Director of Indian Veterinary Research Institute, Izatnagar–243122 (U.P.) and former Animal Husbandry Commissioner of India and Deputy Director General (Animal Science) of Indian Council of Agricultural Research, Ministry of Agriculture, Krishi Bhawan, New Delhi. The basic idea of the series is provide first rate text books to students and scholars in developing countries based on the experiences of developing countries themselves with special focus on Southern Asia in conformity with standards laid out by regulatory agencies in India (VCI, ICAR, UGC, AICTE) and similar agencies in other developing countries.

The titles to be brought out in present series are given below:

1. Goat Production
2. Dairy co-operatives in India
3. Sheep Production
4. Buffalo Production
5. Dairy cattle production
6. Pig Production
7. Poultry Production
8. Cross breeding of cattle for improved milk production in tropics

9. Camel Production
10. Yak Production
11. Mithun Production
12. Rabbit Production
13. Laboratory Animal Production
14. Dog Management, Breeding and Health
15. Animal Biodiversity
16. Livestock Statistics
17. Livestock Economics
18. Livestock Extension
19. Breeding and Health of Equine
20. Animal Nutrition
21. Animal Physiology

It is hoped that in the next two years, these books will be available for the benefit of student, teachers and professionals in the area and fill the gap which is currently wide.

Prof (Dr) Pushkar Nath Bhat

ABOUT THE AUTHORS

Dr B.U. Khan is graduate of Veterinary College, Mathura followed by M.V.Sc. and Ph.D from Indian Veterinary Research Institute. He has been involved in research, teaching on sheep and goats for more than 35 years and retired as Director of Central Sheep and Wool Research Institute at Avikanagar, Rajasthan. He was Head of Division of Goat Management and Production at Central Institute for Research on Goat, Makhdoom, Mathura before becoming the Director. He has published 150 research publications, 4 Books and bulletins. He was advisor with FAO and reviewed the goat/sheep systems in Iran and Pakistan.

Prof (Dr) P.N. Bhat got his Ph.D in population genetics from Institute of Population Genetics Purdue University, West Laffayette, Indiana, USA followed by several post doctoral assignments and visiting professorship in Genetics, biotechnology, livestock production systems. He returned to India and joined Punjab Agricultural University at Hisar – Ludhiana followed by several professional assignments he joined as Head, Division of Animal Genetics at Indian Veterinary Research Institute in 1971. He was responsible for establishing coordinated projects on cattle, sheep, goat, pigs and poultry during 1970–74. He was founder Director of Central Institute for Research on Goats and in 1984 he became Vice Chancellor and Director of IVRI. He joined as Deputy Director General (Animal Science) in 1992 and Animal Husbandry Commissioner in December 1992. He is fellow of several National and International Science Academies.

PREFACE

Goats are multi-purpose animals, producing meat, milk, skin and hair. In India, it is generally known as poor man's cow, in temperate countries milk has become of greater importance; skins are a valuable by-product, especially in those countries with large goat population. Goat meat is relished in all countries of Asia, Africa and Middle East where there is a tradition for meat consumption from both sheep and goats in some countries, such as plains of India, the goat is the major supplier of meat, there is social resistance to change. In Rajasthan, in the desert conditions of the Luni catchment area, the preference for meat from goats to that from sheep was so great that the owners were unwilling to replace their goats with sheep, although sheep were more economic (Bose, 1963). In parts of Africa, such as Ghana (Jollans, 1959), there is a preference for goat meat, and this also applies to parts of the Middle East, South-East Asia and the West Indies.

In India, 95% of goat meat produced is consumed locally and the per capita availability is far below the requirement. Thus, there is considerable potential for developing goat production not only for meat for internal consumption but also for export, for quality leather production in which India ranks high among the goat skin exporting countries, the opportunities are unlimited.

Goat rearing occupies an important place in the economy of desert districts of India as it provides livelihood to millions of goat keepers. The goats are known to be especially useful to people in semi-arid zones, where goats can sustain themselves on sparse forage and extreme climate.

While goat's place among the livestock species is an important one, the research and development in this species is meagre and information available among the species is small. The editorial board decided that the first project should be a Text Book on Goat, so that a basic text is available to the students of veterinary and animal science colleges as supplement to their college notes.

A number of personal have contributed to the development of this text. The basic draft was revised by Dr Chaman Lal Arora, former Assistant Director General (Animal Production & Breeding) at Indian Council of Agricultural Research Headquarters, which was then worked over by Dr B.U. Khan, former Director of Central Sheep and Wool Research Institute, Avikanagar, Rajasthan, who has been Head, Division of Goat Breeding and Management at Central Institute for Research on Goats, Farah, Mathura.

We are grateful to Mr R.C. Garg and Dr G.K. Gaur of Project Directorate on Cattle for carefully going through the manuscript and preparation of content list and for making several new suggestions which have improved the text. The advantage of having outstanding colleagues and friends like Dr R.M. Acharya and Dr N.K. Bhattacharyya for referral discussion is acknowledged.

Dr J.N. Govil, Publishing Director and Managing Editor, Researchco Books & Periodicals Pvt. Ltd., Daryaganj, New Delhi, who is the brain behind this initiative deserves special thanks for making it possible to see that this volume is brought out in time and to the expected standards. Mr Anil Jain and Mr Shray Jain, proprietors of Studium Press (India) Pvt. Ltd. need to be complemented for sustained support.

The hard work put up by Dr H.P.S. Arya and Dr Stella Esther for proof reading the manuscript is gratefully acknowledged. The coordination work of Mr G.P. Gangadharan Pillai, Executive Assistant to Chairman, in preparation of the draft manuscript and typing by Mr S.K. Chowdhry and Mr Pius Joseph is gratefully acknowledged.

New Delhi
21st November 2008

Prof (Dr) Pushkar Nath Bhat
Dr B.U. Khan
Authors

CONTENTS

About the Series v
Preface ix

Chapter 1 Introduction 1

Chapter 2 Classification, Origin and Domestication 4
- 2.0 History and origin 4
- 2.1 Taxonomy 4
- 2.2 History of domestication 6
- 2.3 Theories of domestication 6

Chapter 3 Ecology, Habitat and Distribution 8
- 3.1 Geographical distribution 8
- 3.2 Rationation 9
- 3.3 Population distribution and trends in the states 14

Chapter 4 Goat Breeds and their Description 18
- 4.1 Description of goat breeds 18
  - 4.1.1 Temperate himalayan region 18
    - 4.1.1.1 Gaddi 18
    - 4.1.1.2 Chegu 19
    - 4.1.1.3 Changthangi 19
  - 4.1.2 North-western region 20
    - 4.1.2.1 Jamunapari 20
    - 4.1.2.2 Beetal 21
    - 4.1.2.3 Sirohi 22
    - 4.1.2.4 Jakhrana 23
    - 4.1.2.5 Marwari 23
    - 4.1.2.6 Zalawadi 24
    - 4.1.2.7 Gohilwadi 25
    - 4.1.2.8 Mehsana 25
    - 4.1.2.9 Surti 26
    - 4.1.2.10 Kutchi 26
    - 4.1.2.11 Barbari 27
  - 4.1.3 Southern region 28
    - 4.1.3.1 Malabari 28
    - 4.1.3.2 Osmanabadi 29
    - 4.1.3.3 Sangamneri 29
    - 4.1.3.4 Kanni Adu 30
  - 4.1.4 Eastern region 30
    - 4.1.4.1 Black Bengal 30
    - 4.1.4.2 Ganjam 31

Chapter 5 Production Performance 39
- 5.1 Growth 39
  - 5.1.1 Birth weight 39

		5.1.2 Weaning weight	40
		5.1.3 Yearling weight	40
		5.1.4 Weight gain	40
	5.2	Carcass characteristics	40
	5.3	Milk yield	41
	5.4	Fibre production	42
	5.5	Genetic evaluation of production performance	42
	5.6	Performance of native goats in tropics and sub-tropics	42
	5.7	Performance of european breeds in tropical and sub-tropical countries	43
	5.8	Performance of crossbreds	43
		5.8.1 Age at first kidding	44
		5.8.2 Heritability of milk yield	44
	5.9	Performance of Angora under purebreeding and crossbreeding	46
		5.9.1 Mohair production and quality	46
		5.9.2 Angora breeding in India	47
	5.10	Review of work done	48
Chapter 6	**Genetics**		59
	6.1	Chromosomal profile	59
	6.2	Biochemical polymorphism	60
		6.2.1 Milk protein polymorphism	60
		6.2.2 Serum lysozyme-marker for immunocompetence	62
	6.3	Genetic variability of Indian goats by microsatellite analysis	63
	6.4	Inheritance of qualitative traits	65
		6.4.1. Coat colour	65
		6.4.2. Horn pattern	65
		6.4.3. Conformation	65
		6.4.4 Wool/hair types	66
		6.4.5. Sweat glands	67
	6.5	Inheritance of quantitative traits	67
		6.5.1 Growth rate	67
Chapter 7	**Breeding**		70
	7.1	Systems of breeding	70
	7.2	Inbreeding	71
		7.2.1 Coefficient of inbreeding	71
		7.2.2 Reasons for inbreeding	72
		7.2.3 Prepotency	73
	7.3	Outbreeding	74
	7.4	Crossbreeding	75
	7.5	New breeds from crossbreeds	76
	7.6	Outcrossing	76
	7.7	Top crossing	76
	7.8	Back crossing	77
	7.9	Grading-up	77
	7.10	Species hybridization	77
	7.11	Heterosis or hybrid vigour	78

Chapter 8 Reproduction 80

8.1 Male reproduction 80
8.1.1 Reproductive organs 80
8.2 Types of breeding 80
8.2.1 Natural breeding 80
8.2.2 Artificial insemination (AI) 81
8.2.3 Factors affecting semen quality 83
8.2.4 Semen volume of different breeds and semen quality traits 84
8.3 Anatomy of male reproductive organs 87
8.4 Female reproduction 89
8.4.1 Anatomy of female reproductive organs 89
8.4.2 Breeding season 92
8.4.3 Puberty in does 92
8.4.4 Oestrous cycle profile 93
8.4.5 Ovulation characteristics 93
8.4.6 Gestation period 94
8.4.7 Parturition process 94
8.4.8 Age at first kidding 95
8.4.9 Involution of uterus 95
8.4.10 Post-partum interval 95
8.4.11 Kidding interval 95

Chapter 9 Nutrition and Feeding of Goats 97

9.1 Feeding for maintenance 97
9.2 Energy requirement 98
9.3 Protein requirement 98
9.4 Mineral and vitamin requirement 98
9.5 Components of goat nutrition 99
9.6 Traditional grazing practices 99
9.7 Grazing 100
9.8 Top feed resources 101
9.9 Cultivated fodders 101
9.10 Nutrient requirement 104
9.10.1 Based on EUN and MFN trials 104
9.10.2 Based on metabolism trials and balance studies 108
9.10.3 Based on DM/IBN intake 108
9.10.4 Based on basal calories 109
9.10.5 Work at RBS College, Bichpuri (Agra) 109
9.10.5.1 Based on metabolism trails and balance studies 109
9.10.5.2 Based on DM/TDN intake 110
9.10.5.3 Based on basal calories 110
9.10.5.4 Based on EUN and MFN trials 110
9.10.5.5 Based on energy: protein ratio 111
9.10.5.6 Requirements for maintenance 111
9.10.5.7 Requirement for milk production/lactation 112
9.10.5.8 Requirements for growth 113
9.10.5.9 Requirements for pregnancy 113

9.10.5.10 Feeding of young stock 114
9.10.5.11 Mineral mixture 115
9.10.5.12 Common salt 115
9.10.6 Work at NDRI, Karnal 116
9.10.7 Nutrient intake and digestibility on silvi-pastures 118
9.10.8 Water requirement 118
9.10.9 Requirement of vitamins and antibiotics 119
9.10.10 Ration formulation 119
9.11 Practical feeding systems for goat production 120
9.11.1 Subsistence system of production 120
9.11.2 Extensive system of production 120
9.11.3 Intensive system of production 120
9.11.4 Semi-intensive system of production 121
9.11.5 Special characteristics of goat nutrition 121
9.12 Preparation of concentrate mixtures for feeding to goats 121
9.13 Rumen microbes 122
9.14 Tips for feeding goats 130

Chapter 10 Management and Housing 132
10.1 Systems of management 132
10.1.1 Traditional system 132
10.1.2 Extensive system 133
10.1.3 Semi intensive system 134
10.1.4 Intensive system 135
10.2 Handling and care of animals 135
10.2.1 Kidding does 135
10.2.2 Kids 136
10.2.3 Pregnant does 136
10.2.4 Dry does 136
10.2.5 Fattening stock 136
10.2.6 Male kids 137
10.2.7 Teaser bucks 138
10.2.8 Breeding bucks 138
10.3 Grazing and feeding systems 138
10.3.1 Rotational grazing 138
10.3.2 Silvipasture grazing 139
10.4 Supplementary feeding 139
10.4.1 Fattening of kids 139
10.4.2 Dry does 139
10.4.3 Pregnant and lactating does 140
10.4.4 Bucks 140
10.5 Breeding management 140
10.5.1 Natural breeding (individual, pen and flock mating) 140
10.5.2 Artificial insemination 141
10.5.3 Heat detection methods 142
10.5.4 Insemination procedures 142
10.6 Handling and management of goat products 143
10.6.1 Wool/hair 143
10.6.2 Meat 144

10.6.3 Milk 145
10.6.4 Edible and non-edible offal 146
10.6.5 Skins and hides 146
10.7 Goat housing 147
10.7.1 Orientation of sheds 147
10.7.2 Ventilation 148
10.7.3 Roofing material 148
10.7.4 Height and shape of roof 149
10.7.5 Floor type and space 149
10.7.6 Shelter surroundings 150
10.7.7 Optimum floor space requirement per goat 150
10.7.8 Kidding pens 151
10.7.9 Lean to type shed 151
10.7.10 Shelter for buck 151
10.7.11 Housing for pregnant does and kids 151
10.7.12 Quarantine/segregation shed 152
10.7.13 Exercising yard for stall-fed goats 152
10.8 Essential appliances 152
10.8.1 Feeders 152
10.8.2 Water troughs 152
10.8.3 Hay racks 153
10.8.4 Shed hygiene 153
10.9 Grazing schedule and supplementary feeding 153
10.10 Health management 153
10.11 Feeding management 154

Chapter 11 Goat Production and Land Degradation 155

11.1 Land degradation and goats 155
11.2 Desertification and goats 156

Chapter 12 Meat Production 158

12.1 Effect of live weight on meat production 159
12.2 Dressing percentage 165

Chapter 13 Milk Production 168

13.1 All India Coordinated Research Project (AICRP) on goats 169
13.1.1 National Dairy Research Institute (NDRI), Karnal 169
13.1.2 Kerala Agricultural University, Mannuthy, Trichur 169
13.1.3 Bikaner (Rajasthan) centre 170
13.1.4 Central Institute for Research on Goats (CIRG) 170
13.1.5 Indo-Swiss goat development and fodder production project, Ajmer (Rajasthan) 171
13.2 Special attributes of goat's milk 171
13.3 Variation in yield 172
13.4 Weaning practice 174
13.5 Influence of age 176
13.6 Repeatability and heritability of milk production characters 178
13.7 Efficiency of milk production 179
13.8 Crossbreeding for improved milk production 181

Chapter 14 Utilization of By-products 183

14.1 Hides/skins 183
14.2 Dead carcass utilization 184
14.3 Slaughter by-products 184
14.3.1 Bones 184
14.3.2 Intestines 185
14.3.3 Intestinal mucosa 186
14.3.4 Rumen contents 186
14.3.5 Animal fats 186
14.3.6 Blood 186
14.3.7 Horns and hoofs 188
14.3.8 Meat unfit for human consumption 188
14.3.9 Hair 188
14.3.10 Glands and organs 188
14.4 Goats manure and composting 189

Chapter 15 Goat Records 190

15.1 Various records 191
15.1.1 Livestock inventory 191
15.1.2 Wool/Hair weight and quality 191
15.1.3 Growth records 192
15.1.4 Prolificacy 192
15.1.5 Health 192
15.2 Guidelines for recording of data 192
15.2.1 Body weights 192
15.2.2 Body measurements 193
15.2.3 Wool/hair yield and quality 193
15.2.4 Culling and disposal 193

Chapter 16 Goat Diseases 195

16.1 Abortions and embryonic losses 195
16.2 Neonatal kid mortality 196
16.3 Mortality in kids 197
16.4 Chronic diseases in adult goats 206
16.5 Diagnostic methods 208
16.5.1 Diagnosis by clinical symptoms 208
16.5.2 Diagnosis of faecal examination 208
16.5.3 Diagnosis by blood smear examination 210
16.5.4 Immunodiagonostic tests 211
16.6 Slow-growing viruses 211
16.7 Contamination of goat meat 211
16.8 Zoonotic disease factors 212
16.9 Problems and constraints for preventive health 212
16.10 Diagnostic facilities and immuno biologicals 212
16.11 Vaccination schedule 212

Chapter 17 Economics of Goat Rearing 214

17.1 Scope for goat rearing and its national importance 214
17.2 Economics of goat rearing 215
17.3 Financial assistance available from banks/NABARD for goat rearing 216

	17.4	Scheme formulation	217
	17.5	Requirement of a good project	217
	17.6	Sanction of bank loan and its disbursement	225
	17.7	Unit cost	225
	17.8	Margin money	225
		17.8.1 Interest rate for ultimate borrowers	226
		17.8.2 Security	226
		17.8.3 Repayment period of loan	226
		17.8.4 Insurance	226
	17.9	Package of common management practices for goat rearing	226
Chapter 18	**Marketing of Goats and their Products**		230
	18.1	Marketing agencies	231
		18.1.1 Private enterprises	231
		18.1.2 Cooperative associations	231
	18.2	Marketing of live goats and goats meat	231
		18.2.1 Marketing of live goats	231
	18.3	Marketing of goat meat	233
	18.4	Export of live goats and goat meat	236
		18.4.1 Export of live goats	236
		18.4.2 Export of goat meat	236
Chapter 19	**Goat Hair Processing and Technology**		238
	19.1	Skin follicular characters	239
	19.2	Goat shearing	239
Chapter 20	**Biotechnological Approaches for Enhancing Goat Production**		245
	20.1	Introduction	245
	20.2	Breeding practices and biotechonology	246
	20.3	Multiple ovulation embryo transfer and open nucleus breeding system	246
	20.4	Reproductive biotechnologies	250
	20.5	Improving nutrition through biotechnology	251
	20.6	Health management and biotechnology	254
	20.7	Detection of nucleic acid	255
	20.8	Vaccines	256
Chapter 21	**Conservation of Goat Genetic Resources**		259
	21.1	Introduction	259
	21.2	Goat genetic resources	259
	21.3	Production systems	260
	21.4	Evaluation, conservation and management	261
		21.4.1 Evaluation	262
	21.5	Management	264
	21.6	Impact assessment and technology transfer	264
	21.7	Future thrusts	265
References			266–290
Index			291–294

List of Colour Figures

Plate 1: Various Indian Goat Breeds (a–f) 33

Plate 2: Various Indian Goat Breeds (a–f) 34

Plate 3: Various Indian Goat Breeds (a–f) 35

Plate 4: Various Indian Goat Breeds (a–f) 36

Plate 5: Various Indian Goat Breeds (a–f) 37

Plate 6: Various Indian Goat Breeds (a–f) 38

List of Tables

Chapter 3	**Ecology, Habitat and Distribution**	
	Table 3.1	8
	Table 3.2	10–13
	Table 3.3	15–16
	Table 3.4	17
Chapter 5	**Production Performance**	
	Table 5.1	17
	Table 5.2	39
	Table 5.3	49
	Table 5.4	50
	Table 5.5	51
	Table 5.6	52
	Table 5.7	52
	Table 5.8	53
	Table 5.9	54
	Table 5.10	55–56
Chapter 8	**Reproduction**	
	Table 8.1	57
	Table 8.2	58
	Table 8.3	80
	Table 8.4	83
	Table 8.5	84–85
	Table 8.6	86
Chapter 9	**Nutrition and Feeding of Goats**	
	Table 9.1	86
	Table 9.2	87
	Table 9.3	96
	Table 9.4	97
	Table 9.5	100
	Table 9.6	100
	Table 9.7	102–103
	Table 9.8	103
	Table 9.9	105–106
Chapter 10	**Management and Housing**	
	Table 10.1	114
Chapter 12	**Meat Production**	
	Table 12.1	117–118
	Table 12.2	127
	Table 12.3	131
	Table 12.4	132
	Table 12.5	150
	Table 12.6	158
	Table 12.7	160
		161
		162
		163
		164
		166
		167

Chapter 13	**Milk Production**	168
	Table 13.1	177
	Table 13.2	179
	Table 13.3	180
Chapter 14	**Utilization of By-products**	183
	Table 14.1	185
	Table 14.2	187
Chapter 16	**Goat Diseases**	195
	Table 16.1	196
	Table 16.2	196
	Table 16.3	198
	Table 16.4	207
	Table 16.5	213
Chapter 17	**Economics of Goat Rearing**	214
	Table 17.1	218–219
	Table 17.2	219–220
	Table 17.3	220–221
	Table 17.4	221–222
	Table 17.5	222
	Table 17.6	222
	Table 17.7	223
	Table 17.8	223
	Table 17.9	223–224
	Table 17.10	224
	Table 17.11	225
Chapter 18	**Marketing of Goats and their Products**	230
	Table 18.1	237
Chapter 19	**Goat Hair Processing and Technology**	238
	Table 19.1	240
	Table 19.2	241
	Table 19.3	241–242
	Table 19.4	242–243
	Table 19.5	243
	Table 19.6	244
Chapter 21	**Conservation of Goat Genetic Resources**	259
	Table 21.1	260

CHAPTER 1

INTRODUCTION

Goats are multi-purpose animals, producing meat, milk, skin and hair. Their primary function is meat production, although in temperate countries milk has become of greater importance; skins are a valuable by-product, especially in countries with large goat population. Goat meat is relished in all countries of Asia, Africa and Middle East where there is a tradition for meat consumption from both sheep and goats. In some countries and in some specific areas which are major supplier of goat, like plains of India, there is social resistance to change and shift to any other species, be it economical. In Rajasthan, in the desert conditions of the Luni catchment area, the preference for meat from goats to that from sheep was so great that the owners were unwilling to replace their goats with sheep, although sheep were more economic (Bose, 1963). In parts of Africa, such as Ghana (Jollans, 1959), there is a preference for goat meat, and this also applies to parts of the Middle East, South-East Asia and the West Indies.

In India, 95% of goat meat produced is consumed locally and the per capita availability is far below the requirement. Thus, there is considerable potential for developing goat production not only for meat for internal consumption but also for export, for quality leather production in which India ranks high among the goat skin exporting countries.

Goat rearing occupies an important place in the economy of desert districts of India as it provides livelihood to millions of goat keepers. The goats are known to be especially useful to people in semi-arid zones, where goats can sustain themselves on sparse forage and extreme climate.

Goats constitute an important species of livestock in Asia and contribute greatly to food, rural employment and Gross Domestic Product.

Goat raising is one of the important agricultural enterprise particularly in rural parts of this country and have proved very useful to man throughout the ages, largely because of their adaptability to varying environmental conditions under which the breeds and strain types have evolved and in which they are maintained. They have tremendous ability to survive and often thrive on sparse vegetation unsuitable for feeding of other livestock. Goats can be profitably raised with low investment under intensive and most extensive forms of nomadic grazing. The vast majority of the poorer section of rural population depends on goat rearing for income and certain amount of meat and milk for home consumption. Goat rearing requires low cost and hence is suited to landless labourers, marginal farmers and industrial workers.

The goat is a versatile animal. It is known as the 'poor man's cow' in India and as 'wet nurse' of infants in Europe. Goats can be kept with little expense. Marginal or undulating lands, unsuitable for other types of livestock, may be used and any inexpensive shelter will suffice. Goat milk is cheap, wholesome, easily digestible and nutritious. It is recommended for use in dyspepsia, peptic ulcer and pyloric stenosis. It is preferred to cow milk in liver dysfunction, jaundice, biliary disorders, acidosis and insomnia.

Goats provide a dependable source of income to 40% of the rural population below the poverty line in India and to many who do not possess any land. Contrary to the scientific information, an inbuilt bias and concern among soil conservationists and foresters have left the goat as a species to be neglected in this country for many decades.

There has been a lot of controversy over the role of goats in ecological degradation and in desertification. Of the domesticated animals, no animal has suffered from so much abuse as the goat due to this rather wrong conception. In many instances the criticism has been extreme and has resulted in curbs on breeding and development of goats. Contrary to cattle and sheep, goats are mainly browsers and would seldom go for grown up trees (Lu, 1988). They dislike the leaves of timber trees. They also do not prevent the establishment and spread of grasses which are so essential for soil conservation. Acharya *et al.* (1980) found that goats spend more than 90% time on browsing and hardly graze for 10% time on surface vegetation and that too only when bushy plants are not available. In several behavioural studies conducted at the Central Arid Zone Research Institute (CAZRI), Jodhpur (India), goats have been found to reclaim saline soils by consuming salt-laden leaves of range plants (Shankaranarayanan *et al.*, 1985) and contribute fertility to soil by even distribution of essential manure on the lands they graze. Cattle exploit 3–5 times more of phytomass than goats. Prajapati *et al.* (1988) reported that intensity of grazing of 2 to 4 goats/ha had no effect on run off and soil loss in hot arid regions of India in normal rainfall years. The vast majority of the poorer section of the rural population depends on the goat rearing for subsistence and to meet the house-hold occasional needs for meat and milk. In spite of about 42% annual slaughter of goats, the population continues to increase at an average rate of about 3.4% per year. The socio-economic impact of goat farming is, perhaps, evident in the sharp increase of their population during the post-independence period. The direct contribution of goats to the Indian economy is, however, estimated at Rs. 59,741.16 million annually. The slaughter by-products, skin and fiber provide raw material to industries such as leather and textile.

CHAPTER 2

CLASSIFICATION, ORIGIN AND DOMESTICATION

2.0 History and Origin

2.1 Taxonomy

Archaeological studies provide sufficient evidence that goat was the earliest ruminant and probably the first animal after dog to be domesticated by man. The goat belongs to the family *Bovidae* (hollow-horned ruminants) under order *Artiodactyl* with genus, *C. hircus* and *C. falconeri*. Domesticated goat (*Capra hircus*) are descendants of the Pasang (*C. aegagrus*) represented in Europe by the Cretan and Cyclades races. The East was probably their original home, the earliest recorded being the Persian race. All the species of *Capra* possesses 60 chromosomes and they cross and breed between different races/ strains producing fertile offsprings (Gray, 1972).

Goat is associated with human beings along with the progress of civilization since 70 B.C. The origin and development of goat breeds in Indian sub-continent is not clearly known, although the remains of goat along with other animals like dog, buffalo, sheep and pig have been reported from the Neolithic site excavation in the districts of Hoshnagabad. The ancestors of present breed of goat are *Capra hircus*. The Harappa toys and seals of Mohenjodaro represent the wild goat of 4000 B.C. with elegantly curved horns.

Harris (1962) has reviewed evidence suggesting that the main ancestor of these five possible wild ancestors of domestic goats is the Bezoar of South West Asia and that the Markhor of North West India-Pak sub-continent may have given rise to certain breeds in India, Pakistan and the near East. Mackenzie (1970) considers that the Mediterranian breeds probably owe some of their distinctive characteristics to ancestral influence of the extinct *Capra prisca*.

Methods of Classification

At least four methods of classifying domestic goats have been used, namely:

(i) Body size (Mason, 1951; Epstein, 1953)
(ii) Ear shape and length (Mason and Maule, 1960)
(iii) Function (Williamson and Payne, 1978)
(iv) Origin (Webster and Wilson, 1966)

Each of the above methods has its special attributes and limitations. Epstein (1953) identifies the dwarf goats as a separate group altogether. The classification based on ear shape (lop or prick ears) has been successfully applied to the goats of Eastern and Southern Africa (Mason and Maule, 1960). The goats in West Africa have been arbitrarily divided in terms of size by Mason (1951). The classification of Williamson and Payne (1965) on the basis of function (meat, milk and wool) and that of Webster and Wilson (1966) in terms of origin (European, Oriental, Asiatic and African) are somewhat broader in scope and both have their own limitations. This is particularly true in areas where goats are unspecialised all purpose animals (which is often the case) and their origin lacks documentation.

1.	**Body size and weight**			
	(i)	Large size	:	Jamunapari, Beetal, Jakhrana, Sirohi
	(ii)	Medium size	:	Marwari, Kutchi, Surti, Barbari, Mehsana, Gohilwadi, Kannai Adu, Malabari, Sangamneri, Osmanabadi, Ganjam, Gaddi, Chegu, and Changthangi
	(iii)	Small size	:	Black Bengal
2.	**Production type**			
	(i)	Milk/meat	:	Jakhrana, Jamunapari, Beetal, Barbari, Sirohi, Surti
	(ii)	Meat/milk	:	Sangamneri, Kutchi, Zalawadi, Gohilwadi, Mehsana, Osmanabadi
	(iii)	Meat	:	Bengal, Ganjam, Malabari, Kanni Adu
	(iv)	Fibre/meat	:	Gaddi, Marwari, Changthangi, Chegu

The Indian goat breeds have been classified based on their body size/body weight and also based on production type (Khan and Rai, 2000).

2.2 History of Domestication

The goat was domesticated first among all ruminants during Neolithic age along with the cultivation of cereals sometime before 7000 B.C. at the slopes of the Zagros mountains on the borders of presently Iran and Iraq. According to Mason (1988), the ancestor of the present breeds of goat is *Capra hircus* and *C. aegagrus* – the Bezoar goat. This sub species is distributed from Sind in the East through Iran and Asia Minor to Crete and the Cyclades in the West. The Harappan toys demonstrate the representation of a goat and two seals from Mohenjodaro show the wild Bezoar goat with enormous curled horns, and a bearded domestic male goat with side-spreading horns. Devendra and Nozawa (1976) have suggested routes of migration of domesticated goats from the Western Asia to the Eastern Region. The other is the Northern route via Afghanistan and Turkistan to Mongolia and Northern China. Yet another route is through the Khyber pass to the Indian sub-continent and to the South East Asia besides a sea route.

The goats of *hircus* species reached Indian sub-continent from the Western Asia by Northern-Western and North Eastern routes between 7500–7000 B.C. But it is strongly believed that the present breeds have evolved from the crossings of *hircus* species brought to India along with Muslim invasions up to the 14th Century. Crossbreeding with European breeds have been sporadic in different pockets of India and claims have been made for development of a breed or strain. But the populations of these crosses were so small that sooner or later it got lost in the genetic mosaic already existing in Indian goats.

2.3 Theories of Domestication

The wild goat *Capra hircus*, from which the various breeds of domestic goats have been derived, is found in the barren hills of Baluchistan and Western Sind. In the North-East of Quetta, Markhor (*Capra falconeri* Wanger) is found. This species is also recorded in Turkestan, Afghanistan, Baluchistan and Kashmir. By far, the most important breed is the Bezoar goat (*Capra hircus aegagrus*), which ranges from Sind in the East, through Iran and Asia Minor to Crete and the Cyclades in the West. The goat was the earliest ruminant to be domesticated (long before 6700 B.C.), probably in Palestine or Iran. The Harappa toys contain representation of a goat (Bezoar); these could have been domesticated in Harappan period (Randhawa, 1982).

The breeding of goats under domestication was based on natural instinct within and between the species of genus *Capra*. The retention of male kids on the basis of faster growth and larger body size seems to play an important role. The choice of does were mainly on the reproductive efficiency but all the conditions were finally dominated by the disease resistance and survival of the fittest. Archaeological studies are scanty with regard to the evidence of the goat husbandry practices.

CHAPTER 3

ECOLOGY, HABITAT AND DISTRIBUTION

3.1 Geographical Distribution

The goat was domesticated first among all ruminants sometime before 7000 B.C. at the slopes of the Zagros mountains on the borders of presently Iran and Iraq. This sub species is distributed from Sind in the East through Iran and Asia Minor to Crete and the Cyclades in the West. The Harappan toys demonstrate representations of a goat and two seals from Mohenjodaro show the wild Bezoar goat with enormous curled horns, and a bearded domestic male goat with side spreading horns. Devendra (1979) have suggested routes of migration of domesticated goats from the Western Asia to the Eastern Region. The other is the Northern route via Afghanistan and Turkestan to Mongolia and Northern China. Yet another route is through the Khyber pass to the Indian sub-continent and to the South East Asia, besides a sea route.

India has vast genetic resources of goats having 22 defined breeds and a vast majority of goats are non-descript or intermixture of a number of breeds. They are adopted to harsh climate, long migration, poor vegetation and drinking water resources. India is a vast country with variable agro-climatic zones. The breeds of goats adopted to these regions also show wide variations in their size, weight and carcass quality.

Most of these breeds have been evolved over ages through geographical isolation and natural selection to suit the specific agro-ecological conditions. The concerted effort to tailor the breeds based on selection criteria for commercial gains have been meagre. These breeds are, therefore, adapted more to the harsh climate, long migration, tropical diseases, poor nutrition, and shortages of drinking water. A majority (about 75%) of the Indian goats are considered as mixed and non-descript in different States (Table 3.1). There is obvious proneness to a large intermixture of germplasm amongst the recognized breeds and non-descripts in different agro-climatic regions, particularly where two or more breeds are being raised on the contiguous breeding tracts. There are yet no viable Breeders' Societies or Agencies, established to register the animals of a particular breed for maintaining the Flock Book and to ensure the conservation of a breed or type. A systematic effort has been initiated lately by the National Bureau of Animal Genetic Resources (NBAGR), Karnal to develop the breed descriptors and evaluate the goat breeds. A few breeding farms belonging to the Central and State Government currently continue to maintain flocks of indigenous breeds for production of superior bucks and distribution to the farmers.

Numerically, Marwari in the Western dry region of Rajasthan, Kutchi in the Gujarat Plains and Black Bengal in the Lower Gangetic Plains of West Bengal are the important breeds for the purpose of meat production. Black Bengal, Barbari (in the Upper Gangetic Plains) and Surti have, of course, more market demand for their quality of meat. The Jamunapari in Upper Gangetic, Beetal in Trans-Gangetic, Jakhrana in the Central Plateau and Osmanabadi in the Western Plateau and Hill regions have been the breeds of choice for higher milk production. Hilly goats in the foot-hills of Himalayas have long hairs which are rough and poor in quality. They have the potential to produce annual fibre like Mohair in addition to milk and meat. At the high altitude in the Himalayas especially in Ladakh (J&K) and Himachal Pradesh, Pashmina goats produce Pashmina Cashmere-don fibre called 'Pashim'.

3.2 Rationation

Planning Commission has divided the country into 15 different agroclimatic zones depending upon the agro-ecological conditions. In respect of types of goats available, the country can be divided into 4 different regions (i) North-Western and Central Arid and Semi-Arid region, (ii) the Southern Penninsular region, (iii) the Eastern region and (iv) the Northern Temperate region.

In the North Western dry region of Rajasthan and Gujarat, Marwari, Surti and Kutchi are main meat breeds. In the upper Gangetic plain (central region), Barbari, Jamunapari and Jakhrana are the breeds of choice for milk production. In the East Gangetic Plain, Black Bengal and in Southern Plateau, Osmanabadi is the breed. In the Northern Temperate region, Gaddi, Chegu and Changthangi are the main meat and Pasmina breeds.

Table 3.1 Regional Distribution of Estimated Goat Population, Recognized Breeds and Breeding Policy (1992)

Agro-climate	State/Union territory	Climate annual rainfall (mm) temperature (°C)	Estimated population $\times 10^6$	Recognized breeds	Breeding policy
Western Himalayan	J&K, HP, Northern Punjab and Hill Dists. of UP and Uttarakhand	Ann. R.f. 250–280 Temp. Max 30, Min 40	6.1	Bakharwal, Changthangi, Chegu, Gaddi, Kangan, White Himalayan	Generally unorganised Crossing of dairy goats with Angora has been adopted by HP and J&K. No improvement programme of Changthangi, Chegu and Gaddi breeds
Eastern Himalayan	Sikkim, Arunachal Pradesh, Meghalaya, Nagaland, Manipur, Tripura,Mizoram, Assam and Jalpaiguri, Coochbihar, Districts of WB, Bihar, Orissa and Assam.	Ann. R.f. 1000–3300 Temp. Max 38, Min 2	5.0	Black and Grey Bengal and Khasi	Not defined
Lower Gangetic Plains	West Bengal	Ann. R.f. 1100–1925 Temp. Max 41, Min 10	12.4	Black Bengal (syn. Bengal)	Not defined
Middle Gangetic Plains	North Bihar and Eastern UP	Ann. R.f. 803–1470 Temp. Max 49, Min 4	16.1	Non-Descript	Not defined

Table 3.1 (Contd.)

Agro-climate	State/Union territory	Climate annual rainfall (mm) temperature (°C)	Estimated population $\times 10^6$	Recognized breeds	Breeding policy
Upper Gangetic Plains	UP excluding Hill districts & UP and Bharatpur Dist. in Rajasthan	Ann. R.f. 668–1400 Temp. Max 45, Min 4	5.0	Jamunapari, Barbari	Buck distribution programme improved strain
Trans Gangetic Plains	Punjab, Haryana, Delhi, Chandigarh & Sriganganagar, Sirohi, Ajmer, Bhilwara, Tonk, Jaipur, Districts of Rajasthan	Ann. R.f. 200–1000 Temp. Max 49, Min 2	4.6	Beetal	Not defined
Eastern Plateau	South Bihar Chattisgarh & Bastar Plateau area of M.P., Eastern Vidarbha zone of Maharashtra & Plateau zone of Eastern Ghat, Western Undulating zone & West and Mild Central Tableland area of Orissa and A.P.	Ann. R.f. 960–1700	11.4	Ganjam	Not well defined
Central Plateau amd Hill	Kymore Plateau, Satpuda Hill & Plateau, Vindhya	Ann. R.f. 531–1500 Temp.Max 45, Min 1	10.1	Barbari, Jhakhrana, Sirohi	Sirohi cross breeding with Swiss-Alpine has been followed up

Table 3.1 (Contd.)

Agro-climate	State/Union territory	Climate annual rainfall (mm) temperature (°C)	Estimated population × 10^6	Recognized breeds	Breeding policy
	Plateau, Central Narmada Valley, Grid & Bundalkhand area of M.P., Semi–arid Eastern Plain zone, Flood prone Eastern Plain, Subhumid zone and Aravalli Hills zone of and of Rajasthan and Bundalkhand area of Maharashtra				with ONB and selection of breeding bucks through a network programmes of ICAR
Western Plateau and Hill	Malwa Plateau, Nimar Valley and Jhabua Hill area of MP. Western Ghat, Sub-Montane zone, Plain and Plateau zone.Scarcity and Central Vidarbha zone of Maharashtra	Ann. R.F. 600-2,500 Temp. Max. 41 Min.5	10.8	Osmanabadi Sangamneri	Not well defined
Southern Plateau and Hill	Telangana & Scarce zone of Rayalaseema of Andhra Pradesh, Karnataka and North Western zone and Western zone of Tamil Nadu	Ann. R.F. 456-1,303 Temp. Max. 42 Min.10	12.1	Kanni Adu	Not well defined

Table 3.1 (Contd.)

Agro-climate	State/Union territory	Climate annual rainfall (mm) temperature (°C)	Estimated population × 10^6	Recognized breeds	Breeding policy
East Coast Plains & Hills	Krishna Godavari zone Hill Altitude and Tribal zone of AP, Coastal Zone and Eastern Ghat of Orissa, North Eastern zone, Cauvery Delta zone and Southern zone of Tamil Nadu & Pondicherry	Ann. R.f. 700–12597 Temp. Max 39, Min. 11	8.1	Bengal	Not well defined
West Coast Plains and Ghat	Coastal zone of Karnataka, Konkan, Coastal zone of Maharashtra, High Rainfall and North Konkan, Coastal zone of Tamil Nadu	Ann. R.f. 904–4,694 Temp. Max 36, Min 4	2.9	Malabari	Crossing with Exotic milch breeds Saanen and Alpine tested
Gujarat Plains & Hill	Gujarat, *i.e.* Saurashtra and Coastal zone,	Ann. R.f. 700–2200 Temp. Max 40, Min 13	3.7	Gohilwadi, Kutchi, Mehsana Surti, Zalawadi	Not well defined
Westerm Dry	Arid Western Plain, Transitional Plain zone of Inland Drainage and Lumi Basin	Ann. R.f. 100–490 Temp. Max 45, Min 10	6.0	Marwari	Not well defined
Island Region	Andaman & Nicobar Islands Lakshadweep, Minicoy and Amindivi	Ann. R.f. 700–1600 Temp. Max 36, Min 4	0.1	Andamani wild goats	Not defined

Source: Handbook of Animal Husbandry (3rd Edition, 2002–ICAR, Nerw Delhi)

The goats are also classified according to major produce associated with them. Although goats are main meat animals, good milk producing breeds like Jakhrana and Jamunapari and fibre producing like Chegu and Changthangi that produce elite Pashmina fibres and Angora (imported to India) that produce mohair fibre are available. Goat hair also fetches good price and there has been demand for goat hair of breeds like Marwari of Rajasthan.

3.3 Population Distribution and Trends in the States

The goat population in India (Table 3.2) has increased unabated since independence in spite of prejudices, negligence and depletion through slaughter. The average annual growth rate of goat population in India between 1972 and 1987 has been of the order of 3.1%. The increase in 1982 over 1977 was 24.4% and in 1987 over 1982 was 4.3% despite huge slaughter between 1982 and 1987 (Table 3.3).

According to 2003 livestock census, India possesses 124.4 million goats, out of which 38.740 million (33.6%) were under one year of age and 76.539 million (66.4%) were more than one year of age. Sex-wise 32.218 million were male and 83.060 million females. About 94.9% of goats are maintained in rural areas and 5.1% in the urban areas (Table 3.4).

In terms of goat population of the world, India is having 17.7% of goats during 2001 and thus ranks second in the world after China. As per Livestock census (2003) after cattle (185.18 M), goats are second in number (124.36 M) in the country. Maximum increase has been in the state of Assam followed by Maharashtra, U.P. and West Bengal.

Table 3.2 Goat Population in Different States/Union Territories of India (in '000)

State/Union territory	1972 Census	1977 Census	1982 Census	1987 Census	1992 Census	1997 Census	2003 Census
Andhra Pradesh	4380	4364	5559	4876	4328	5213	6277
Arunachal Pradesh	*	*	76	108	125	154	231
Assam	1258	1657	1729	2134	3454	2717	2987
Bihar	7364	9925	12221	15032	17459	20229	9490
Chhatisgarh	–	–	–	–	–	2154	2336
Gujarat	3210	3084	3300	3584	4241	4386	4541
Goa	*	**	-	17@@	14	13	11
Haryana	479	519	608	673	800	968	460
Himachal Pradesh	906	1035	1060	1119	1118	1168	1125
Jammu & Kashmir	569	692	1004	1396	1767	1864	2055
Jharkhand	–	–	–	–	–	0	5031
Karnataka	3726	3388	4547	3888	6287	4875	4484
Kerala	1468	1683	2004	1581	1848	1598	1213
Madhya Pradesh	6167	6725	7572	7751	8370	6470	8142
Maharashtra	5911	7563	7705	9191	9943	11434	10684
Manipur	*	**	42	44	38	33	33
Meghalaya	*	**	186	194	195	280	327
Mizoram	*	**	28	20	21	15	17
Nagaland	*	**	62	72	151	161	175
Orissa	2884	3417	4931	4804	4943	5772	5803
Punjab	801	722	700	537	544	414	278
Rajasthan	12163	12307	15479	12579	15309	16971	16809
Sikkim	*	89	96	98	114	86	124

Table 3.2 (Contd.)

State/Union territory	1972 Census	1977 Census	1982 Census	1987 Census	1992 Census	1997 Census	2003 Census
Tamil Nadu	3954	4202	5246	5921	6343	6416	8177
Tripura	147	199	343	441	429	639	472
Uttar Pradesh	6610	8462	9686	11320	13110	11784	12941
Uttarakhand	–	–	–	–	–	1070	1158
West Bengal	5211	5211	10916	22694	14170	15648	18774
A. & N. Islands	*	**	34	44	56	71	64
Chandigarh	*	**	1	1	–	1	1
D. & N. Haveli	*	**	14	19	19	20	21
Delhi	*	**	28	22	17	25	17
Lakshadweep	*	**	10	15	17	26	47
Pondicherry	*	**	53	31	44	41	48
Daman & Diu	*	**	15@	–	5	5	4
All India	67518	75602	95255	110207	115279	122721	124358

@ including Goa; @@ including Daman & Diu; Total of other territories (*) for which break-up is not available = 310000 Total of other territories (**) for which break-up is not available = 358000

Source: Ministry of Agriculture – Directorate of Economics and Statistics, New Delhi (Indian Livestock Census – 1972, 1977, 1982, 1987, 1992,1997, 2003)

Table 3.3 Annual Growth Rate (%) of Goat Population in India

Year	Population	Period	Annual growth rate (%)
1951	47.020	1951–56	3.26
1956	55.040	1956–61	1.91
1961	60.090	1961–66	1.19
1966	64.060	1966–72	0.88
1972	67.518	1972–77	2.29
1977	75.602	1977–82	4.73
1982	95.225	1982–87	2.96
1987	110.207	1987–92	0.99
1992	115.028	1992–97	1.26
1997	122.072	1997–03	0.22
2003	124.036		

Source: Ministry of Agriculture – Directorate of Economics and Statistics, New Delhi (Indian Livestock Census – 1972, 1977, 1982, 1987, 1992,1997, 2003)

Table 3.4 Goat Population in India (2003 Census) (in '000)

	Rural	Urban	Total
Male	33,791	2,023	35,815
Female	83,688	4,855	88,543
Total	**117,479**	**6,878**	**124,358**

Source: Indian Livestock Census – 2003

CHAPTER 4

GOAT BREEDS AND THEIR DESCRIPTION

4.1 Description of Goat Breeds

India possesses 20 recognized breeds of goat, which constitute 20–25% of the total goat population and remaining are non-descript with mixed features. Country is divided into four eco-regions (Temperate Himalayan, North-Western, Southern and Eastern region) and region wise description of goat breeds is given as under:

4.1.1 Temperate Himalayan Region

4.1.1.1 *Gaddi*

The breed derives its name from the nomadic tribe "Gaddi" who normally keep this breed. The breed is also known as Chamba in some of the dwellings. It is found in Kangra and Kullu valley, Chamba, Sirmur, Shimla and Lahul and Spiti in Himachal Pradesh, hilly districts of Tehri Garhwal and Chamoli in Uttarakhand and in parts of Jammu hills. Flocks size vary from 20 to 500.

Gaddi is a medium sized breed, well built, hardy with long hairs measuring about 20–25 cm (*Plate 1a and b*). The coat colour is mostly white but some animals with black, brown or tan makings are also found. The face is small with convex nose line and alert eyes. Both sexes have long spiral horns directed upward and backward with pointed tips. The ears are long and drooping. The average body weight of adult male and female is 27.45 ± 0.41 and 24.72 ± 0.51 kg respectively. The average body length, body height and heart girth is 69.5 ± 0.84, 61.3 ± 0.84 and 72.2 ± 0.68 cm in male and 65.2 ± 1.18, 58.1 ± 1.02 and 69.3 ± 0.48 cm in females respectively. Prolificacy is very less and twinning ranges from 10 to 15%. The average age at first kidding ranges from 350 to 390 days. The udder is small and well set with conical teats. They are poor milkers and produce 200–300 gm milk/day.

The mortality varies form season to season and ranges from 10–15% in kids and 5–8% in adult. Pneumonia is the main cause of death. Smaller flocks are stationary but most of the flocks migrate in summer to Alpine pastures on high altitude and remain in foothills and valleys during winter. During winter they are mostly fed with stored feed and fodder. The Gaddi goats are valued for their hairs, which are used for making rugs, ropes and other items of domestic utility. They are also used for meat production and as pack animals particularly males for transportation of mercantile in hilly terrain.

4.1.1.2 *Chegu*

Chegu is a cashmere "Pashmina" producing goat and found in Lahul, Spiti Valley of Himachal Pradesh and hilly regions of Uttar Kashi, Chamoli and Pithoragarh of Uttarakhand mainly the areas bordering Tibet region.

It is a medium sized breed with compact body and short legs. The coat colour is predominantly white (*Plate 1c and d*). Animals having dark gray, and mixed colours are also found. Face is slightly longer with straight profile. Both the sexes have twisted long horns directed upwards, backward and outward with pointed tips. Ears are short and kept laterally. They have long lustrous hairs with a fine under coat of cashmere, which is harvested through combing once in a year. The average cashmere yield/goat is about 120 gm with a fibre diameter of 12 m. The average body weight of a adult male and female is 39.90 kg and 27.8 kg respectively. The average body length, body height and heart girth is 72.8, 66.2 and 88.7 cm in males and 65.5, 58.1 and 73.9 cm in females respectively. The age at first kidding and kidding interval is 615.8 ± 23.03 and 272.8 ± 8.4 days respectively. Twinning is very less. Udder is small and round with small teats. They are poor milkers and yield 69 kg milk in a lactation of 187 days.

Normally holding size vary from few goats to larger flocks which migrate to upper ranges in summer and return back at the onset of winter to their dwellings in the valleys and foothills. During this period, stored fodder including dry grasses, tree leaves and straw are fed.

4.1.1.3 *Changthangi*

Changthangi is cashmere (Pashmina) producing goat, found in cold and arid Leh region of Kashmir and adjoining areas of Himachal Pradesh.

Animals are medium in size and hardy. Coat colour is generally white but gray and brown animals are also found (*Plate 1e and f*). The breed is adapted to cold conditions and larger flocks inhabit the cold desert areas of "Changthang" located at the height of 3500 to 5000 m ASL. They are sturdy with deep body and short but strong legs. The ears are small pointed and kept laterally. Both sexes have horns, the size and shape of horns vary but mostly directed upward, backward and outward inward. Body is covered with long and lustrous hairs with a fine undercoat of cashmere, which is harvested mostly through combing once in a year leaving the long hairs to be clipped later. The cashmere is very soft and warmest of the animal fibres, which is used for manufacturing high quality fabrics and shawls fetching very high market price. The long hairs are used for making ropes, rugs and many other items of domestic utility. The goats are also used for meat and for transporting mercantile at high altitude. The average body weight of adult male and female is 20.37 ± 0.24 and 19.75 ± 0.16 kg respectively. The average body length, body height and heart girth is 49.8 ± 0.36, 49.0 ± 0.29 and 63.0 ± 0.44 cm in males and 52.4 ± 0.23, 51.6 ± 0.20 and 65.2 ± 0.29 cm in females respectively. The age at first kidding vary from 540 to 600 days and twinning is very less. Udder is small and round with small teats. The average yield of Pashmina is 214 gm with a fibre diameter of 13.86 μ.

The mortality is high in young kids under field conditions raging from 25 to 35%. Adult mortality is recorded varying form 5 to 15%. Smaller flocks are mostly stationary ranging from 10 to 15 goats, but larger flocks having 200 to 500 goats are migratory.

4.1.2 North-Western Region

4.1.2.1 *Jamunapari*

The breed derives its name from the location of its native habitat in the area across the river Jamuna (Jamuna Par) in the Etawah district of Uttar Pradesh. The breed is mainly found in the Chakarnagar block comprising of the villages of Nagala Kathori, Nagala Jor, Jagtoli, Kola, Rampura, Aheria and Neemdera etc. of Etawah district, where it is found in its most pure form. Breed is also spread in the adjoining districts of Bhind, Morena and Shivpuri in Madhya Pradesh. The population of Jamunapari goats is decreasing over the years and the number is estimated to about 9000 goats and need active consideration and action for the conservation of such a valuable germplasm. The flock size ranged from 2–25 goats.

Jamunapari is a large sized animal with a majestic getup and look. The coat colour is predominantly white with brown patches on the ears and head (*Plate 2a and b*). The face is convex which is very peculiar to this breed and called Roman nose. The body is long with a tuft of hair on the backside of the thighs. In most of the animals lower jaw is longer giving a parrot mouth appearance. Ears are long and pendulous. Both male and female have horns. The breed is known for its outstanding ability for milk production. The udder is well developed, round with long conical teats. The breed has been used for upgrading the native goats in many south and South-East asian countries like Sri Lanka, Malayasia, Indonesia and Vietnam. Jamunapari breed was also used to evolve the famous Anglo-Nubian breed of England. The body weight of adult male and female ranged form 45–60 kg and 35–40 kg respectively. The average body length, body height and heart girth is 77.4 ± 1.23, 78.2 ± 1.25 and 79.5 ± 1.20 cm respectively in adult males and 75.2 ± 0.46, 75.2 ± 0.38 and 76.1 ± 0.38 cm respectively in adult females. The average age of first kidding and kidding interval is 786 and 390 days respectively with 30% twinning. A twinning of 55% and 4% triplets was recorded under high plane of feeding. The average lactation yield is 113.57 ± 2.41 kg in a lactation period of 148 days under farm conditions. The milk yield of 2–3 kg is not common in the farmer's flocks.

The mortality in village flocks is very low, ranging from 4–6% in young kids and 3–4% in adult goats. All the flocks are stationary and maintained on extensive grazing however, supplementary feeding with grains and also with the tree leaves is provided to the pregnant and lactating does and breeding bucks. The animals are grazed from early morning to late evening. Thatched huts commonly used for the housing of the goats. They are also kept in open paddocks fenced with thorny bushes. Smaller flocks or 2–3 goats are kept in the house itself where the owner lives. The dam's milk yield and phenotype of animal is the only consideration for the selection of the bucks in the villages.

4.1.2.2 *Beetal*

Beetal breed of goat is predominantly found in the districts of Gurdaspur, Amritsar and Firozpur in Punjab in its purest form, however, it has spread in other parts of Punjab and adjoining parts of Haryana. The population of Beetal goats has reduced to a greater extent due to intensive agriculture and shrinkage of grazing land in Punjab. Normally smaller flocks are found and vary from 2–10 goats. Only 1% of the total flock have goats exceeding 20 but not more than 50. Beetal is an outstanding dairy breed and carry easy adaptation capability to varying environmental conditions.

It is a large sized tall breed with well-set body. Coat colour is mainly black or brown with white patches on the head or body (*Plate 2c and d*). Face profile is convex with long and flat drooping ears. The long ears and Roman nose indicate common ancestry with Jamunapari breed. Both sexes have horns. Males generally carry beard. Udder is large and well set with long conical teats. The body weight of adult male and female is 59.07 ± 2.42 kg and 34.97 ± 0.52 kg respectively. The body length, body height and girth is 85.5 ± 1.41, 91.60 ± 1.97 and 86.0 ± 1.20 cm in adult males and 70.4 ± 0.88, 77.13 ± 0.46 and 73.7 ± 0.70 cm respectively in adult females. The average age at kidding and kidding interval is 560 days and 355 days, respectively. Prolificacy is not different to Jamunapari breed. The milk yield is 156.9 ± 6.80 kg in a lactation period of 186 days.

The mortality is very less (not more than 5%) under field conditions. Under farm conditions it is as high as 25% in young kids and 13% in adults.

Flocks are stationary and reared under extensive grazing conditions. The potent source of forage is from tree leaves and browsing of bushes and shrubs on the verges of roads and canals. Supplementary feeding with the grains is occasional.

4.1.2.3 *Sirohi*

The Sirohi breed is native of Sirohi district in Rajasthan from where it derives its name. It is distributed over a wider area of Ajmer, Bhilwara, Tonk, Jaipur districts and nearby areas of Rajasthan and also adjoining region in Northern Gujarat. The flock size varies from place to place and ranges from 10 to 200 goats. Sirohi is very potent breed and many large flocks are seen in central Uttar Pradesh also which are raised for sale of milk by traditional shepherds community who are in this business over the generations.

It is a medium to large sized breed having compact body and strong legs. The coat colour is predominantly brown with dark brown or tan patches of different shape and size all over the body (*Plate 2e and f*). Face profile is straight or slightly raised. Ears are medium in length, leafy and drooping. Both sexes have horns, which are curved upward and backward with pointed tips. Some have wattles. Sirohi are good milkers and udder is well developed and round with long conical teats. The average body weight of a adult male and female is 50.4 ± 2.52 and 22.5 ± 0.17 kg respectively. The average body length, body height and heart girth is 80.0 ± 1.02, 85.6 ± 1.40 and 80.3 ± 1.00 cm in males and 60.3 ± 0.20, 68.4 ± 0.20 and 62.4 ± 0.20 cm in females respectively. The average age at first kidding and kidding interval is 661 and 469 days respectively with twinning as 12.5%. The average lactation yield is 122.30 ± 4.61 kg in a lactation period of 198 days.

The mortality under field conditions is very less ranging from 2 to 5% in young kids and 1 to 3% in adult goats. Normally goats are maintained under extensive grazing system and most of the flocks are stationary. Large sized flocks migrate seasonally to nearby areas for want of forage and water.

4.1.2.4 *Jakhrana*

The breed derives its name from the name of a village "Jakhrana" where it is found in the most pure form. The breed is distributed in the area around Jakhrana village near Behror in Alwar district of Rajasthan. The population of the breed is small which is localized in few villages only. According to a rough estimate the number of goats is not more than 6000. Most of the flocks are small in size having 1–25 goats but few flocks having 100–200 goats are also found.

It is a large sized breed with compact body and long legs. The coat colour is predominantly black with a typical white speckles on the ears (*Plate 3a and b*). The coat is short and lustrous. Face is straight and forehead is narrow and raised. Both sexes have strong and thick stumpy horns. Ears are of medium length and drooping. Wattles are found in some animals. Udder is large and well developed with long conical teats. The average body weight of adult male and female is 43.50 ± 1.16 and 39.29 ± 0.40 kg respectively. The average body length, body height and heart girth is 84.1 ± 2.11, 90.40 ± 1.61 and 86.0 ± 1.91 cm respectively in male and 77.7 ± 79, 79.1 ± 0.29 and 79.1 ± 0.31 cm respectively in females. Average age at first kidding and kidding interval is 574 and 319 days respectively with a twinning rate of 41.0%. Triplets are 2–3%. This breed is famous for its milk production potential and used at many places as an improver breed for increasing the milk production. The lactation yield is 121.8 ± 8.8 kg in a lactation period of 115 days with a daily average of 1060 gm.

The death rate is very small under village conditions ranging to 2–3% in kids and 1–2% in adult goats. The goats are raised under extensive conditions and flocks are stationary. Supplementary feeding with grains is provided to pregnant and lactating goats and farmers take special care of their goats, which is very peculiar to this breed.

4.1.2.5 *Marwari*

Breed derives its name after the region "Marwar" which is the natural habitat of the breed. It is found in an extensive area in Western Rajasthan comprising of Barmer, Bikaner, Jaisalmer, Jalore, Jodhpur, Nagaur and Pali districts. The breed can also be seen in adjoining area of Northern Gujarat. The total goat population of this region is 3.53 million and Marwari breed alone accounts for 28% of the total goat population of the state (Livestock census 1988). The flock size varies from few goat (2 to 5) to larger flock (100 to 500) but the migratory flocks are always larger. The flock holding depends on the resources available to the owner besides his financial capabilities.

Marwari is a medium sized breed with compact body and strong legs. Coat colour is predominantly black but animals with brown and white markings are also found (*Plate 3c and d*). The coat is shaggy and dull in appearance and contains long hairs. The ears are long, flappy and drooping. Both the sexes possess horns, which are medium in length, directed upward and backward and pointed at tips. Males have longer and stronger horns than the females. Wattles are present in some animals. One third of the goats possess beard. The average weight of adult male and female is 46.6 ± 2.63 and 30.0 ± 0.25 kg respectively. The body length, body height and heart girth is 70.97 ± 1.65, 74.74 ± 1.61 and 71.68 ± 1.41 cm in adult males and 63.51 ± 0.44, 69.29 ± 0.22 and 68.60 ± 1.60 cm in adult females respectively. The average age at kidding and kidding interval is 563.1 ± 17.1 and 375.5 ± 11.1 days respectively. Twinning varies from 10 to 13.4%. The milk yield is 110.03 ± 5.66 kg in a lactation period of 196 days.

The overall flock mortality under field conditions is very low and ranged from 3.3 to 4.1% in kids. The adult mortality is 3.5%. The smaller flocks are stationary but the migration of larger flocks is the routine practice during the lean period. They start migrating in the month of March-April in many directions through their well established routes year after year and return back to their dwellings in the month of July. All the time animals are kept on extensive grazing system.

4.1.2.6 *Zalawadi*

Zalawadi is one of the major goat breed in Gujarat. Goats of this breed are reared by traditional shepherd community known locally as "Rabaris" and "Bharwads" in the semi-arid area of Sourashtra region of the state where rainfall is erratic and low. It is believed that the breed originated in the erstwhile "Zalawad" region now known as Suredranagar district and part of Zalawad falling in Rajkot district. It has spread in the adjoining area of Jamnagar and Ahmedabad districts also. Out of the total goat population of the state Zalawadi alone constitute 27.85% as per the bulletin of Animal Husbandry, Poultry and Dairying Statistics of Gujarat (1997–98). Flock size ranges from 15–200.

It is a medium to large sized breed of dual utility with long legs. The coat is predominantly black with lustrous, long shinning hairs (*Plate 3e and f*). Some animals have black and white mixed coat. Face is slightly raised. Both male and females have twisted corkscrew type horns directed upward and backward with pointed tips. Ears are long, leafy and drooping and are invariably white speckled. The ears are so wide loose and leafy that the shepherds traditionally trim the ears and sometimes split the ears to avoid injuries from thorny bushes while browsing. Udder is large and well developed with long conical teats. The average body weight of adult male and female is 38.84 ± 1.46 and 32.90 ± 0.32 kg respectively. The average body length, body height and heart girth is

75.6 ± 1.05, 83.3 ± 0.8 and 76.8 ± 1.1 cm in male and 71.8 ± 0.30, 78.5 ± 0.20 and 74.2 ± 2.3 cm in females respectively. The average daily milk yield is 1.76 kg ranging from 1.5 to 2.0 kg. Goat produces 154 kg of milk in a lactation period of 150 days under village conditions. The prolificacy is high with 55% twinning and 2% triplets. The mortality is high in young kids and low in adults.

Flocks are stationary except few flocks which migrate to Ahmedabad and Khera district in summer and return back in rainy season. The flocks are maintained under extensive grazing management. No specific housing is provided and animals are kept in open area fenced with thorny bushes.

4.1.2.7 *Gohilwadi*

The breed is a native of Gujarat state and found in the district of Bhavanagar. Amreli and Junagarh in its pure form. It is also distributed thinly in neighbouring districts. The flock size ranged from 5 to 80 goats.

It is a medium to large in size and body is covered with long hairs. The coat colour is predominantly black with white marking on the ears. The face is long and lightly convex. The ears are leafy and drooping. Both sexes have long lightly twisted horn, which directs upwards and backwards with pointed tips. Hairs are long rough and thick. The body weight of adult male and female is 37.10 ± 1.42 and 36.03 ± 0.38 kg respectively. The average body length, body height and heart girth is 73.4 ± 1.3, 81.2 ± 1.2 and 74.7 ± 0.9 cm in adult male and 72.4 ± 0.3, 79.5 ± 0.3 and 75.2 ± 0.2 cm in adult female respectively. The age at first kidding ranges from 650 to 700 days with twinning of 15 to 20% under village conditions. The udder is well developed with long conical teats. The average daily milk yield is 1.71 ± 0.145 kg in village goats.

The mortality is very low and ranges from 6 to 10% in kids and 4 to 5% in adult goats. Mostly flocks are stationary and reared on extensive grazing. Few flocks migrate to nearby areas in scarcity period.

4.1.2.8 *Mehsana*

Breed derives its name from the place of its origin called "Mehsana" in Gujarat where it is found in its most pure form. It is widely distributed in Banaskantha, Palanpur, Gandhi Nagar, Ahmedabad and nearby areas in varying intensity. The flock size vary from 15 to 300. Many smaller flocks of 2–5 goats are also found.

The breed is large in size with convex face profile. The coat colour is grayish black with long and coarse hairs. Ears are white with black markings, leafy and drooping. Both sexes are horned which have one or two twists and curved upward and backward with pointed tips. Usually they possess beard. They are good milkers and udder is well developed and capacious with long conical teats. Tail is short and kept upward. The average body weight of adult male and female is 37.14 ± 1.51 and 32.39 ± 0.38 kg respectively. The average body length, body height and heart girth measured as 71.2 ± 1.0, 80.4 ± 1.2 and 76.9 ± 1.2 cm in males and 68.0 ± 03, 74.3 ± 0.2 and 73.0 ± 0.3 cm in females respectively. The age at first kidding vary from 600 to 650 days. Twinning is low and ranged from 10 to 15%. The average daily yield is 1.32 ± 0.13 kg.

Mortality in young kids ranged form 10–15% and in adults 4–5% under village conditions. Most of the flocks are stationary and reared on extensive grazing and browsing except few flocks, which migrate seasonally to nearby places. No specific housing is provided.

4.1.2.9 *Surti*

The breed derives its name from the name of the place called "Surat" in Gujarat where it is found in most pure form. It is widely distributed in adjoining areas around Surat and Nasik in Maharashtra. It is very popular in Bombay area due to its dairy potential. It is believed that Surti breed is derived from Arabian milch goats. The Surti goats are very popular and maintained in small flocks ranging from 2 to 15 goats. The breed is most suited and performs well under stall fed conditions.

It is a small to medium sized goat with compact body. The coat colour is predominantly white having short and lustrous hairs (*Plate 4a and b*). Ears are medium sized and kept dropping. Face profile is slightly raised, forehead prominent. Both sexes have horns of medium size directed upward and backward. Udder is well developed with large conical teats. The average body weight of a adult male and female is ranging from 25–30 kg and 22–25 kg respectively. The age at kidding vary from 400–500 days. The average daily milk yield vary from 1.2–2.0 kg under village conditions.

They are maintained on extensive grazing but small holders keep them under intensive conditions and fed in stall. Tethering is common with the farmers who keep one or two goats.

4.1.2.10 *Kutchi*

The breed derives its name after the region "Kutch" in Gujarat, which is the natural habitat of the breed. It is found in Kutch region in northern Gujarat and spread in the adjoining areas of southern Rajasthan. The flock size ranges from small to large having 5 to 300 goats.

Kutchi is a medium sized goat of dual utility (milk and meat) with a compact body and long legs. The coat colour is predominantly black, few animals with white markings are also found (*Plate 4c and d*). Ears are medium in size, flappy and drooping with typical white markings. Coat is shaggy and dull in appearance and contain medium to long coarse hairs. Both the sexes have horns, males have long and strong horn while the females have flat and weak horns. Face is slightly raised. Udder is reasonably developed with well-placed long and conical teats. The average adult body weight of male and female is 43.50 ± 1.16 and 39.29 ± 0.40 kg respectively. The average body length, body height and heart girth is 77.5 ± 1.2, 86.4 ± 0.7 and 78.4 ± 0.7 cm in males and 75.0 ± 0.3, 82.4 ± 0.3 and 76.1 ± 0.2 cm in females respectively. Age at first kidding and kidding intervals is 777.5 ± 20.4 and 464.3 ± 20.9 days respectively. Twinning is 10%. The lactational milk yield is 112.56 ± 5.65 kg in a lactation period of 202 days under farm conditions.

The annual mortality in young kids vary from 5 to 15% however the mortality in adult flock averages 5% under village conditions. Most of the flocks are stationary but they migrate for a limited period to adjoining areas during the period of scarcity. Flocks are maintained exclusively under extensive grazing round the year.

4.1.2.11 *Barbari*

It is believed that breed derives its name from the place of its origin in Barbara in East Africa. The route of their migration to India has not been known but in all its probability the traders of medieval period might have brought them through sea and land routes during their business entourage to India. The breed is found in Etah, Aligarh, Agra and Mathura Districts of Uttar Pradesh and adjoining district of Bharatpur in Rajasthan. This has been adopted and extensively used in many states of the country under their goat development programmes and also for commercial raising.

Barbari is medium in size with stout and compact body. They look very alert and attractive. The coat colour is white with light to dark brown spots of different size and shape all over the body (*Plate 4e and f*). The orbital bone is prominent and eyes appear as protruding. Ears are short and erect. Both sexes have twisted horns directed upwards, backward and outward. Males have longer and stronger horns than the females. Males have beard and some of the animals have wattle. Udder is well set and round with conical teats of good length. The breed is very versatile and popular in cities because of its size and adaptability to stall-feeding. They are very prolific breeders and usually kid twice in 14 to 16 months with a high rate of prolificacy. The average body weight of adult male and female is 37.85 ± 1.96 and 22.56 ± 0.32 kg respectively. The carcass weight ranges from 8 to 10 kg when the kids are reared under intensive feeding and slaughtered at the age of 9 months with a dressing of 55.15% on empty live weight basis. Average body length, body height and heart girth 64.3 ± 0.4, 70.7 ± 0.7 and 75.5 ± 1.3 cm in adult males and 58.9 ± 04, 56.2 ± 0.4 and 64.3 ± 0.4 cm in adult females respectively. The average age at first kidding and kidding interval is 491 and 290 days respectively with 52% twins and 4% triplets. Lactation yield of 82.2 ± 3.16 kg is recorded in a lactation period of 130 days.

4.1.3 Southern Region

4.1.3.1 *Malabari*

The breed derives its name from its native habitat of "Malabar" area of northern Kerala where they are densely populated. The breed is widely distributed in the districts of Kasargod, Trichur, Kannur, Kozikkot and Malapuram and mainly concentrated in and around Tellicherry district and therefore sometimes called as Tellicherry breed. Malabari breed is believed to have Surti blood. Flock size is small ranging from 2–10 goats. Few flocks have 15–20 goats.

It is a medium sized animal of dual utility and constitutes only 10% of the total goat population of the state. They do not have any uniform body colour (*Plate 5a and b*). Majority of the goats are white or black but goats with mixed colours of black, white and brown are also found. They have medium sized head with straight face, some have slightly raised face also. Males mostly have beard and 10–15% females also have beard. Wattles are present in about 25% of the goats. Both sexes have twisted horns of small to medium size directed sideward and downward. Most of the goats have small shinning hairs while other have long hairs. Ears are medium in size directed outward and downward. Tail is small, thin and kept upwards. Udder is small with conical teats and medium size. The average body weight of adult male and female is 35 and 28 kg respectively. The body length, body height and heart girth is 70, 72 and 74 cm in males and 64, 63 and 67 cm in females respectively. The breed is very prolific with 45% twins and 4.5% triplets. The litter size at birth varied from 1.25 to 1.98. The age at first kidding and kidding interval was 581 and 307 days respectively. The average milk yield is recorded as 65 kg in a lactation of 172 days.

The morality in young kids is high (20%) but in adult it is low (5%). The main cause of death is pneumonia. The flocks are stationary and maintained on extensive system. Goats are grazed for an average period of 6 hours a day. They are also fed with the jack leaves and coconut oil cakes. Shortage of quality bucks is common feature in the breeding tract.

4.1.3.2 *Osmanabadi*

The breed derives its name form habitat, Osmanabad in Maharashtra where it is found in most pure form. The breed is mainly concentrated and found in Latur, Parbhani, Ahmednagar and Solanpur districts of Maharashtra besides Osmanabad. It has spread over a wide range of agro-climatic conditions in Vidharva, Marathwada, Western Maharshtra and adjoining parts of Karnataka and Andhra Pradesh.

Osmanabadi is a medium sized breed with comparatively long body and long legs. The body colour is mostly black but some animals with white patches on ears, neck and sometimes on body are also found (*Plate 5c and d*). The hair coat is short and shinning. The face profile is straight, ears are of medium size and dropping. Most of the males are horned but about 5% of the females are polled. Wattles are found in many animals. The adult males and females weigh 33.66 ± 2.73 and 32.36 ± 055 kg respectively. The average body length, body height and heart girth is 69.12 ± 1.93, 77.87 ± 1.68 and 72.06 ± 2.04 cm respectively in male and 67.51 ± 0.38, 74.79 ± 0.30 and 72.04 ± 0.04 cm in females respectively. The average age at first kidding and kidding interval is 523 and 214 days respectively. Goats of this breed have very efficient reproduction and in well-managed flocks kidding is noticed twice in 14–15 months with 30% twinning and 2% triplets. The daily milk yield ranged form 700 gm to 1500 gm under well managed village flocks with lactation length of 130–150 days. The meat is fairly of good quality and it preferred by majority of the rural and urban population.

Mortality ranges from 10–18% in kids and 5–8% in adult goats under field conditions. Goats are widely reared either by grazing in harvested field/roadside bushes/hillocks or semi-grazing system is followed. In sole grazing system, many times the kids are born in the field. No elaborate housing is provided. Goats are kept in thatched huts fenced by thorny bushes. Kids are also sent for grazing for long distances along with the mothers. The extensive walking results in heavy kid losses.

4.1.3.3 *Sangamneri*

Sangamneri goat is a dual purpose breed found in "Sangamener" Tehsil of Ahmednagar district in Maharshtra. The breed is also found in adjoining districts of Pune, Solapur, Nasik and Dhule. The breed is found in irrigated areas and generally maintained in small flocks of 2–10 goats. Few larger flocks of 50–100 goats also seen.

It is medium in size with straight face. Body is long with, long legs. Coat colour is white in a majority of the cases but black or brown coloured animals are also not uncommon (*Plate 5e and f*). Some have spots of different colours. Ears are of medium size and drooping. Both sexes have horns directed backward and upward. Tail is thin, short and kept upward. Hair coat is short and coarse, some animals have wattles. The udder is small, round with conical teats. The body weight of adult male and female ranges from 35 to 55 kg and 25 to 36 kg respectively. The maximum weight recorded was 67 kg in males and 52 kg in females. Reproductive efficiency is high with 30–40% twinning and 3–5% triplets. The average age at first kidding and kidding interval is 420 and 340 days respectively. Earliest age at first conception is recorded as 240 days. The milk yield ranges form 103–130 kg in a lactation period of 160 days under village conditions.

The mortality among the kids ranged from 5–10% and in the adults it was less than 10%. The flocks are stationary and maintained under extensive grazing of 5–6 hours a day. They are also fed tree leaves and wilted fodder along with small quantity of concentrate and cake.

4.1.3.4 *Kanni Adu*

The breed is known for its fairly large size, vigour, disease resistance and heat tolerance capability under humid conditions. Generally this breed is reared for meat and skin. The breed is found in the districts of Ramanathapuram and Tirunelveli in Tamil Nadu. The flock size varies from 2–40 goats.

They are tall having straight face profile. Coat colour is predominantly black or black with white spots of different sizes (*Plate 6a and b*). Legs below knee and hocks are white. Peculiar white strips are found on the face. Ears are long and pendulous. Males having strong horns and females are polled. Tail is of medium size, thin and kept up. Udder is small and round with small conical teats. The body weight of adult male and female varies from 32 to 38 kg and 28 to 31 kg respectively. The average body length, body height and heart girth measured as 71.83 ± 1.36, 84.12 ± 1.74 and 77.53 ± 1.55 cm for males and 67.30 ± 0.38, 76.15 ± 0.46 and 70.83 ± 0.44 cm for females respectively. Prolificacy is low with the twinning of 10%. Age at first kidding is recorded as 650 days. Goats of this breed are poor milkers.

4.1.4 Eastern Region

4.1.4.1 *Black Bengal*

The origin of Bengal goats is not definitely known but it is considered that their native habitat is in Bengal (now West Bengal in India) and it is found in central and south parts of West Bengal and distributed in adjoining states of Bihar, Orissa, and Assam but is found in most pure form in the district of Murshidabad and surrounding places. Flock size is generally very small ranging form 2–20 goats. This breed is famous for the quality of meat and produces superior quality of skins.

It is a dwarf breed having short legs, straight back and deep body. The breed is found in three coat colours namely black, white and brown but black colour is most predominant and majority of goats are of this type (*Plate 6c and d*). Some times white markings on head and abdomen are also seen. The hair coat is short and lustrous. The face is small with straight or slightly depressed profile. Both sexes have small and stumpy horns directed upwards and backward. Beard is found in both sexes. The ears are short, flat and horizontally placed. The average body weight of adult male and female is 32.37 ± 2.74 and 20.38 ± 0.16 kg respectively. Average age at first kidding and kidding interval is 430 and 260 days, respectively. Bengal goats are very prolific and litter is 5.3% twins and 4.4% triplets and quadruplets. They are poor milkers and yield 25.24 ± 1.03 kg of milk in a lactation of 96 days.

The mortality in black Bengal goats particularly in young kids is high as compared to other breeds and ranged from 20 to 40%. This is due to smaller birth weight of the litter and low milk yield of the dams. Mortality among the adults ranged from 10–15%. Flocks are stationary because of small holding size. Goats are mainly maintained on extensive grazing and tethering is very common where one or two goats are reared. Tree leaves or cultivated fodder is also fed to these goats at home.

4.1.4.2 *Ganjam*

The breed derives its name from the name of the place "Ganjam" in Orissa. It is widely distributed in south Orissa and adjoining areas of coastal Andhra Pradesh and Jagdalpur of Madhya Pradesh. Goats are mainly raised by the tribals and also peoples living in rural areas. The size of flock is small; some are large but not exceeding 200 goats.

It is a tall and leggy animal with body of medium size. The breed is primarily valued for meat. The coat colours are predominantly black but animal with white spots are also seen (*Plate 6e and f*). The hairs are short and lustrous. Face is straight and forehead is prominent. Both the sexes are horned. Horns are long, flat, directed upward and backward some times straight. Ears are medium in size and kept straight or horizontal with pointed tips. Males usually have beard and some animals have wattle. The average adult body weight of male and female is 44.05 ± 0.13 and 31.87 ± 0.37 kg respectively. The average body height, body length and heart girth is 76.2 ± 0.1, 84.5 ± 0.7 and 83.1 ± 0.9 cm in males and 67.6 ± 0.3, 77.1 ± 0.3 and 74.6 ± 0.3 cm in females respectively. It is a prolific breed and two kidding in 14–15 months are common with twinning of 45%. The age at first kidding ranges from 700 to 750 days and kidding interval is 376 days. The udder is small with small teats. Ganjam goats are poor milker and the daily milk yield ranges from 300 to 400 gm under village conditions. The lactation period varies form 130 to 150 days.

The mortality among young kids is high and ranged from 20 to 30%. The adult mortality is less and not exceeding 5%. Animals are kept on extensive grazing and the goats move mostly along with the nomadic tribals who mostly rear goats.

Plate 1. Various Indian Goat Breeds

(a) Gaddi (Male)

(b) Gaddi (Female)

(c) Chegu (Male)

(d) Chegu (Female)

(e) Changthangi (Male)

(f) Changthangi (Female)

(*Contd.*)

Plate 2. Various Indian Goat Breeds

(a) Jamunapari (Male)

(b) Jamunapari (Female)

(c) Beetal (Male)

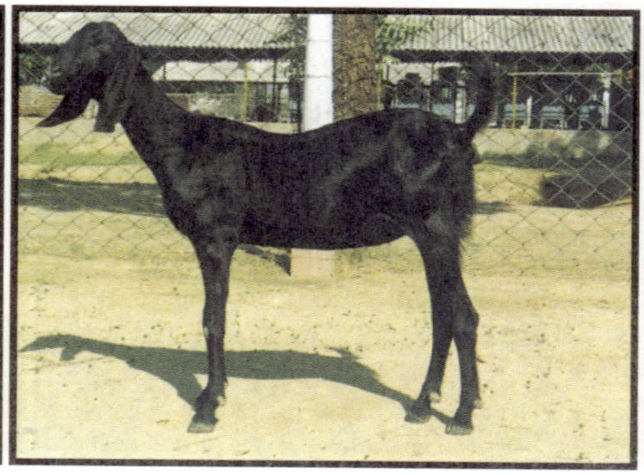

(d) Beetal (Female)

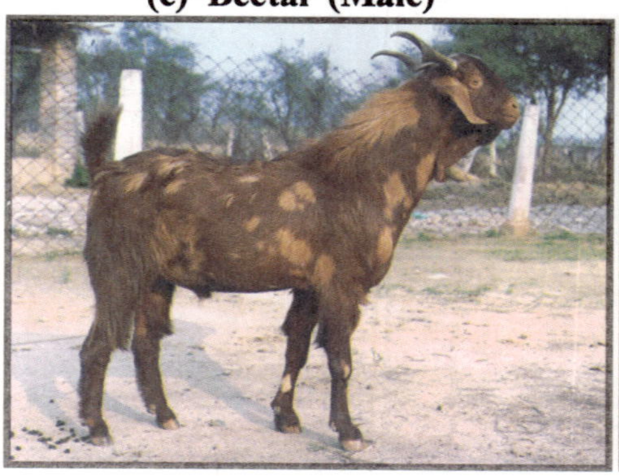

(e) Sirohi (Male)

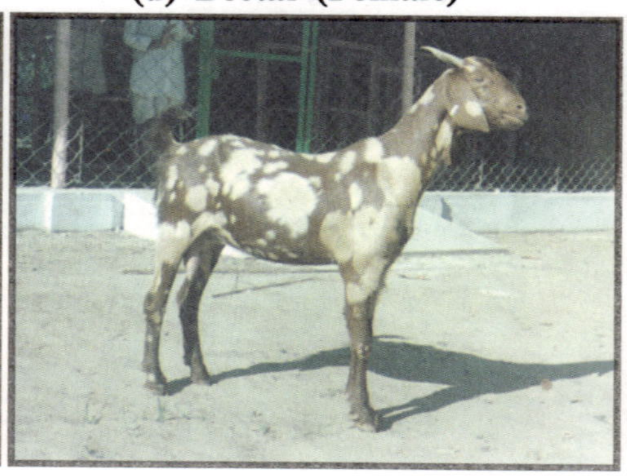

(f) Sirhoi (Female)

(Contd.)

Plate 3. Various Indian Goat Breeds

(a) Jakhrana (Male)

(b) Jakhrana (Female)

(c) Marwari (Male)

(d) Marwari (Female)

(e) Zalawadi (Male)

(f) Zalawadi (Female)

(*Contd.*)

Plate 4. Various Indian Goat Breeds

(a) Surti (Male)

(b) Surti (Female)

(c) Kutchi (Male)

(d) Kutchi (Female)

(e) Barbari (Male)

(f) Barbari (Female)

(Contd.)

Plate 5. Various Indian Goat Breeds

(a) Malabari (Male)

(b) Malabari (Female)

(c) Osmanabadi (Male)

(d) Osmanabadi (Female)

(e) Sangamneri (Male)

(f) Sangamneri (Female)

(*Contd.*)

Plate 6. Various Indian Goat Breeds

(a) Kanni Adu (Male)

(b) Kanni Adu (Female)

(c) Black Bengal (Male)

(d) Black Bengal (Female)

(e) Ganjam (Male)

(f) Ganjam Flock (Female)

CHAPTER 5

PRODUCTION PERFORMANCE

The production performance in terms of adult body weight, growth rate, milk yield and fibre production of Indian breeds is lower than many exotic breeds developed for specialized production *e.g.* meat, dairy and fibre. The fecundity of several Chinese small and dwarf breeds excels the Indian breeds. It is believed that the perfect genetic expression of the production potential in Indian breeds remains un-exploited due to in-adequate nutrition and sub-optimal environment.

5.1 Growth

Growth is one of the traits of great economic importance because of its relationship with market weight, survival and milk production. The growth in Indian breeds of goats has been studied in terms of body weights at different ages.

5.1.1 Birth weight

The average birth weight in different Indian goat breeds ranged from 1.15 ± 0.25 kg to 3.62 ± 0.46 kg (Bhat *et al.*, 1981). Smaller breeds have lighter weight at birth. The average birth weight in Jamunapari kids ranged from 2.20 to 4.80 kg in males and 1.40 to 4.00 kg in females (Khan *et al.,* 1979). Single born kids weighed significantly heavier than twins (Dutt, 1968; Singh, 1973). Seth *et al.* (1968) Moulick and Systrad (1970) reported significant effect of season and sex on birth weight. A low estimate of heritability was reported in Beetal goats (0.19 ± 0.17) by Amble *et al.* (1964) in Black Bengal goats (0.067 ± 0.006) by Guha *et al.* (1968) indicating that birth weight in goats is not highly heritable.

5.1.2 Weaning weight

Body weight at weaning (90 days) was highest (11.5 kg) in Jamunapari kids and lowest (4.8 kg) in black Bengal goats (Bhat *et al.,* 1981). Mukundan (1979) and Kumar (1972) reported a significant effect of sex on weaning weight. It was found that male kids weighed heavier than the female kids.

5.1.3 Yearling weight

The yearling body weight varies from 22.9 kg in larger breeds to 11.8 kg in smaller breeds (Bhat *et al.*, 1981). The body weight at 12 months in crossbred goats was not higher than the purebred goats. The variation in yearling weight was reported to be significant due to season of birth and type of birth in Jamunapari goats (Khan,1980) and in Malabari and its Saanen half breds (Mukundan, 1979). Mittal and Pandey (1978) did not observe any effect of twinning on the yearling body weight in Barbari goats.

Average body weights of some of the Indian breeds at different stages of growth (Table 5.1) under intensive (feedlot) management system indicate diversity among breeds and superiority of Jamunapari over the other breeds with respect to body weight at 12 months and average daily weight gain (107.0 g) between 6 and 9 months of age.

5.1.4 Weight gain

Average daily body weight gains between 3–12 months under range/semi extensive system of management in Beetal and Jamunapari kids were similar (Table 5.2). The Jamunapari and Beetal breeds are, therefore, evidently preferred to, as improver breeds, for contributing growth rate and body weight traits among small and dwarf breeds under range grazing and management.

5.2 Carcass Characteristics

The goat is an important meat animal in India and contributes about 36% of the total meat produced in the country except meat from poultry. Srivastava *et al.* (1968), Johri and Talpatra (1971), Singh and Sengar (1978) and Khan (1980) have reported the dressing percent varying from 44.2 in Jamunapari to 49.5 in Barbari.

The information on the carcass characteristics of different breeds is meagre. Available data from different sources and experiments (Table 5.3) indicate that the dressing percentage ranges between 39.16 and 52.10. The bone percentage ranges between 9.7 and 14.3 in various breeds.

5.3 Milk Yield

Average daily milk yield of some of the Indian breeds (Table 5.4) varied between 35.20 kg (for Bengal-Meat Breed) and 154.6 kg (for Surti-Milch Breed). Jamunapari, Beetal and Surti excelled other breeds in milk production followed by Kutchi, Jakhrana, Sirohi and Marwari. Average lactation length (150–243 days) of milch breeds varies considerably. Khan and Sahni (1982) reported overall means for lactation yield, lactation length and daily yield in Jamunapari goats as 70.30 kg, 106.27 days and 0.686 kg respectively. Lall and Singh (1949) reported that milk production of Jamunapari goats maintained at Etawah farm averaged 0.68 kg/day with a maximum yield of 1.36 kg/day. Similar yields in Jamunapari goats have been reported by other workers (Sahni *et al.*, 1973; Rai and Chorey, 1965). Malabari and Beetal had an average lactation yield of 39 and 105 kg (Mukundan, 1979 and Bhatnagar *et al.,* 1973).

Effect of various non-genetic factors on lactation yield was studied by Khan and Sahni (1982) and reported a significant effect of year. Singh *et al.* (1970b) also reported similar year effect. Highly significant effect of season of kidding on lactation length was reported by Barhat and Chowdhary (1978) in Rajasthani goats and on lactation yield by Singh *et al.* (1970) and Parkash *et al.* (1971) in Beetal goats. It was observed that with the increase in the lactation order, milk yield increased significantly linearly as reported by Prakash *et al.* (1971) in Beetal goats. Rathore (1970) demonstrated in Saanen goats that milk yield increases from 751 kg to 1201 kg from 1 to 3 years of age but it decreased as the age advanced beyond 3 years. Saanen, Alpine and Anglo Nubians were imported for cross breeding of native goats. The performance of their crossbreeds was better than the natives (Bhatnagar *et al.*, 1973). A significant improvement was found by the introduction of genes from these breeds. Repeatability estimates for milk production and lactation length were reported to range from 0.18 to 0.35 and 0.04 to 0.36 respectively (Amble *et al.*, 1964; Kumar *et al.*, 1962). The heritability estimate for first lactation yield and first lactation length was reported as 0.29 and 0.04 respectively (Acharya, 1979).

The milk production performance of crossbreeds (Table 5.5) in the country showed that the Alpine × Beetal crosses were superior to Sannen × Beetal crosses. Alpine × Beetal Halfbreds produced 309.60 kg in a lactation of 240 days. Average milk yield per day in crossbreds with Alpine × Beetal inheritance was 1.28 kg as against 1.03 kg in Jamunapari. Superior Jamunapari does are, however, capable of producing 150–200 kg milk in 90 days. At the Central Institute for Research on Goats (CIRG), Makhdoom, the highest record is 362 kg in a lactation period of 225 days (Roy, 1991) which is still lower by about 100 kg claimed by the breeders in the home tract (Chakarnagar, Etawah, U.P.) in respect of exceptional does.

5.4 Fibre Production

Indian goats produce two of the three types of fibres, *viz.* hair, pashmina and mohair. All the hairy goats produce hair fibre, which is utilized for preparation of ropes and gunny bags. The finest natural fibre, Pashmina (Syn. cashmere, pashm, siflit, liftik, tibit and sivit) is produced by Chegu and Changthangi breeds in India. The pashmina produced by the Indian goats is either gray or white in colour. White Pashmina is preferred to gray Pashmina due to its lustre. The average Pashmina yield per year of Indian goats varies from 111 to 149 g during second to fourth year of production as compared to an average yield of 400 g claimed in case of Soviet Don breeds. The quality of Pashmina of Indian goats is finer (8–14μ diameter) but shorter in staple length (Table 5.6). The shawl prepared out of such commercially superior fibres is claimed to be so fine that it can pass through a finger ring. The goats in the hills such as Gaddi produce hair which is used for making ropes and hair patties. ½ to 1 kg of coarse hair with stale length of 25 cm is obtained from each goat each year. Crossbreeding of the goats with Angora yields Mohair, halfbreds and three quarter breds producing 0.306 and 0.745 kg per animal respectively (Bhat, 1973).

5.5 Genetic Evaluation of Production Performance

Most of the research has been limited in this country, to study the factors affecting the growth and production performances. Singh and Acharya (1982) estimated the genetic and environmental trend in the milk production in a closed flock of Beetal goats. The selection was primarily on the milk yield and breeding bucks were selected on the basis of dam's milk yield. Genetic gain was 1.05% to 1.02% in the first lactation milk yield and lactation per day. The heritability estimates for 90 and 140 days milk yield, total lactation yield and lactation length on the flock of Jamunapari goats maintained at the Central Institute for Research on Goats (CIRG), Makhdoom were 0.34 ± 0.22, 0.38 ± 0.36, 0.30 ± 0.21 and 0.09 ± 0.17 respectively. Genetic and phenotypic correlations among the production traits were medium to high. Improvement in 90 days milk yield in Jamunapari goats in the first and second generations were to the tune of 8.25 and 7.19% respectively. The breeding bucks were primarily selected on the basis of dam's 90 days milk-yield. For maximum gain of genetic potential, the improved management conditions are required and therefore, it is necessary that there should be simultaneous improvement in the level of management so that the genetic improvement brought forth can be fully supported for expression.

5.6 Performance of Native Goats in Tropics and Sub-Tropics

Milk yields of native breeds of tropical and sub-tropical countries (Table 5.7) vary from 15–22 kg in Black Bengal (Bangladesh) to 350–450 kg in Mamber

(Israel). In general native types of tropics are characterized by low yield over shorter lactation period. The fat varied from 4.1 to 5.65%. Most of the breeds in the tropics have not been thoroughly exploited, so that in many instances the true productive potential of individual breeds has not been adequately documented. As a general rule, there is very little controlled breeding but rather, uncontrolled breeding is very common in 18 countries of APHCA Region (Devendra, 1979). Lack of detailed description of the breeds and their genetic potential, has limited the development programmes. The goats of the tropics are multi-purpose animals and are used for meat, milk, fibre and hides. Their varied roles have led to scant attention as high milk producers as is the case in Europe (Sands and McDowell, 1978).

5.7 Performance of European Breeds in Tropical and Sub-Tropical Countries

Four famous and recognized milk breeds, *viz.* Saanen, Alpine, Anglo-Nubian and Toggenberg, of temperate zone have been introduced to the tropical countries. Their yield is given in Tables 5.8 and 5.9. Milk yield in these went down by 25 to 50%, but the average lactation period was almost the same. Import of below average germplasm and environment and nutritional stress may be the major cause of their low milk production in most of the tropical countries. The average lactation length of Alpine and Saanen maintained at National Dairy Research Institute (NDRI), Karnal was 13 to 16% less than that reported from the country of origin, whereas their milk yield tended to decrease from 42 to 44%.

5.8 Performance of Crossbreds

Crossbreeding programmes have been initiated in some of the tropical countries for increasing the milk production of native goats. Milk yield of native and crossbreds is summarized in Table 5.10. Crossbreeding of Malabari with Saanen and Alpine gave good results at Trichur, Kerala.

At NDRI, Karnal, excellent results were achieved from crossbreeding of Beetal (B) goats to Saanen (S) and Alpine (A) under stall-fed condition. Crossbreds produced 65 to 130% more milk than Beetal. Lactation period of crossbred was significantly higher than native. A small number of Alpine × Barbari (9) and Saanen × Barbari (5) crossbred females born at NDRI, Karnal, produced 117 ± 13 and 164 ± 13 kg milk in first lactation over 172 ± 12 and 187 ± 14 days respectively. Milk yield was 75 and 145% higher than contemporary Barbari while lactation period was 33 and 45% respectively.

Comparative performance of Beetal and its exotic crossbred goats which had completed first three lactations at Institute's Farm from 1972 to 1980 showed that first 150 days milk yield of AB, SB, SAB and ASB goats was 67, 112, 60 and 73 kg higher than contemporary Beetal in first lactation; 76, 92, 79 and 89 kg in second lactation and 22, 52, 60 and 155 kg in third lactation respectively. First total lactation yield of crossbreds was 50% (AB) and 122% (SB) higher than of Beetal while their lactation length was 18.5 (ASB) and 33.3% (SB) higher. First 150 days and total lactation yield of first, second and third lactations of crossbreds were significantly higher than of contemporary native goats. Saanen × Beetal goats produced more milk than AB in the first three lactations while ASB goat produced more milk than SAB except in first total lactation yield. Differences were significant. First 150 days milk yield increased from first to third in Beetal, Alpine, Saanen and ASB while in the other crossbreds increase was from first to second lactation.

5.8.1 Age at first kidding

Average age at first kidding of indigenous breeds varied from 430 days (Assam Hill) to 1036 days (Barbari). Substantial variation in age at first kidding among breeds and even within breed has been reported (Misra, 1976; Acharya, 1979).

Age at first kidding of most of the temperate breeds varies from 390 days to 495 days (Rako, 1950 and Bouillon and Ricordeau, 1975). However, European does in the tropics are usually older 19.0 to 25.4 months at the first kidding (Khattar, 1975; Mishra *et al.*, 1976; Gill and Dev, 1972; NDRI, 1980).

Genes of improved dairy breeds are very effective in lowering the age at first kidding in crossbreds. Crosses of exotic (Alpine and Saanen) with indigenous (Beetal and Malabari) does have 15 to 25% lower age at first kidding (Nair, 1978; Mishra *et al.*, 1976; NDRI, 1979).

Heritability estimates for age at first kidding are reported high (0.51) for European goats (Bouillon and Ricordeau, 1975), and from 0.35 ± 0.34 (Mehla, 1979) to 0.54 ± 0.12 (Singh *et al.,* 1970a) in indigenous breeds. The value reported by Mehla (1979) was based on pooled data of Beetal, Alpine × Beetal and Saanen × Beetal. High h^2 estimate suggested that AFK could be changed through selection or by grading up with improved breeds.

5.8.2 Heritability of Milk Yield

For Beetal goats, heritability estimates for first lactation yield, age at first kidding and first kidding interval were 0.25 ± 0.08, 0.54 ± 0.12 and 0.15 ± 0.09 (Singh *et al.*, 1970a). Prakash *et al.* (1971) gave estimates for milk yield in first five lactations as 0.32, 0.29, 0.32, 0.28 and 0.16. Both the estimates were based on fairly large breeding data of 1432 goats progeny of 114 bucks. Amble *et al.* (1964) reported 0.06 to 0.13 h^2 from intro-sire regression of daughter on dam for using part of the data from Hissar flock.

In general, heritability values for milk yield are similar in temperate and tropical goat breeds. Although the heritability estimates for goats vary with sample size and method of calculation, they fall within the range reported for cattle (McDowell, 1972).

Mishra *et al*. (1976) estimated hypothetical and real heterosis for milk yield and lactation length based on small records of Alpine × Beetal crossbreds. The values were 37.7 and 4.9% for milk yield and 18.2 and 5.2% for lactation period respectively.

Singh *et al*. (1970b) constructed selection indices combining age at first kidding, first lactation yield and first kidding interval and concluded that indices based on age at first kidding and first lactation yield gave highest correlation with aggregate breeding value (R = 0.86). It will provide maximum genetic gain in the aggregate genetic worth (H = 48.3). They suggested that selection based on this index will provide higher gains in all the individual characters compared to be selected based on any one of them. Garg *et al.* (1980) studied the impact of selection in increasing the milk production in Beetal goat and observed that pooled milk yield would increase at the rate of 2.8 kg/year. Acharya (1979) reported genetic gain of 1.62 ± 0.94 kg in first lactation yield and 0.009 ± 0.20 kg in yield per day of lactation from the same flock. Genetic gain as percentage of mean was 1.05.

Breed average milk yield may be expected when the total digestible nutrients (TDN) of the diet provide energy equivalent to near 2.5 multiples of maintenance. When feeding corresponds to 1.8 times maintenance, lactation milk yield will be reduced approximately by 45% (McDowell *et al.*, 1975). They suggested that milk production performance of crossbreds (290 kg) is 54% of the expected level of production (800 kg exotic + 290 kg native) which indicated that feeding levels at the farms was equivalent to approximately 1.8 × maintenance. Low level of feeding deteriorates the annual milk production potentiality of an animal as given below:

Influence of feeding level multiple of maintenance	Performance of F_1 crosses estimated milk yield (kg) per lactation
2.45	535
1.80	305
1.50	198
1.15–1.20	50–60

McDowell (1974) estimated that feeds available to animals on small farms varied with season but the overall feeding level during the year was equivalent to 1.15 to 1.20 maintenance requirements. With these low level of feeding, the predicted milk yield for crossbreds or high grades of exotic breed was likely to be on the order of 50 to 60 kg/lactation.

The famous milch breeds of India have not been explored thoroughly at organized farms to identify their genetic potentialities for milk production and for reproductive and production traits. For this, bold and ambitious research and development efforts are required to promote the value of this species.

The impact of selection on increasing the milk production in mature goat is unattractive. If our projections are to increase the milk production from present 150 to 300 kg/lactation through selection in native breeds, assuming the annual expected genetic gain in lactation yield as 1% per year, this would take 100 years. This type of slow improvement programme with the increasing human population would not be encouraging. Therefore alternative way to increase the milk production from present to double in short duration is through the introduction of improved breeds of temperate region for crossbreeding purpose.

For improving the milk production of native milch breeds, farms should be established for maintaining and multiplying bucks at places where environmental constraints are limited. For exploiting the genetic worth of the buck, large scale AI in goats may be practiced. Crossbred male kids of Alpine × Beetal, Saanen × Beetal and three breed crosses (SAB/ASB) which are easily available at NDRI, Karnal, would definitely be superior in increasing the milk yield than most of the Beetal and other breeds. It is suggested that these crossbred kids may be procured by agencies for raising as bucks, which could be used for improving the milk yield of non-descript and poor yielding goats. Moreover these crossbred male kids being genetically superior in body weight and large in size will be useful for increasing the meat production of small size breeds.

5.9 Performance of Angora under purebreeding and crossbreeding

5.9.1 Mohair production and quality

A few Angora goats were imported by the State Governments and introduced in Maharashtra, Himachal Pradesh and Uttar Pradesh in 50's. The results in Maharashtra and Himachal Pradesh were quite encouraging. The crosses of $^3/_4$ bred Angora with local hairy breeds yielded lustrous Mohair. The flocks at Pipalkoti in Uttar Pradesh could not survive apparently due to poor adaptability. In a subsequent experiment, $^3/_4$ and $^7/_8$ crosses of Angora × Deccani, produced 0.183 and 0.454 kg of mohair respectively. Pant (1968) reported 0.475, 0.105, 0.182 and 0.369 kg of mohair from pure bred Angora, and $^1/_2$, $^3/_4$ and $^7/_8$ Angora × Gaddi crosses, respectively, at Himachal Pradesh. Half breds of Angora × Sangamneri, yielded no mohair under the AICRP on Goats at Rahuri Maharashtra (Mishra, 1985), but 75% crosses produced finer fibres and at $^7/_8$ level, the total yield being 1.1 kg/clip as against 1.5 kg/clip in contemporary purebred Angoras. The mohair production was also tried in the arid-regional station of the CSWRI at Bikaner, Rajasthan, by crossing Angora with white Marwari goats (Singh, 1986). The results were similar to those at Rahuri. The mohair yield from the $^3/_4$ and $^7/_8$ crosses was 0.955 and 1.335 kg/annum respectively. The problem of low fertility and high mortality among higher crosses quite discouraged all the efforts to synthesize the Indian mohair strains.

5.9.2 Angora Breeding in India

More than 100 years ago, there was news in Times of India on 1 August 1876 drawing attention to the subject of introducing the Angora breed into this country. However, the first attempt of introducing Angora breed in this country was done by the then Imperial Council of Agricultural Research in 1939. W.S. Read was deputed to South Africa for 6 months to study all aspects of mohair industry and to investigate the possibility of introducing Angora in India. He identified regions, particularly North West India, which would provide environments suitable for profitable farming of mohair-producing goats (Read, 1940 a, b).

A proposal was then put up to import 10 bucks and 150 does as the nucleus of what may prove to be a profitable in New Agro-industrial Enterprise of India. Unfortunately this could not be effected due to shipping difficulties. However, preliminary crossbreeding work was started in Punjab in 1941 with the purebred bucks brought by Capt W.S. Read from South Africa (Nanda, 1942). The progress of crossbreeding was hindered during the war because purebred bucks could not be imported from abroad. During this period, only three good specimens of Angora arrived from South Africa (Handa, 1946).

The Angora breeding scheme was shifted from Cangra district of Punjab to Uttar Pradesh at Pipalkoti as there was a feeling amongst some workers that size of the crossbred goats had gone down in Cangra district (Lall, 1968). These animals were then shifted to Gwaldom located at altitude of 1900 m and had rainfall of 250 to 260 cm. This scheme was terminated on 31 March 1956 by the Indian Council of Agricultural Research, and was handed over to the State Department of Animal Husbandry, Uttar Pradesh.

The crossbreeding work of crossing Zonal Deccani doe's with Angora bucks was also discarded from l956 to 1965 at Pune, Maharashtra. All these experiments suffered due to non-availability of sufficient Angora bucks and ended with fragmentary picture of crossbreeding results.

5.10 Review of Work Done

The extensive work done in various countries on Angora was reviewed by Pahl, 1978. However, the salient features relating to the crossbreeding work with Angora is only reviewed and the same is given below:

The crossbreeding of local goat with Angora was started in South Africa as early as 1838 (Marincowitz, 1958) and took more than 21 years for the farmers to evolve a new breed of Angora by applying judicious selection. Read (1941) reported that the marketable crossbred hair with pronounced mohair characteristic was obtainable in the second generation. Mohair from the third crossing commanded good price. In general, crossbreeding takes about generations to produce Angora goat.

The performance of the crossbreds (Gaddi × Angora) at Gwaldom was studied in detail by Pant and Kapri (1968), Pant (1968, 1969) and Lall (1968), and at Poona (Deccani × Angora) and at Gwaldom was reviewed by Khera (1965). The observations were however based on small number of observations. The findings indicated that halfbred Angora did not yield mohair. On the contrary, the halfbred from hairy breed yielded less hair than indigenous hair goat. There was some improvement in the mohair yield of $^{3}/_{4}$th Angora, *i.e.* 0.250 to 0.350 kg/year. The yield was further increased to a greater extent in $^{7}/_{8}$th Angora, *i.e.* 0.500 to 0.900 kg. There was no further improvement in mohair yield in $^{15}/_{16}$th Angora. The medullation percentage was 56.43, 33.57, 19.76, 7.41, 5.82 and 4.21% in Gaddi, $^{1}/_{2}$ Angora, $^{3}/_{4}$ Angora, $^{7}/_{8}$th Angora, *inter-se* $^{7}/_{8}$ Angora and Angora respectively.

The s/p ratio improved to a great extent of halfbred Angora, *i.e.* 0.187 in Gaddi to 5.58 in halfbreds. The s/p ratio graduall1y increased with increase in Angora inheritance in further crosses. At Poona centre, the s/p ratio in halfbred Angora did not improve to the extent observed at Gwaldom. The s/p ratio observed was 1.6, 2.6 and 6 in $^{1}/_{2}$, $^{3}/_{4}$ and $^{7}/_{8}$th Angora respectively.

On *inter-se* mating between $^{7}/_{8}$th Angora at Gwaldom the quality of mohair was as good as in $^{7}/_{8}$th; however, the s/p ratio of 8.30 in $^{7}/_{8}$th Angora declined to 6.78 in interbreds. Unfortunately, the further work remained to be done.

Table 5.1 Body Weight and Weight Gain of Some Indian Breeds under Intensive System of Management (Feed-lot Experiment on Males)

Breed	Body weight (kg)				Av. daily weight gain (g)				Source
	3 M	6 M	9 M	12 M	3-6 M	6-9 M	9-12 M	3-12 M	
Jamunapari	12.60 ± 0.54(11)	18.51 ± 0.73(11)	28.15 ± 1.19(11)	35.71 ± 1.54 (16)	66.0	107.0	84.0	86.0	Annual reports, CIRG(1985–91)
Barbari	8.16 ± 6.37 (21)	14.40 ± 0.61(21)	20.58 ± 0.65 (21)	24.60 ± 0.70(21)	69.0	69.0	45.0	61.0	Annual reports, CIRG (1985–91)
Black Bengal	5.40 ± 0.90	8.96 ± 1.48	11.50 ± 1.77	–	39.6	28.6	–	–	Mishra (1985)
Sirohi	17.24 ± 0.75(22)	23.95 ± 0.84 (23)	28.71 ± 1.47(7)	33.42 ± 2.05(6)	75.0	53.0	52.0	60.0	Mishra (1985)
Marwari	15.15 ± 1.01(14)	23.24 ± 1.05(11)	28.25 ± 4.82(2)	32.00 ± 4.50(2)	90.0	56.0	42.0	62.0	Mishra (1985)
Kutchi	16.21 ± 0.96(14)	23.80 ± 1.00(14)	28.33 ± 1.65(6)	32.83 ± 1.87(6)	85.0	50.0	50.0	62.0	Mishra (1985)

M = Month

Figures within parentheses indicate number of observations.

Table 5.2 Body Weight and Weight Gain of Some Indian Breeds under Range/Semi–Extensive System of Management

Breed	Body weight (kg)				Av. daily weight gain (g)				Source
	3 M	6 M	9 M	12 M	3–6 M	6–9 M	9–12 M	3–12 M	
Black Bengal	4.74 ± 0.11	6.81 ± 0.21	9.30 ± 0.36	12.08 ± 0.18	22.88	–27.66	30.88	27.14	Singh (1992)
Jamunapari	9.11 ± 0.15	12.69 ± 0.28	17.54 ± 0.51	22.48 ± 0.64	39.77	53.28	54.88	49.51	Annual reports
Beetal	8.49 ± 0.11	12.45 ± 0.19	16.26 ± 0.43	22.47 ± 0.65	44.00	42.30	69.00	51.77	Singh (1992)
Sangamneri	7.28	10.06 ± 0.39	13.45 ± 0.15	17.33 ± 2.19	30.88	37.77	43.11	37.22	Singh (1992)
Malabari	6.05 ± 0.09	8.85 ± 0.12	11.62 ± 0.15	14.39 ± 0.20	31.11	30.77	30.77	30.88	Singh (1992)
Jakhrana	8.88 ± 2.09	14.60 ± 0.49	19.97 ± 0.52	24.68 ± 0.75	64.00	60.00	52.00	59.00	Tyagi *et al.* (1992)
Osmanabadi	7.34 ± 0.05	11.07 ± 0.05	15.12 ± 0.24	–	41.44	45.00	–	–	Singh (1992)
Chegu	7.52 ± 0.11	10.67 ± 0.17	13.06 ± 0.17	16.89 ± 0.28	35.00	26.55	42.5	23.03	Singh (1992)
Changthangi	8.44 ± 0.30	10.22 ± 0.20	12.72 ± 0.33	–	19.77	27.77	–	–	Misra (1985)

M=Month

Table 5.3 Meat Production Profile in Different Indian Breeds at 9 Months of Age (Male)

Breed	No. of observations	Pre-slaughter weight (kg)	Hot carcass weight (kg)	Dressing (%) on Pre-slaughter weight	Bone (%)	Source
Sirohi	4	37.40 ± 0.62	21.80 ± 0.10	52.10 ± 0.80	13.80	Mishra (1989)
Kutchi	3	23.10 ± 1.34	11.70 ± 0.69	50.50 ± 0.70	14.30	Mishra (1989)
Sangamneri	5	11.50 ± 1.46	5.08 ± 0.74	44.2	–	Mishra (1989)
Beetal	6	18.20 ± 1.02	7.94 ± 0.51	43.56 ± 1.06	–	Singh and Sengar (1990)
Black Bengal	17	10.00 ± 0.49	4.47 ± 0.24	44.90 ± 0.58	–	Singh and Sengar (1990)
Barbari	51	14.42 ± 0.30	6.55 ± 0.38	45.4 ± 0.41	10.60	Prasad *et al.* (1992)
Jakharana	4	19.60 ± 0.64	8.99 ± 0.30	45.9 ± 1.24	9.7	Prasad *et al.* (1992)
Jamunapari	2	14.9 ± 0.45	6.47 ± 0.43	39.16 ± 0.84	11.4	Prasad *et al.* (1992)

Table 5.4 Milk Production in Some of the Indian Breeds

Breeds	Lactation yield (kg)	Lactation length (days)	Average milk yield/day (kg)	Reference
Jamunapari	144.54 ± 7.34 (48)	153.08 ± 8.65 (48)	1.03	Annual reports CIRG (1985-91)
Beetal	150.00 ± 1.56 (206)	169.00 ± 6.0 (206)	0.88	PC Report (1989)
Surti	154.6 ± 6.1	150.0	1.02	PC Report (1989)
Jakhrana	119.32 ± 13.91 (6)	245.27 ± 3.08 (6)	0.48	PC Report (1989)
Sirohi	113.62 ± 2.43	193.92 ± 2.33	0.58	PC Report (1989)
Kutchi	124.06 ± 2.84 (162)	95.20 ± 2.36 (162)	0.63	PC Report (1989)
Marwari	90.02 ± 4.10 (94)	178.17 ± 4.39 (94)	0.50	PC Report (1989)
Barbari	84.18 ± 1.0 (87)	111.23 ± 5.0 (87)	0.75	Annual reports CIRG (1989)
Bengal	35.20 ± 1.36 (69)	111.20 ± 2.25 (69)	0.31	PC Report (1989)

Table 5.5 Milk Yield in Crossbreds

Genetic Group	Lactation yield (kg)	Lactation length (days)	Average milk yield/ day (kg)	Reference
Jamunapari × Black Bengal	46.15 ± 1.24 (226)	124.10 ± 2.7 (226)	0.37	PC Report (1989)
Beetal × Black Bengal	49.96 ± 1.58 (131)	127.10 ± 3.4 (131)	0.39	PC Report (1989)
Saanen × Beetal	309.60 ± 13.5 (462)	240.80 ± 0.80 (462)	1.28	Acharya (1992)
Alpine × Beetal	257.10 ± 6.4 (462)	229.90 ± 0.37 (462)	1.12	Acharya (1992)
Saanen × Malabari	127.27 ± 12.42	190.50 ± 6.5	0.66	Acharya (1992)
Alpine × Malabari	98.01 ± 10.57	186.90 ± 4.2	0.52	Acharya (1992)

*140 days milk yield.
Figures in parentheses indicate the number of observations.

Table 5.6 Fibre Production of Indian Pashmina Goats and Angora × Sangamneri Crosses

Characters	Pashmina annual yield (g)			Mohair annual yield (kg)		
	Changthangi	Chegu		Angora	¾ Angora	7/8 Angora × Sangamneri
First year	67 (104)	86 (413)		–	–	–
Second year	142 (112)	98 (399)	Greasy overall	1.323 ± 0.04 (724)	0.423 ± 0.02 (3796)	0.996 ± 0.05 (1271)
Third year	154 (135)	114 (289)	–	–	–	–
Fourth year	152	121 (166)	–	–	–	–
Staple length (mm)	40.00 ± 0.60 (59)	54.91 ± 0.60 (324)	103.2 ± 1.9 (173)	61.2 ± 2.1 (1070)	94.1 ± 3.3 (382)	–
Fibre diameter (μ)	12.72 ± 0.17 (59)	12.14 ± 0.08 (324)	21.19 ± 0.25 (173)	17.95 ± 0.28 (1070)	21.03 ± 0.44 (382)	–
Medulation (%)	9.6	24.0	2.72 ± 0.22 (166)	12.05 ± 0.63 (1056)	4.80 ± 0.60 (360)	–
Clean yield (%)	63.8	76.0	84.98 ± 0.66 (173)	89.78 ± 0.95 (809)	93.0 ± 0.60 (382)	–

Source: Terminal Report (1985); Patil (1992)

Figures within parentheses indicate the number of observations.

Table 5.7 Milk Production Performance of Native Goats in Tropics and Sub-Tropics

Breed	Location	Lactation yield	Lactation length	Lactation	Fat	Source
Mamber	Israel	Mat.	350–450	–	–	Mon. J. Brit GoatSci. (1972)
Boer	S. Africa	–	160	120	5.65	Ukecker mann *et. al.* (1974)
Native	Greece	–	93	210	–	Vlachos (1975)
Native	Korea	–	91	–	–	Lee *et. al.* (1974)
Criollo	Venezuela	–	40–60	200	–	Garcia *et. al.* (1972)
Crioilo	Puerto Bico	1st	129	218	–	Sanfiorenzo (1957)
	Puorto Rico	Mat.	190	272	–	Sanfiorenzo (1957)
Kambing	Katiang Malaysia	–	90	126	–	Mahmud and Devendra (1966)
Native	Niger	–	100	–	–	FAO (1977)
Native	Nigeria	–	110	–	–	FAO (1977)
Beetal	Pakistan	–	147	186	4.9	Devendra (1979)
Barbari	Pakistan	–	100	100	41	Devendra (1979)
Bihaneri	Pakistan	–	75	100	5.1	Devendra (1979)
Chopper	Pakistan	–	75	105	5.2	Devendra (1979)
Dera Dean	Pakistan	–				
Panah	Pakistan	–	205	130	4.9	Devendra (1979)
Demani	Pakistan	–	107	105	–	Devendra (1979)
Kamori	Pakistan	–	227	120	4.7	Devendra (1979)
Local Hill Goat	Nepal	–	59	106	4.8	Pradhan (1979)
Terai Goat	Nepal	–	87	126	4.0	Pradhan (1979)
Black Bengal	Bangladesh	–	15–22	90	5.5	Pradhan (1979)
Neldi	Iran	–	248	249	–	Beheshi & Saleh (1979)
Native	Philippine	–	66	187	–	Chamehey *et. al.* (1989)

Table 5.8 Milk Production Performance of European Breeds in Tropical and Sub–Tropical Area

Breed	Location	Lactation yield	Lactation length	Lactation	Fat	Source
Saanen	France	Ist	430	260	3.30	Ricordeau (1963)
	France	Mat.	527	279	3.30	Ricordeau (1963)
	France	All	575	241	–	Patre (1966)
	France	–	530	270	–	Ricordeau (1964)
	France	–	700–900	300	–	Gall (1975)
	Switzerland	Ist	530	264	–	French (1970)
	Switzerland	Mat.	684	273	–	French (1970)
	England	–	1277	365	–	Devendra & Burns (1970)
	United States	All	921	276–305	3.60	Dickinson & King (1977)
	Britain	–	650	300	–	Great Britain (1970)
Alpine	France	–	600	270	–	Ricordeau (1964)
	France	Ist	470	260	3.37	Ricordeau (1963)
	France	Mat.	583	279	3.37	Ricordeau (1963)
	United States	All	916	276–305	3.50	Dickinson and King (1977)
Anglo-Nubian	England	–	989	365	–	Devendra and Burns (1970)
	United States	–	752	276–305	4.50	Dickinson and King (1977)
Toggenberg	Switzerland	Ist	468	266	–	French (1970)
	Switzerland	Mat.	605	281	–	French (1970)
	United States	–	878	276–305	4.0	Dickinson and King (1977)
	The Netherland	All	1043	294	–	
La Mancha	United States	All	800	276–305	4.0	Dickinson and King (1977)
White Polled	Czechoslavakia	All	800	290	3.80	Jurik (1971)
	Czechoslavakia	All	997	300	–	Horak & Pindak (1969)

Table 5.8 (Contd.)

Breed	Location	Lactation yield	Lactation length	Lactation	Fat	Source
Banat	Romania	–	532	246	–	Ciolca *et al.* (1959)
White German	Germany	Mat.	702	282	4.18	Altenkirch (1957)
improved	Germany	Mat.	910	–	3.35	Helmke (1952)
Fawn German	Germany	Mat.	616	270	2.95	Aitenkirch (1957)
improved	Germany	Mat.	771	–	3.63	Helmke (1952)
Thruingian	W. Germany	–	658	381	–	Aitenkirch (1957)
Dole	Norway	All	548	7250	3.70	Ronningen (1965)
	Norway	–	560	–	3.38	Trodahl (1975)
Granda	Spain	–	600–700	275	–	Gall (1975)
Malaga	Spain	Mat.	309–360	180	–	Saraza Ortiz (1952)
Damascus	Cyprus	All	558	263	–	Louca *et al.* (1975)
	Lebanon	All	392	238	–	Choueiri (1973)
Kilis	Turkey	–	281	260	4.70	Yarkin and Eker (1961)
	Turkey	–	250–300	150–210	–	Yarkin and Sonmez (1962)
Poitou	France	–	600	270	–	Rocordeau (1964)
Local	Britain	–	907	365	–	Great Britain (1961)
Local	Hungary	–	664	265	–	Bodo (1961)
Local (4961)	France	–	575	241	–	Patre (1966)
Local (2148)	Bohemia and Moravia (France)	–	1080	299	–	Harak (1969)
Local	France	–	450	260	–	Disset *et al.* (1969)

Table 5.9 Milk Production Performance of European Breeds in Tropical and Sub–Tropical Area

Breed	Location	Lactation yield (kg)	Lactation length (days)	Source
Sannen	Israel	650	200–335	Gall (1975)
	Israel	761	240	Mon. J. Brit. Goat. Sci. (1972)
	Israel (Mat.)	990	336	Mon. J. Brit. Goat. Sci. (1972)
	South Africa	1011	263	Hofmayer (1972)
	Puerto Rico	292	270	Sanfiorenzo (1962)
	Venezuela	295	278	Garcia *et al*. (1972)
	Cyprus	536	240–300	Richards (1959)
	Australia	538	300	Pegg (1957)
	India	287	244	NDRI (1980)
Alpine	Venezuela	232	264	Garcia *et al.* (1972)
	Trinidad	274	209	Devendra *et al*. (1969)
	South Africa	904	263	Hofmayer (1972)
	India	310	239	Gill and Dev (1972)
	India	319	252	NDRI (1980)
Anglo-Nubian	Trinidad	143	124	Devendra *et al.* (1969)
	Malaysia	250–300	300	Devendra (1962)
	Philippines	167	294	Villegas (1932)
	Venezuela	155	224	Garcia *et al.* (1970)
	Mauritius	191–285	210	Maule (1966)
	India	289	240	Gill and Dev (1972)
	Egypt	60–70	60	Gall (1975)
Toggenberg	Venezuela	283	283	Garcia *et al*. (1970)
	Tanzania	250	212	Maule (1966)
	South Africa	532	263	Hofmayer (1972)

Table 5.10 Milk Production Performance of Native and Crossbreds in Tropics

Location	Types of mating	Lactation milk yield (kg) level of exotic inheritance: Native	1/2 (F1)	1/2 (F2)	3/4	7/8	Source
Korea	Sannen × Native	91	288	–	355	373	Lee *et al.* (1974)
Puberto-Rico	Saanen × Native	188	245	–	285	273	Sanfiorenzo (1957, 1962)
Turkey	Saanen × Kills	261	710/705	–	718/305	–	Eker *et al.* (1976)
Malaysia	Anglo × Nubian Kambing Katjand	90	296	237	–	–	Mahmud &Devendra (1966)
India	Saanen × Beetal (NDRI)	164 ± 9/185 ± 7	306 ± 16/239 ± 10	–	372 ± 24/330 ± 18	–	NDRI (1980)
India	Alpine × Beetal (NDRI)	164 ± 9/185 ± 7	306 ± 16/239 ± 10	–	372 ± 24/330 ± 18	–	NDRI (1980)
India	Saanen(Alpine × Beetal)	164 ± 9/185 ± 7	–	–	268 ± 32/231 ± 21	–	NDRI (1980)
India	Alpine (Saanen Beetal)	164 ± 9/185 ± 7	–	–	289 ± 16/231 ± 7	–	NDRI (1980)
	Alpine × Barbari	67/129	177/172	–	–	–	NDRI (1978)
	Saanen × Beetal	67/129	164/187	–	–	–	NDRI (1978)
India	Alpine × Beetal	163 ± 7/186 ± 6	205	–	–	–	Khattar (1975)
India	Saanen × Malabari	38/101	129/241	–	–	–	Nair (1978)
India	Alpine × Malabari	38/101	155/241	–	–	–	Nair (1978)
West Germany	Saanen × Carpathian	247/280	532/248	–	–	–	Ciolca *et al.* (1959)
Bulgaria	Saanen × Local	330/238	303/237	–	–	–	Kaditski (1957)
West Germany	West Germ Improved	247/280	532/247	–	–	–	Ciolca *et al.* (1959)

CHAPTER 6

GENETICS

6.1 Chromosomal Profile

Indian goats exhibit a model diploid number of 60 chromosomes similar to that reported in Saanen (Evans *et al.*, 1973) and Korean native (Yeo, 1984).

The karyotype consists of 29 pairs of acrocentric autosomes and 1 pair of acrocentric X-chromosomes in female animals. The acrocentric nature of X-chromosome was similar to previously published observations in various breeds of goats (Evans *et al.*, 1973; Hansen, 1973; Bhatia and Shanker, 1989, 1990).

Biarmed as well as acrocentric Y-chromosomes were discerned in male animals. The male component in goats contained a small biarmed Y-chromosome which varied from submetacentric (Khavary, 1973) to metacentric in its morphology (Ford *et al.*, 1980; Berardino *et al.*, 1987; Bhatia and Shanker, 1991). Similar cases of both types of Y-chromosome being present in Malawi zebu cattle and cattle in Brazil were reported by Meyer (1984) and Pinheiro *et al.* (1980) respectively. Polymorphism of Chianina and Brahman crossbred steers was reported by Eldridge and Blazak (1977).

Polymorphism of Y-chromosomes is considered to occur through 2 cytogenetical events, *viz.* pericentric inversion or acquisition of a heterochromatic short arm, whose respective influences on the size of the Y-chromosome differ considerably (Pathak and Kiefer, 1979). The overall length of an element following a pericentric inversion is expected to remain constant while the addition of heterochromatin increases the total length of the chromosome. To ascertain the role of either of these cytogenetical events in the present morphological change of Y-chromosome, comparison was made in relative lengths of acrocentric and biarmed Y-chromosomes. It revealed marked constancy in the size of the dimorphic Y-chromosomes ($p<0.01$). Change in the morphology of the Y is entirely because of pericentric inversion (Pathak and Kiefer, 1979). The improved resolution of the G-banding pattern of the Y-chromosome would hopefully facilitate the strengthening of this observation in future studies.

Little information is available on the karyotypic characteristics of Indian goats. The present study compares the cytogenetic parameters of Jamunapari, Barbari, Beetal, Black Bengal, Sirohi, Jakhrana, Marwari and Kutchi breeds (all horned). Somatic chromosome preparations were obtained from lectin stimulated blood leukocyte cultures of 306 animals of these breeds. The diploid chromosme number in all the breeds is 60 (29 pairs of acrocentric autosomes and a pair of sex chromosomes). The X-chromosome is the largest acrocentric element while the Y-chromosme is the smallest and the only submetacentric element in the karyotype. Analysis of relative lengths and ideograms revealed no significant differences between sexes and among breeds. X-chromosome contributed over 5% and the Y-chromosome about 1% to the total genome length. Comparison of C and G banding patterns indicated no apparent band variation between breeds. NORs were localized on telomeres of 4 pairs of large and 1 pair of small autosomes in all breeds.

6.2 Biochemical Polymorphism

There are three haemoglobin phenotypes *viz* Hb AA, Hb AB and Hb BB attributed by two alleles at the locus controlling haemoglobin synthesis. Freqeuncy of haemoglobin types AA in Indian breeds and their crosses with Alpine and Anglo-Nubian has been observed to be AA types in 90% females and which had significantly lower age at first kidding.

There is clear cut bimodality in respect of potassium concentration in Barbari goats and no bimodality has been reported in Jamunapuri goats. There are two transferrin variants and three phenotypes (Tf AA, Tf AB and Tf BB) in Beetal, Barbari and their crosses. The gene frequency of TfA and TfB were 0.473, 0.527, 0.125, and 0.875, 0404 and 0.596 in Barbari, Beetal and their crosses respectively. The TfB was more frequent than TfA.

There are two alkaline-phosphatase phenotypes designated as fast (FA) and slow (FO) in Jamunapari, Barbari amd Black Bengal. There are two variants in Barbari and Jamunapari goats.

6.2.1 Milk protein polymorphism

Milk samples were collected from Barbari (34) and Jamunapari (24) goats to study a lactoalbumin (a-La) and b-lactoglobulin (b-Lg) polymorphism. Polyacrylamide-gel electrophoresis was carried out for fractionation of proteins from the milk. a-La in both breeds was influenced by 3 co-dominant alleles *i.e.* a-LaA, a-LaB and a-LaC with 5 different genotypes AA, AB, AC, BB and BC. The genotypic frequencies of different genotypes were 11.76, 20.53, 32.35, 20.59 and 14.70% in Barbari, and 4.17, 37.50, 16.67, 16.67 and 25.0% in Jamunapari goats respectively. b-lactoglobulin was governed by 4 alleles *i.e.* b-LgA, b-LgB, b-LgC and a-LgD. Six different genotypes AA, AB, AC, AD, BB and CD were common to both the breeds with genotypic frequencies as 8.82, 29.41, 14.70, 5.88, 5.88 and 8.82% in Barbari, and 4.17, 33.33, 29.17, 25.00, 4.17 and 4.17% in Jamunapari respectively. Three additional genotypes observed in Barbari goats were BC, BD and CC with the gene frequencies of 5.88, 17.65 and 2.95% respectively.

Milk samples from lactating does of Jamunapari and Barbari breeds, maintained at this Institute, were collected and casein was separated from the whole milk to study the components of casein using the high performance liquid chromatography (HPLC). Milk casein was separated into 5 components in a single run using gel-permeation phase of HPLC in Jamunapari and Barbari goat breeds. Milk casein was analyzed at the flow rate of 0.5 ml/min with the chart speed of 2 mm/min and 0.04 sensitivity in both the breeds. The casein sample injected was 1.5 ml in Jamunapari and 1.0 ml for Barbari breed. The elution time was recorded at 280μm absorbance. The main 5 isolated components under HPLC were aS-Cn, aS-Cn, aS-Cn, b-Cn and k-Cn. These components were compared with corresponding bands in polyacrylamide-gel electrophoressis (PAGE), although the resolution of bands was not so clear in their appearance. The subcomponents of a–Cn fused in a single band as observed on PAGE could be clearly distinguished into 3 peaks on HPLC, *i.e.* aS-Cn, aS_1-Cn and aS_2-Cn. Remaining b-Cn and g-Cn bands were correspondingly identical to the last peak in similar conditions. The elution time differs in both breeds showing the difference in their molecular weight.

The whey proteins (a-La) and (b-Lg) in milk of Jamunapari and Barbari goats were analysed with high performance liquid chromatography (HPLC) under the gel-permeation system. Flow rate, chart speed and sensitivity were 1 ml/min, 2 mm/min and 0.04, respectively; elution time was recorded at 280 μm absorbance. Samples were injected with the concentration of 100 for a-La; for b-Lg the sample concentration was adjusted at 0.5 ml and 0.25 ml in Jamunapari and Barbari respectively and 3 distinct peaks were noticed in both breeds.

The genetic variability of milk proteins in six Indian goat breeds was studied. The milk samples were collected from individual Jamunapari, Barbari, Marwari, Sirohi, Beetal and Jakhrana goats respectively, from the natural habitat of each breed. The main variation in electrophoretic patterns was observed in a S-C_n, k-case in and aS_2-C_n loci between breeds. The other loci β-casein, β-lactoglobulin, a-Lactoalbumin showed little variation between breeds. The relationship of total casein, total protein and calcium with milk protein variants were also analysed in all these Indian goat breeds. Breed had significant ($p<0.01$) effect on calcium and total protein yield. The aS-C_n AA and AF genotypes had lowest casein, calcium and protein contents. Least squares analysis of variance revealed that the aS_1-C_n genotype had significant ($p<0.05$) effect on casein and protein contents and k-Cn locus had significant ($p<0.05$) effect on calcium content of milk.

The other loci β-casein, k-casein, β-Lg, a-La were without much variation between breeds. The aS_1-C_n A allele was dominant in all breeds and the majority of goats were homozygous AA. The frequency of the aS_1-C_n A allele (which was directly related to higher casein yield) was 0.90, 0.90, 0.93, 0.78 and 0.86 in Jamunapari, Barbari, Marwari, Sirohi and Jakhrana goats, respectively and was highest in Marwari and lowest in Sirohi goats. The F allele was found as heterozygous AF in all breeds except Marwari. The F allele was also observed as homozygous FF (which was directly related to the low production of casein content in goat milk) in two Jamunapari goat. In the aS_2-C_n locus, the most prevalent allele was the A type in goat breeds and mostly as homozygous AA. The genotypic frequency of aS_2-C_n was similar in Barbari, Marwari and Sirohi goats but less frequent in Jamunapari (0.64) and Jakhrana goats (0.57). For Marwari goats the β-C_n fraction was observed in all 100 tested individual milk sample. A faster migrating allele in the β-C_n locus was observed in Jamunapari goats indicating the existence of an additional β-C_n allele (β-C_n B) in the electrophoretic pattern with a frequency of 0.0036. Monomorphic patterns were observed in the k-C_n locus and found as homozygous AA in all five Indian breeds. The k-C_n locus was absent in some individuals. All populations were at H.W equilibrium at the aS_1-C_n and β-Lg loci.

6.2.2 Serum lysozyme-marker for immunocompetence

Serum lysozyme activity reflects the homeostatic expression of reticuloendothelial system, which is one of the most fundamental defense mechanism against infection. The significant role of lysozyme action in the body's defense against bacterial infection has been confirmed in different studies (Lie, 1980). The hydrolytic activity of lysozyme damages the cell wall of gram (+ve) bacteria and to some extent on gram (–ve) bacteria in cooperation with antibody/complement system. The resulting hydrolytic products further activate and stimulate cells of immune system (Jolles and Jolles, 1984). A distinct bimodal distribution of serum lysozyme activity has been observed in some cattle population (Lie and Solube, 1983) suggesting that the activity of this enzyme is influenced by at least one major gene. Therefore the serum lysozyme pattern and its genetic control in goats was analyzed with an aim to enhance general disease resistance. All animals analysed were clinically healthy and exhibited normal behaviour. About 128 individuals of 6 month of age belonging to 12 half sib families of Jamunapari goats were characterized for lysozyme activity and all the sires had at least 7 offspring in each family. Lysozyme activity was measured in blood serum using Modified Lysoplate assay (Lie *et al.,* 1986). High and low lysozyme phenotypes were recognized keeping the 0.7mg/ml as dividing point. A fixed model for estimation of least squares were applied to test the effect of sire, sex on lysozyme activity on a within sire family basis.

Blood samples from Barbari and Jamunapari were collected and DNA was isolated from each sample using standard freeze fraction method. Samples were digested with the *HaeIII*, *Hinfl*, *BglI* and *BstNI* at the recommended temperature and electrophoresed using 1% agarose gel. Size fractioned DNA samples were transferred on to Hybond N+ nylon membrane and hybridized with a P32 labeled Bkm 2(8) at 60°C for 16 hours. Blots were washed in 2XSSC + 0.1% SDS and exposed on to phosphor imager and X-ray filter subsequently. The distinct bands were distributed over the range of 0.5–23.0kb. All the bands showing similar sizes and intensities were considered to be identical. DNA fingerprints were scanned and fragments were calibrated for size by using λ *HindIII* marker. The study revealed that the average number of total bands in Barbari and Jamunapari goats were 19.7, 21.8 in *HinfI* profiles and 22.2 and 21.4 in *Hae III* profiles and 14.87 and 15.38 in *BstNI* profiles respectively. The analysis of fingerprint data demonstrated that band distribution depends on sex, individual and restriction enzyme. Analysis of the finger printing data also demonstrated that most of the bands were highly conserved. The maximum number of bands observed in the *HaeIII* DNA profiles of the two goat breeds, were almost evenly distributed along the length of DNA fingerprint. The goat as a species also showed species-specific distribution of high intensity major bands in their genetic profiles at about 0.8 kb. The *HinfI* digested DNA-profiles showed female specific band in both the breeds at about 23 kb size. A 5.5 kb species specific band is also observed in this DNA profile. In *BstNI* profile, a specific band was observed in Jamunapari female at about 23 kb range. Moreover, a common band at about 800 bp was observed as species specific band for all goats. Conserved bands of high intensities in the fingerprint of *BstNI* profile was also observed in this two geographically distinct breed of different region. The *Bgl I* profile of both Jamunapari and Barbari breed showed female specific bands at higher molecular size. However, two Jamunapari males showed some specific bands. Similarly, female specific bands in both the breeds was also observed in *Hae III* digested DNA profile and a species specific band was also observed in this case at about 12 kb size. The similarity index was calculated and observed that 0.44, 0.56 and 0.30 in *HinfI*, *BstNI* and *HaeIII* digested DNA profile, respectively.

6.3 Genetic Variability of Indian Goats by Microsatellite Analysis

Microsatellite analysis of samples of Jamunapari, Black Bengal, Pashmina, Sirohi,

Local, Jakhrana, Barbari, Marwari, Kutchi and Osmanabadi breeds were carried out with 23 markers. Out of 23 markers tested in Indian goats, 4 markers did not show amplification in Indian goat breeds. The total 303 samples of 8 goat breeds were completed for all the markers. Analysis was made on samples belonging to 8 Indian goat breeds namely Kutchi, Jakhrana, Sirohi, Jamunapari, Barbari and Marwari, Black Bengal and Pashmina using 19 microsatellite markers. Out of all microsatellites analysed, 2 of them showed monomorphic patterns and remaining markers showed polymorphism. Among the polymorphic markers, BM4621 showed 22 alleles, Jamunapari goats showed highest 15 alleles at this locus. Allele 132 of the locus BM4621 was observed with high frequency. Rare allele of 106 bp was observed in Jamunapari goat and 148 bp was observed in Sirohi. NRAMP marker showed 15 alleles of which, Jakhrana showed 12 alleles, and Jamunapari showed 10 alleles. Allele 234 was found to be more frequent than others. A rare allele of 248 bp was observed in Jamunapari goats. In ILSTS005 marker showed 12 alleles, of which Sirohi and Black Bengal showed 10 alleles each. Allele size of 181 was found to be predominant. About 12 alleles were observed for the locus OarAE101, of which Barbari and Marwari showed 10 alleles each. Allele 98 and 102 were observed with high frequency. A rare allele of 140 bp size was observed in Marwari goats. A total of 12 alleles were observed for the locus ETH225 of which Jakhrana showed 10 alleles. Allele 150 was found to be predominant. A rare allele of 138 bp was observed in Black Bengal goat. Locus Oar HH56 showed a total of 14 alleles. Jakhrana showed 12 alleles. Allele size of 162 bp was found to be the most frequent. About 15 alleles were observed in the locus IDVGA7. Of which, Jakhrana showed 11 alleles, Sirohi showed 13 alleles, Marwari showed 10 alleles, Jamunapari showed 15 alleles, and Barbari showed 9 alleles. A total of 15 alleles were observed in BM6526, of which 12 and 13 alleles were observed in Marwari and Sirohi, respectively. Allele size 180 bp was a rare allele in Sirohi goats. Locus INRABERN 192 showed 16 alleles, of which Jamunapari showed 10 alleles and Marwari showed 8 alleles. Allele size of bp 206 and 208 were rare allele in Jamunapari goats. A total of 15 alleles were found at TGLA 40 locus of which 11 alleles were observed in Barbari and Jakhrana goats each. Allele size 200 bp was rare allele in Barbari goats. FCB 48 locus showed 9 alleles, of which Barbari showed 9 alleles and Jamunapari, Black Bengal, Jakhrana, Marwari and Sirohi showed 8 alleles each. Allele of bp 168 was rare allele in Barbari goats. About 13 alleles were observed at locus BM143, of which 11 alleles were observed in Jakhrana goats and 11 allele were observed in Sirohi goats. Locus SRCRSP5 showed 19 alleles, of which 12 alleles were observed in Black Bengal, Jakhrana and Sirohi goats each. Allele size of 150 bp was rare allele in Sirohi goats. About 12 alleles were observed at SRCRSP 6 locus, of which 10 alleles were observed in Jamunapari, Jakhrana and Sirohi goats. SRCRSP 9 locus showed 22 alleles, of which Jakhrana showed 15 alleles, Black Bengal showed 12 alleles. Allele size of 112 and 116 were rare allele in Jakhrana goats. A total of 9 alleles were observed at SRCRSP 10 locus, of which Jamunapari and Black Bengal showed 9 alleles each, Jakhrana, Marwari and Sirohi showed 8 alleles each. Oar HH 62 locus showed 12 alleles, of which Black Bengal showed 10 alleles. A rare allele of 130 bp size was observed in Black Bengal. Lysozyme and MILSTS076 were monomorphic showing sizes of 237 bp and 120 bp respectively.

In total, 243 allele were observed from 19 loci surveyed. The number of allele ranged from 9 to 23. The mean expected hetrozygosity (HE) ranged from 0.74 in the Barbari breed to 0.79 in Jakhrana population. The highest polymorphism values were obtained at SRCRSP locus (0.85 to 0.89) and BM 4621 locus (0.76 to 0.88). The least polymorphism was observed in ILSTS 005 locus. Breed specific alleles were observed at different loci for different breeds.

6.4 Inheritance of Qualitative Traits

6.4.1 Coat colour

The local goats were usually either red and white or black and white, occasionally grey and white. First cross to Angora were pale beige or pale grey, and all later crosses were pure white, white is dominant (although it is evident from his description that dominance is incomplete), which agrees with the conclusion of Asdell and Buchanan Smith (1928) on the dominance of white in British milch goats.

6.4.2 Horn pattern

Horn Pattern are inherited in the Mendelian fashion. In most of the breeds the males have horns whereas females are polled but there may be some exceptions also.

6.4.3 Conformation

Height at withers is less affected than weight by temporary environmental fluctuations. Goats are classified according to height into three categories: large (over 65 cm), small (51–65 cm) and dwarf breeds (under 50 cm). The large breeds weigh anything from 20 to 63 kg, the small breeds 19–37 kg and the very small goats from 18–25 kg. The variation in body weight is therefore very considerable and between-breed comparisons indicate that some breeds can weigh as much as two or three times than the lighter animals although they have approximately the same height at withers. The large breeds appear to be more dual purpose in function than the smaller breeds. The dwarf goats are kept exclusively for meat production.

Since a relationship between height at withers and body conformation has been reported in some breed of goats is useful in indicating the value of some of the breeds for meat production, although the larger breeds are mainly milk or dual purpose types. In some countries variations in body weight are partly due to the many types of goat that have been improved, resulting in heterogeneous crossbreeds unselected for any standard weight.

Within-breed differences are also considerable although they are not as large as between-breed differences. Within any one evironment the within-breed variability can be ascribed to fluctuations in feeding and management and other environmental influences that elicit a favourable response in some animals and not in others.

6.4.4 Wool/hair types

The fleece of Cashmere goats in desert and semi-desert areas consists of an outer layer of thin, long coarse wool, and in inner layer of soft and thin fine wool. Percentage of coarse and fine wool in quantity are 17.11 and 82.89% and in weight 50.13 and 49.87%. The content of coarse wool in population fleece is little, and fine-wool yield is higher. The productive cashmere wool is of long fibre, pleasing lustre, strong intensity, high whiteness and good handle. The annual cashmere wool yield averages 259.29 ± 19.44g. The fineness of cashmere wool amount to 15.21 ± 2.69 μ, the natural length is 4.30 ± 0.56 cm, strength 5.05 ± 0.64g and stretched length 40.39%. It is the best of all natural fibres.

The wool of Cashmere goats in the desert and semi-desert areas where it is dry and cold has an ideal fineness, but the yield is low. In areas where it is warm and moist the situation is contrary. In desert areas the cashmere wool yield might have increased because of selective breeding of native breed. On the edge of desert and semi-desert areas where the moisture is 0.05–0.15 and rainfall is above 250 mm for rapidly increasing the wool yield the introductive crossing may be properly conducted.

Pashmina fibre biometry of Chegu goats, born during 1972–84 and adapted to medium altitude (2,350 m), temperature and humid climate, was studied for 6 annual collections. The mean values for the various annual collections were 5.36–6.08 cm for fibre length and 11.58–12.58 μm for fibre diameter. Least-squares analysis of variance showed a significant effect of period of birth on fibre length in the first and third collections and on fibre diameter in the first, third, fifth and sixth collections. Season of birth had no effect on these traits. Sex of goat significantly influenced fibre length in the third and fifth and fibre diameter in the first, second, third and fourth collections with males having longer and coarser fibre. Influence of generation on fibre length and fibre diameter was not evident. Inter-action effect of period × season, period × season, period × sex and sex × generation were variable and significant on the traits in some collections. Regression on body weight was significant for length in the first, second and sixth collections and non-significant for fibre diameter. The study showed that Indian Pashmina was a fibre of fine quality, being variably influenced by non genetic sources of variation.

6.4.5 Sweat glands

Sweating in temperate breeds of goat and sheep is in the form of discrete discharges of moisture on to the body surface (Bligh, 1967; Robertshaw, 1968; Allen and Bligh, 1969) and a prominent feature of rapid decline in amount of sweat produced at each discharge is considered as fatigue (Robertshaw, 1968; Jenkinson and Robertshaw, 1971). The rate of sweating in Black Bedouin goats (Brout *et al*., 1979; Dmiel *et al.,* 1980) and Indian goats (Rai *et al*., 1979; Singh and Saxena, 1991) is higher even than that of temperate breeds of cattle.

Consistency in sweating rate in bucks of Jamunapari and Barbari breeds of hot semi-arid region of India was measured during hot dry period. Animals (10), 5 of each of 2 breeds, were exposed from 800 to 1700 hr to direct sun on clear day at the Central Institute for Research on Goats, Makhdoom (27° 10'N, 78° 02'E) where the prevailing daily maximum temperature was above 40°C. Concentrate mixture at the rate of 300g/head was fed in the morning at 800 hr. Roughage and water were available ad lib. throughout the period of exposure. Sweating rate was measured by cobalt chloride-disc method at sacral region at 900, 1300, 1400, 1500, 1600 and 1700 hr. The ambient temperature recorded simultaneously was 39.5, 44.0, 46.0, 46.5, 46.0 and 44.5°C respectively. The vapour pressure varied from 12 to 13 mm Hg. The respective mean values for sweating rate (g/m^2/hr) in Barbari were 274.4 ± 20.45, 320.8 ± 25.59, 349.5 ± 31.76, 286.2 ± 27.79, 199.4 ± 15.64 and 256.6 ± 34.42; and in Jamunapari 240.0 ± 25.89, 288.0 ± 24.42, 300.2 ± 22.73, 289.9 ± 25.27, 172.0 ± 20.47 and 197.2 ± 16.24.

Cutaneous evaporative cooling in both the breeds seems to be the important channel of heat dissipation at high ambient temperature. Barbari bucks showed higher sweating rate than Jamunapari bucks. Not only the sweating rate was high but also it was also consistent throughout the 8 hr exposure to high temperature. This indicated that the phenomenon of muscle fatigue reported in sweat glands of temperate breeds of goats on prolonged exposure to high temperature does not hold true for the breeds of hot semi-arid zone of India.

6.5 Inheritance of Quantitative Traits

6.5.1 Growth rate

Records of 124 male and 93 female Barbari kids kept individually in feed lots were analyzed to study the body weight at 3, 6, 9 and 12 months of age. Year and season of birth and weight of dam at kidding had a highly significant effect on body weight at all ages. Effect of sex was significant at 6, 9 and 12 months whereas type of birth significantly influenced the body weight at 3, 6 and 9 months of age. Sex × type of birth interaction affected body weight except at 9 months. Sex × year of birth interaction had no effect. The estimates of heritability for 3, 6, 9 and 12 months weight were 0.333 ± 0.110, 0.360 ± 0.110, 0.295 ± 0.107 and 0.144 ± 0.149 respectively. Genetic correlations of 6 months body weight with those at 9 and 12 months were 0.731 ± 0.194 and 0.729 ± 0.199, respectively, and corresponding phenotypic correlations were 0.933 ± 0.018 and 0.933 ± 0.019 respectively.

Means for body weight in Sirohi Goats were 12.4 kg at 3 months and 19.1 kg at 6 months of age.

Kids in the Northern part of Ajmer district had a lower body weight than in other parts of Ajmer or Southern districts. This probably reflects a lower weaning age.

Body growth was significantly better in the second year, contrary to what was observed in milk yield. We have no explanation for the differences found.

October-born kids gain better than kids born before or afterwards. This may by because in September there is likely to be an after-effect of illness of the monsoon. Kids born in November or December will be 1–2 months old and more sensitive to diseases during the coldest part of the winter. These kids reach an age of 5–6 months when the less favourable summer starts. Body weights of twin kids remain significantly lower than of single kids during the first 6 months. Male kids are heavier than females.

The performance of Sikkim local goat is poor in terms of growth rate, service period and interkidding interval compared to Barbari goats. The Barbari goats are more adopted to relatively warmer and drier climate. The rainfall and humidity in Sikkim is very high. In order to improve the productive traits of goats, crossbreeding in Sikkim local goats with Barbari goats was started. A 10 weeks growth cum metabolism trial was undertaken using 8 female kids (4 from each group) to study the comparative performance of Sikkim local vs. crossbred kids (Sikkim local × Barbari). The kids were offered concentrate mixture @ 1.5% of body weight and mixed jungle grass *ad lib*. The growth rate in Sikkim local was non-significantly higher (40.12 ± 1.58 g) compared to crossbred kids (34.12 ± 3.45). DM intake per kg metabolic body weight was similar in both the groups. The DM intake per kg gain was higher in crossbred kids (12.17 ± 1.56 kg) compared to Sikkim local kids (10.90 ±

0.63). The results on nutrient utilization revealed that the digestibility coefficients of various nutrients as well as balances of N, Ca and P were similar in both the groups. The studies indicated that growth rate and nutrient utilization did not improve in crossbred kids.

Fifteen weaned Marwari kids grouped into 3, each of 5 were fed differently proportionate concentrate to roughage rations (T1–50:50, T2–35:65, T3–20:80) individually under feed-lot system for 20 weeks. The average values for body weight gain, DM intake/100 kg body weight and DM intake/unit body weight gain under T1 were 82.5 ± 3.72 g, 3.66 ± 0.08 kg and 8.65 ± 0.44 kg respectively. The corresponding figures for T2 and T3 were 82.13 ± 6.22 g, 48.67 ± 5.77 g; 3.78 ± 0.11 kg, 3.05 ± 0.10 kg; 11.56 ± 1.23 kg and 12.55 ± 1.23 kg respectively.

The growth rate between T1 and T2 was at par but different significantly from T3 indicating significant treatment effect. The average values of haemoglobin (g %) (T1–9.16 ± 0.26, T2–9.60 ± 0.38, T3–10.16 ± 0.45), packed-cell volume (%) (T1–28.40 ± 1.5, T2–29.13 ± 1.7, T3–29.03 ± 1.8) and serum protein (g%) (T1–8.90 ± 0.36, T2–9.04 ± 0.39, T3–8.87 ± 0.58) did not show significant effect due to ration but showed significant period (age) effect. The kids on 35% concentrate showed high carcass weight, dressing percentage and low feed cost/kg weight gain in comparison to T1 and T3 suggesting that kids on 35:65 concentrate to roughage ration attain the acceptable market body weight with low fed cost.

CHAPTER 7

BREEDING

7.1 Systems of Breeding

The art of breeding lies in the proper application of principles of heredity to animal improvement. The problem of animal improvement may be approached in two ways:

i. Modification of environment as better feeding, management and disease control.

ii. Genetic improvement, which is permanent, *e.g.* selection and mating systems.

Systems of breeding do not create any new gene. They sort out old genes into new patterns. Success, therefore, depends upon the proportion of favourable genes present in the foundation stock. Genes that are not present in the foundation animals can sometimes be found in other strains or populations and can be introduced through crosses.

Breeding	
Inbreeding	**Outbreeding**
i. Close breeding	i. Cross breeding
ii. Line breeding	ii. Outcrossing
	iii. Top crossing
	iv. Back crossing
	v. Grading
	vi. Species hybridization

7.2 Inbreeding

In breeding is the mating of males and females that are related. We consider animals to be related only when they have one or more ancestors in common in the first 4 to 6 generations of their pedigree. The intensity of inbreeding depends on the degree of relationship, for *e.g.* mating of son to dam or brother sister mating are called close breeding in contrast to cousin matings or those which are not closely related.

Measurement of relationship between individuals help us to understand the intensity of inbreeding. Two animals no nearer related than the average of their breed have a relationship of zero. Two animals with exactly similar genotype have a relationship of 100. They are alike in 100% of the genes, *e.g.* identical twins. The degree of relationship therefore, ranges from 0 to 100.

Relationship may be of two kinds, direct and collateral. You are directly related to your father. That is you and your father have more genes in common (50%) than do unrelated members of the human population. Similarly, one half of your genes are identical with those of your mother. You and your cousins are collateral relatives, because you both have some ancestors in common. Your cousin probably has some identical genes that came to each of you from your common grand parents.

7.2.1 Coefficient of inbreeding

When animals which are related, in other words those having genes in common, are mated, more homozygosity results in the offspring in relation to average animals of the same breed in the foundation stock. Inbreeding, therefore, increases homozygosity or decreases the heterozygosity in individuals.

The average percentage increases in homozygosity or decreases in heterozygosity in an inbreed animal in relation to an average animal of the same breed of the foundation stock is known as the coefficient of inbreeding. It is obtained by multiplying the relationship among parents by ½; since the new generation produced is once further removed from the common ancestors and a further halving of the genetic material occurs. The formula for the coefficient of inbreeding of individuals (Fx) is

$$Fx = \Sigma\ [(1/2)\ n'1 + n+1]$$

n: number of generations or halvings from the sire to the common ancestor.

n' : Number of generations from the dam to the common ancestor.

The coefficients are not absolute but relative measures. It measures the probable similarity of germ cells. It is useful for study of breeds and lines within a breed and analysis and comparison of individuals, groups and breeds for the part inbreeding plays in their respective developments.

Effect of inbreeding

Inbreeding is the mating of animals, which are related or have more number of similar genes. Hence, it increases the likelihood of similar genes becoming paired. In other words it increases the percentage of homozygotes and reduces the proportion of heterozygotes. Inbreeding makes the genes, favourable or unfavourable, homozygous. When the animals are homozygous for a number of traits the regularity of inheritance is assured, *i.e.* it fixes the characteristics. A high degree of homozygosity increases the prepotency of the inbred individuals *i.e.* the ability of a parent to impress its characteristics uniformly on its offspring.

7.2.2 Reasons for inbreeding

1. To promote genetic purity and thereby increase prepotency.
2. To bring undesirable recessives to light and give the breeder an opportunity of culling them from the stock. When a sire is mated to 20 of its daughters, if it does not throw out any recessive character, it may be reasonably stated that the sire is not heterozygous for character under question.
3. To develop inbred lines for nicking ability.
4. For regrouping the genetic material.

It is generally believed that inbreeding reduces vigour. The reasons for this are:–

i. The recessive genes become homozygous during inbreeding.
ii. If overdominance exists, where a2 a2 is superior, inbreeding diminishes the quality of trait and a2 a2 becomes a1 a1 and a2 a2 during inbreeding. To offset the bad affects of inbreeding, it is desirable that it is practiced only in herds that are better than average *i.e.* where frequency of desirable genes is more. It may be practical in herds where an outstanding sire has been used. It is also necessary that the breeder should know the merits and demerits of this system before he practices it. Inbreeding should not be practiced in grades or in commercial herds below average for the sake of economy in a single sire herd.

Line breeding

Line breeding is a form of inbreeding, but so directed as to keep the relationship of the individuals very close to as in admired ancestor. The admired ancestor is usually a male since it can give more offsprings during its life time than a female. When we say that an animal is a line breed, the question immediately arises, line bred to what? It may be line bred to Prince Dixio or Prince Bhima.

When and why **line breed**? When a sire, used on good dams produce offsprings better than their dams, the breeder should line breed at once strongly to this sire, while the animal is yet alive. It can be used on its daughters and grand daughters generation after generation. Often an animal is old or dead before its superiority is recognized. If its sons and daughters are mated to unrelated individuals, within three to four generations, the influence of the outstanding animals is so scattered that no one descendent is like the original individual. So, line breeding holds the expected amount of inheritance from the admired ancestor as a constant level instead of letting it to be halved every generation. If at the time of death, there are no relatives more closely related than 50%, we cannot produce animals more close to it than that, but it may be possible by inbreeding to keep and maintain that level.

Line breeding builds up homozygosity and prepotency. It tends to hold the gain made by selection while attempt is made to make further gain.

Line breeding is specially useful where there is much epistasis, where a desired characteristic depends on a combination of genes and where the combination tends to get scattered at each generation. These genes can be made homozygous in different lines and lines can be crossed for their combining or nicking ability.

When progress by inbreeding comes to a standstill, line breeding makes additional progress possible.

Dangers of line breeding

Line breeding tends to make genes, good or bad, homozygous rapidly. Hence choosing of the ancestor (sire) to line breed is very important. Those that are definitely superior should alone be selected. Besides, rigid selection, culling of the undesirable recessives is highly essential. Line breeding should be practiced only in herds distinctly superior to the general average of the breed.

7.2.3 Prepotency

Prepotency is the ability of an individual to stamp its characteristics on its offsprings to such an extent that they resemble their parents more closely than is usual. It is the property of the characteristic and not the individual breed or sex. When two individuals are mated, one may have more influence than the other on the offspring. Similarly, some lines and breeds are more prepotent than others. However, prepotency cannot be passed on from one generation to another unless it is possessed by both sires and dams.

A high degree of homozygosity and the possession of a high percentage of dominant genes are the inherent qualities that will enable an animal to stamp its own characteristics on majority of its offsprings. A perfectly homozygous animal produces only one kind of gametes and all its offsprings will receive exactly the same gene from it. Any genetic difference between the offsprings would depend entirely on their having received different genes from the other parents. If the parent is homozygous for several dominant genes, all the offsprings will resemble it irrespective of what they received from the other parent. Here, prepotency is the maximum.

Measure of prepotency

Inbreeding and increase of homozygosity is the only means of making animals prepotent for characteristic. The more the animals are inbred the more they become homozygous for a number of genes. The inbreeding coefficient then is the best estimate of an animal's prepotency. Prepotency, however is not transmissible from parent to offspring.

7.3 Outbreeding

Outbreeding is the mating of animals distinctly less closely related to each other than the average of the population, *i.e.* those that have no common ancestors in the preceding 4 to 6 generations of their pedigree. It is just the opposite of inbreeding. It promotes the pairing of unlike genes by mating animals that belong to different families, breeds or species. Thus it increases heterozygosity and variability. The chief reasons for outbreeding are:

i) To bring about an increase in vigour. Vigour includes almost anything that pertains to desirability, *e.g.* rate of gain, efficiency of gain, fertility, general strength etc.
ii) To make full use of dominance of characteristic.
iii) To introduce new genes in a closed population. If a certain breed or family is deficient in a certain trait, the quickest and most certain method of improving that trait is to introduce genes through cross breeding to some stock known to be superior in that trait.
iv) To start new breeds with a broad genetic background.

v) To produce market animals making use of heterosis.

Outbreeding includes:

(a) cross breeding,
(b) outcrossing,
(c) back crossing,
(d) top crossing,
(e) grading and
(f) species hybridization.

7.4 Crossbreeding

Cross breeding is the mating of two animals, which are pure bred but belonging to different breeds. It is widely practiced in swine, sheep, and poultry and less so in cattle and horses. The main purpose in this is to produce commercial stock where the individual merit for economic traits is promoted. The breeding value of the individual, however, is lowered.

When the crosses are used for breeding purposes, their offsprings are more variable than the crossbreds and generally average, somewhat, lower in individual merit, below their purebred grand parents.

Judicious crossing of breeds that complement each other might result in increased vigour. The economy of crossbreeding, therefore, depends upon whether the increase in production is more than enough to balance the possible confusion regarding the breeding value of the crossbred individuals and also increase in cost of replacement of pure bred stock under a cross breeding system. It is more profitable where fertility is highest and females can be kept for long and the cost of their replacement is lowest. Mainly for these reasons, it is mostly practiced in swine, poultry and sheep.

In a cross breeding system, the males are to be discarded because of their lowered breeding value. The heterosis in females can, however, be utilized by crossing it with a third different breed and rotating the same in a systematic manner. This is known as triple crossing, or rotational crossing.

Three breeds are used in this system. The females or crosses are used on sire of pure breeds in rotation. The crossbreds will soon come to have $^4/_7$ of inheritance of the breed of immediate sire, $^2/_7$ from the breed of maternal grand sire and $^1/_7$ of the hereditary material of the other pure breed. Heterosis is thus continuously maintained.

Crisscrossing is another method proposed for utilizing heterosis in dams, without incurring the full decline in average individual merit which usually occurred when crossbreds are mated.

7.5 New Breeds From Crossbreeds

Crossbreeding has been utilized for developing several new breeds of livestock. It offers a broad genetic basis from which by a process of selection and inbreeding, new gene combinations can be made for specific purposes.

Hissardale is a new breed of sheep developed in Punjab by crossing Bikaneri ewes and Marine farms. Columbia is a new breed of sheep produced by crossing Lincoln rams and Rambouillet ewes, by the U.S.D.A. Pannama has been developed privately by James Laidlow of Idaho by crossing Rambouillet rams with Lincoln ewes.

The question of development of new breeds arises only when we are not satisfied with the existing breeds as regards their utility value and when we feel that their value can be enhanced by making new gene combinations from two breeds which is likely to complement each other in different traits.

7.6 Outcrossing

Outcrossing is the mating of the animals, that are members of the same breed, but show difference in the herd. Intense inbreeding makes the genes homozygous and at the same time closes the door for further improvement. It fixes the deleterious genes also. An outcrossing brings into the herd new genes and gives an opportunity for selection and further improvement of the herd.

Outcrossing is a useful procedure where it is desired to change the type of the herd rather drastically, when necessitated by market demands.

7.7 Top Crossing

Top crossing is the mating of a male of a certain family to females of another family of the same breed. It is the same in principle as grading up, except that top crossing is usually applied to different families within a pure breed, whereas grading up is applied to continued use of sires of one pure breed starting with foundation females which are of another breed or mongrel stock.

7.8 Back Crossing

Back crossing is the mating of a crossbred animal back to one of the pure parent races which were used to produce it. It is commonly used in genetic studies, but not widely used by breeders. When one of the parents possesses all or most of the recessive traits, the back cross permits surer analysis of the genetic situation than an F_2 does.

A heterozygous individual of the F_1, when crossed with a member of the homozygous recessive parent race, the offsprings group themselves into a phenotypic ratio of 1:1; on the other hand if the individual of the parent race were to be homozygous dominant, all the offsprings will be phenotypically alike.

7.9 Grading Up

Grading up is the continuous use of purebred sires on females of another breed or mongrel stock, to raise them quickly to the level of the purebred sires. When the purebreds are relatively scarce, this is the only quickest way available for improving the mongrel stock. In grading up, generally, the first cross shows a marked improvement over the original stock. Further improvement by each successive cross is progressively less. In a purebred, which is stationary in level, the mongrel stock by the seventh generation almost reaches that of the purebred.

7.10 Species Hybridization

By crossing two different species sometimes we get good fertile individuals. The mule is a good example of a commercially important species hybrid. Mare × Jack = Mule; She Ass × Stallion = Hinny. Male mules are always sterile as far as is yet known. A few cases of fertile mare mules have, however been reported. But these are very rare. Hinny is generally inferior to mule as work animal. It is also sterile.

Horse having 33 pairs of chromosomes and ass 32 pairs, the mule comes to possess 51 single chromosomes in all. The mare mules have given birth to mule foals and horse foals when bred to jack and stallion respectively. The inference is that the mare rules essential function as mares as far as the genetics of their eggs is concerned.

If all the horse chromosomes were extruded in the polar body, these mules will function genetically as asses. But no case of this sort has been reported. True breeding of mules as such seems also theoretically impossible.

European cattle and American bison when crossed, produce sterile males and fertile females. By back crossing the females to bison and cattle, attempts are being made to form a new breed of cattle, the Cattalo.

7.11 Heterosis or Hybrid Vigour

Heterosis or Hybrid vigour is a phenomenon in which the crosses of unrelated individuals often result in progeny with increased vigour much above their parents. The progeny may be from the crossing of strains, varieties, or species. One of the explanations for this increased vigour is that genes favourable to production are usually dominant over their opposites. As a species or breed develops, it becomes homozygous for some dominant genes. They also have few unfavourable recessive ones. When one breed is crossed with the other one parent supplies a favourable dominant gene to offset the recessive one supplied by the other and vice versa.

The offspring, therefore, has a larger number of dominant genes than does either parent and is likely to be more vigorous.

Another explanation for hybrid vigour is overdominance, where a heterozygous condition is much more superior to any of the homozygous conditions.

Heterosis is much employed to produce commercial stock where the individual merit is promoted, but the breeding value is lowered. The successful exploitation of heterosis depends upon how superior the crosses are over the purebreds and whether it is worth the confusion caused in lowering the breeding value of the individual and the cost of replacement of purebred stock. For these reasons it is more commonly practiced in poultry, swine and sheep where the fertility is high and the cost of replacement of purebred stock is likely to be low.

The breeding policy in respect of Indian goats has not been seriously considered and adhered to in the country. In most of the States, upgrading of the non-descript and dwarf breeds of goats is favoured with larger breeds of the Upper Gangetic or Trans-Gangetic Plains regions. Selection for improvement in the Indian milch and mutton breeds has been advocated widely, though sporadic attempts have been made for crossing with exotic milch and mutton breeds in Rajasthan, Maharashtra and Kerala. Similarly, the attempts have been made to introduce Angora in Himachal Pradesh, Uttar Pradesh and Maharashtra for producing mohair by crossing with local hairy breeds.

The development scheme on Beetal was taken up in the country in 1939 at the Government Livestock Farm, Hissar to improve the milk production through selection. The effort, however, dwindled as there was a decline in the lactation yield, daily yield and yield/day of kidding-interval by 6.8, 19.5 and 21.8%, respectively, and an increase in the age at first kidding, lactation length and kidding interval by 12.6, 22.5 and 12.9%, respectively, over the foundation stock after 5 generations of selection. The decline of performance was ascribed to the poor breeding value of the sires used in the successive generations.

The few State Governments, during the post independent years prior to the organization of the State Agricultural Universities, undertook Research and Development work at the State Farms on Beetal, Jamunapari, Sirohi and Black Bengal flocks. Occasional efforts were also made to produce the exotic crosses with Saanen, Alpine, Anglo-Nubian and Toggenburg. A few Missionaries, Voluntary and Non-Governmental Organizations undertook the sporadic developmental work in some pockets of the country introducing Swiss-Alpine, Israel-Saanen and Boer goats. Such goat development efforts in the different States have been rather meagre, except in Rajasthan, Gujarat, Maharashtra, Uttar Pradesh, Himachal Pradesh and Jammu and Kashmir. The early experience of crossbreeding under field conditions, except in the Western Himalayan and Southern Plateau and Hill regions, is discouraging, for the crossbreds generally failed to sustain the nutritional and environmental (including disease) stress and endured high mortality.

In the absence of reasonable data, on comparisons among the breeds having identical production trait(s) and on compatibility with improved breeds in diverse geoclimate, a few Indian breeds have since been identified as improver breed, fascinated often by the adult body size and milk yield. Jamunapari, Beetal and Barbari have been employed more often as improver breed in India and the other neighbouring countries to upgrade the regional non-descripts, small and dwarf breeds. The valuable traits, like fecundity, disease resistance, grazing habit, meat quality and skin characteristics became secondary though crucial in the development programme of goats. The short-term crossbreeding experiments in the institutional farms really demonstrated the transmissibility of some of the production traits of the Indian improver breeds to the breeds of lower productivity. Jamunapari × Bengal and Beetal × Bengal crosses showed an improvement of 76.3 and 90.8% respectively, in the milk production over purebred Bengal. The improvement was obviously spectacular due to the large genetic diversity between the parental breeds. The effect of the heterosis was rather conspicuous when the high-yielding dam of Jamunapari was crossed with low-yielding sire breeds, *e.g.* Bengal. The reciprocal crosses did not show the heterosis (Misra and Mukundan, 1990). The improvement was imperceptible when the Sirohi breed was crossed with Beetal where the genetic divergence between the two breeds was limited. The situation was similar when Black Bengal was crossed with Barbari though the milk yield was considerably increased in the crosses (Agrawal and Bhattacharyya, 1978) over the Black Bengal.

CHAPTER 8

REPRODUCTION

Reproduction is of fundamental importance to profitable commercial goat production. It is the means by which hereditary material is passed from generation to generation. It also plays a major role in the success or failure of breeding system. In the male, spermatozoa are produced in the testes. The remainder of the system has the purpose of providing for maturation and storage of spermatozoa and for their deposition in the female. The basic female organ is the ovary, in which the female reproductive cells *i.e.* ova are produced. Other ducts and organs exist for the purpose of transporting the ova to the site of fertilization and receiving the sperms and transporting them to the site of fertilization. The uterus provides for the growth and development of the embryo and fetus until it reaches a stage of development compatible with independent life.

8.1 Male Reproduction

8.1.1 Reproductive organs

The male component of reproduction in goats, like other mammals, is distinctly different and independent of the female component. The spermatogenesis, including sperm maturation and epididymal storage, quality semen production and semen technology have been the areas of major interest in recent years as these aspects have direct bearing on the utilization and conservation of meritorious male gametes.

8.2 Types of Breeding

8.2.1 Natural breeding

The natural breeding is done either by flock mating, pen mating or hand mating.

In flock mating system, goat can be bred naturally or through artificial insemination. In flock-mating system, breeding bucks are usually turned out in the flock during the mating season at the rate of 2 to 3% of the does all through day and night. It is most widely practiced in the flocks of all farmers. Semi flock breeding or pen breeding is done to conserve the energy of bucks and give them rest. In this, the bucks are turned out for service with the flock in the pen during night, and confined and stall-fed or grazed separately during the daytime. Hand mating is practiced when exotic purebred sires are used, or when it is considered desirable to extend the services of ram over much larger flocks. Goats in heat do not manifest behavioural symptoms. Hence, the teaser rams are employed for detecting the does in heat. These does are then taken out of the flocks and bred to the designated sire of the flock. In pen and hand-mating systems and when teaser bucks are used for heat detection, some dye mixed in grease or simple linseed oil is smeared on the brisket of the bucks. This makes it possible to record the date when the does are bred and also to remove them from the breeding flock. The colour of the dye should be changed every 16 to 18 days so that the repeaters can be discovered, if the bred does have not been removed from the flock. This is termed as marking of does by breeding buck and marked does are considered as bred.

8.2.2 Artificial insemination (AI)

AI technique in goats is simple, though it demands skilled operation and the practical experience of workers. The technique, elaborately described by Sahni (1967) and Tiwari (1967), has since been widely adopted in research laboratories and limitedly practiced in the field. AI equipments have been developed for easy operations and for application of frozen semen, though the basics remain the same.

Pioneering work on AI in goats was carried out by Guha *et al.* (1951) at the Indian Veterinary Research Institute, Izatnagar. Extensive studies on various aspects of AI were also undertaken later at the Veterinary College, Mathura. 50% conception at first insemination with washed spermatozoa diluted in EYC glycine diluent was achieved (Roy and Gupta, 1959; Roy *et al.*, 1962). Irrespective of the stage of heat, an overall conception rate in Barbari goats, with single insemination each at 3 consecutive oestrous cycles (60 days) was 88%. Tiwari *et al.* (1968) succeeded in getting 79% conception rate with 0.1 ml freshly diluted semen (1:5) containing about 33 million spermatozoa. Results of AI with rural goats using freshly diluted semen on the same day of storage (12 hr) gave 85% conception rate (Sabnis, 1991). AI performed with stored semen (beyond 24 hr) resulted in a significantly lower conception (29.2%) than with freshly diluted semen (Mathai and Nair, 1981). However, 64 and 61% conception rate with 48 hr stored semen of Saanen bucks in milk and EYC diluent was reported by Borgohain *et al* . (1985).

Fertility trials were carried out with frozen buck semen at the CIRG, Makhdoom between 1986 and 1990. Irrespective of the stage of heat, Jamunapari, Barbari and Jakhrana does in oestrus were inseminated twice at an interval of 6–8 hr in the same oestrus by depositing thawed semen (0.25 ml) over the opening of cervix. A higher conception rate was observed in Jamunapari (42.2%) in comparison to Barbari (35.8%) and Jakhrana (37.5%) and variation in the conception rate was observed between breeds within seasons and years (Tiwari and Bhattacharyya, 1988). The conception rate reported by Sinha *et al.* (1987) was 45.16, 44.0 and 42.1 % in Jamunapari, Barbari and Black Bengal does respectively. Washing spermatozoa resulted in higher conception rate (69.35%) than unwashed (66.38%) frozen semen (Deka and Rao, 1989). The conception rate (following deep cervical insemination) with egg yolk-citrate-fructose glycine and tris-egg-yolk-citric acid-fructose-glycine diluents was 50 and 60% respectively (Deshpande and Mehta, 1991). However, Chauhan and Anand (1990) claimed 81% conception with tris diluent in Jamunapari goats. Though the conception rate with frozen buck semen (Table 8.1) using cervical insemination (deposition of semen over opening of cervix) is very encouraging from research point of view, yet there is a strong need to adopt deep cervical and intra-uterine insemination technique for obtaining acceptable fertility (70%) over large goat population. Effective utilization of AI technique with the use of liquid or frozen semen of superior sires could be possible after systematic analysis of socio-economic, technical and organizational problems in the country.

Sahni (1987) identified several factors which should induce the adoption of AI Technology in goats.

- Shortage of males of superior genetic makeup among indigenous as well as exotic milch breeds for natural service.
- Small size of farmer's flock, 1 or 2 goats per family. Farmers cannot afford to bear the cost of maintaining the breeding buck(s) exclusively for their own does.
- Availability of cheap skilled labour in handling goats for AI.
- Ample sunlight for locating the cervix.
- Symptoms of 'heat' are pronounced and it is easy to detect a doe in oestrus as compared to other farm animals.
- Non-seasonal nature of sexual activity both in males and females. AI operations remain continuous throughout the year.
- Semen from bucks is of reasonably good quality and could be used effectively in any season of the year.
- Conception rates are fairly satisfactory with AI technology in goats.
- Goat semen stands freezing better than that of the sheep, buffaloes, horse and pigs.

- Possibilities of natural synchronization of oestrus in goats in tropics exist when sudden association of male (after long period of separation) is ensured daily for a short period (telescoping). In certain breeds (small sized) almost 60 to 70% goats come in heat within the first 7 to 8 days of start of teasing.
- Considerable potential exists in obtaining 2 crops of kids from each doe in 13 to 14 months. The post-partum conception rates with AI are quite satisfactory.

Table 8.1. Conception Rate with Frozen Buck Semen

Conception rate	No. of sperm/ Inseminating dose	Reference
40.0 (131)	–	Sinha *et al.* (1987)
69.0 (130)	1:10	Deka and Rao (1989)
81.0 (26)	18.00×10^7	Chauhan and Anand (1990)
40.0 (374)	60.80×10^6	Tiwari (1991)
55.0 (22)	–	Deshpande and Mehta (1991)

Figures in parentheses indicate number of does inseminated.

8.2.3 Factors affecting semen quality

Puberty in male goats denotes the age at which the young buck just develops the ability to release the spermatozoa, while the sexual maturity indicates the full reproductive potential of the buck. The puberty is influenced by many factors such as birth weight, breed, nutritional level, physical and other environmental conditions. The first sex play has been reported in a number of Indian breeds at the age of even 1 month, though the first ejaculation seems to occur from 267 days onward (Singh and Sengar, 1990). Age of copulation and first ejaculation varies with breeds, seasons, season of birth, plane of nutrition and presence of teaser females (Prasad *et al.*,1970; Singh and Sengar, 1979). The age at first ejaculation has been recorded as 395 days in Jamunapari, 381 days in Jakhrana, 515 days in Sirohi, 511 days in Kutchi and 521 days in Marwari (Tiwari, 1990) and 361 days in Barbari, 371 days in Beetal and 267 days in Black Bengal (Singh and Sengar, 1979). Nutritional deficiency delays the onset of puberty and depresses the production of quality semen. Growing males are more susceptible to the nutritional stress and prolonged Vitamin A and E deficiencies are known to cause irrestorable reproductive disorders.

Several genetical, nutritional, physiological and disease factors are known to affect the seminal attributes in bucks. While other factors could be manoeuvred by rigid selection of bucks, balanced nutrition, disease prevention and physical exercise. The seasonal effect on sperm concentration, sperm motility and percent live count appeared to be economical in all the Indian breeds (Shukla and Bhattacharyya, 1952; Singh, 1961; Patil and Raja, 1978; Mittal, 1987; Singh and Sengar, 1990). The seasonal deterioration in semen quality could be prevented in Barbari, Saanen, Jamunapari and Saanen × Jamunapari bucks by a twice daily collection schedule for 54 days (Tiwari *et al.*, 1968). The age has considerable effect on the semen quality and sex libido. Semen volume and motility were significantly higher in adult Barbari bucks in comparison to young males.

Single service daily from a buck seems to have no adverse effect either on seminal attributes or the health of the animal. Continuous use (3–4 ejaculation in 30 min) of Jamunapari and Barbari bucks for 5 weeks did not affect the semen quality and health adversely (Tiwari *et al.*, 1968; Mittal and Pandey, 1972). The effect of season on the number of ejaculations was statistically not significant in Jamunapari bucks (Mittal, 1985). The semen quality, however, deteriorated when collection was made 10 times at 1 hr interval on a particular day (Singh and Sengar, 1990).

8.2.4 Semen volume of different breeds and semen quality traits

Semen volume and other quality traits in different breeds of goat are presented in Table 8.2.

Reaction time, volume, initial motility, pH, percentage of motile sperm, live sperm and abnormal sperm showed variation in breeds as well as seasons. Similar was the trend in sperm concentration and total number of sperms. Although, there was significant fluctuation in seminal attributes between breeds and seasons, the semen of optimum quality could be obtained throughout the year from mature bucks.

The information on biochemical studies of buck semen is meagre. Patil and Raja (1973b) estimated the biochemical constituents of seminal plasma in Malabari bucks. Fructose level varied from 320 to 866 mg/100ml. Citric and

Table 8.2 Seasonal Variation of the Seminal Attributes in Some Breeds of Indian Goats

Seminal attributes	Rainy	Autumn	Winter	Sprint	Summer
Reaction time (sec)					
Jamunapari	25.5	–	35.8	–	25.4
Jamunapari	30.0	–	33.0	–	21.0
Beetal	15.0	–	21.0	–	21.0
Marwari	18.0	–	30.0	–	27.0
Volume (ml)					
Jamunapari	0.58	–	0.33	–	0.22

Table 8.2 Contd.

Seminal attributes	Rainy	Autumn	Winter	Sprint	Summer
Jamunapari	1.00	–	1.15	–	1.10
Jamunapari	1.18	0.73	0.72	0.84	1.02
Beetal	1.40	–	1.30	–	1.35
Marwari	0.75	–	0.65	–	0.80
Barbari	0.76	0.75	0.61	0.52	0.63
Initial motility (0-5)					
Jamunapari	4.30	3.52	4.02	2.94	3.66
Barbari	3.87	3.98	3.80	3.20	3.62
pH					
Jamunapari	6.65	–	6.82	–	6.75
Marwari	6.82	–	8.50	–	6.78
Motile sperm (%)					
Jamunapari	80.0	–	79.2	–	81.0
Jamunapari	75.6	–	78.4	–	80.0
Beetal	82.0	–	78.0	–	80.0
Marwari	78.0	–	75.0	–	75.0
Live sperm (%)					
Jamunapari	79.40	–	80.40	–	81.80
Jamunapari	70.02	61.63	61.17	55.75	59.34
Beetal	85.80	–	82.70	–	84.70
Marwari	81.70	–	80.50	–	81.00
Barbari	71.62	70.44	70.32	65.29	69.79
Abnormal sperm (%)					
Jamunapari	7.5	–	9.0	–	5.9
Beetal	6.7	–	8.2	–	6.2
Marwari	7.2	–	8.6	–	6.4
Sperm concentration/ml ($\times 10^6$)					
Jamunapari	2952	–	4049	–	4063
Jamunapari	2420	–	2502	–	2475
Jamunapari	1806	1986	3309	2210	1805
Beetal	2700	–	2910	–	2820
Marwari	2470	–	2680	–	2450
Barbari	1880	1608	2410	1840	1810
Total No. of sperms/ejaculate ($\times 10^6$)					
Jamunapari	1712	–	1336	–	894
Jamunapari	2420	–	2877	–	2723
Jamunapari	2082	1576	2513	1627	1352
Beetal	3780	–	3783	–	3807
Marwari	1853	–	1742	–	1960
Barbari	1380	1320	1440	1130	1160

Tabulated from Sinha *et al.* (1981), Mittal (1987), Sahni and Tiwari (1992)

ascorbic acids were reported to be 641.7 and 11.7 mg/100ml respectively. Nair *et al.* (1982) reported the mean value of sodium and potassium in semen as 237 and 239 mg % and observed a positive correlation between the two ions. Markendya and Pargaonkar (1990) compared the biochemical constituents of the first and second ejaculates of the Osmanabadi bucks (Table 8.3).

Table 8.3 Comparison of Biochemical Constituents of First and Second Ejaculates in Osmanabadi Bucks

Constituents	First Ejaculate (Mean)	Second Ejaculate (Mean)	"T" Value
Fructose (mg/100ml)	454.30	513.84	1.032
Protein (g/100ml)	5.32	5.13	0.729
Sodium (mEq/litre)	79.08	76.83	0.309
Potassium (mEq/litre)	72.79	82.20	1.441
Chloride (mg/l00ml)	126.73	118.54	0.829
Calcium (mg/100ml)	8.03	8.85	0.981
Inorganic Phosphorus (mg/100 ml)	10.67	11.18	0.482

Source: Markendya and Pargaonkar (1990)

Information on cytometry of buck spermatozoa indicated that the mean length, width, and width at the base, of the sperm head in Malabari breed were 8.4, 4.4 and 2.5 μ respectively. Length and width of the mid-piece were 11.80 and 1.10 μ respectively. Length of the tail was 41.30 μ (Patil and Raja, 1973a). The cytometric measurements of sperm in Osmanabadi breed (Table 8.3) and its Beetal, Alpine and Saanen crosses, are available (Parendekar *et al.*, 1991; Puranik, 1991). From the scanty data, it appears that there are breed difference in the sperm size.

Table 8.4 Cytometry of Spermatozoa

Measurement specification	First ejaculate (mean ± SE)	Second ejaculate (mean ± SE)
Head length (μ)	8.96 ± 0.05	8.85 ± 0.07
Head width (μ)	4.54 ± 0.0	4.54 ± 0.0
Head area (μ^2)	40.74 ± 0.24	40.02 ± 0.36
Mid-piece length (μ)	12.70 ± 0.18	11.55 ± 0.26
Mid-piece width (μ)	0.75 ± 0.0	0.75 ± 0.0
Mid-piece area (μ)	9.62 ± 0.13	8.74 ± 0.19
Tail length (μ)	36.37 ± 0.94	39.00 ± 0.03
Total length (μ)	58.41 ± 0.87	59.04 ± 0.29

Source : Puranik (1991)

Semen collection once daily from Barbari, Jamunapari, Sannen × Jamunapari crossbred bucks either for 20 or 40 days and twice daily (in quick succession) for 54 days did not produce adverse effect on semen quality (Table 8.5). In view of unwillingness of bucks to donate semen in AV for more than 2 collections within 1 hr, it seems reasonable to suggest that single collection per day from a buck has no adverse effect on seminal attributes up to 60 days on the basis of 50 million spermatozoa required for ensuring fertility and if semen utilization is 100%, the number of goats that can be expected to be bred in a collection schedule of once or twice daily for 60 days from different goat breeds varies from 600 to 3000.

Table 8.5 Approximate Number of Does Which Could be Inseminated under Different Frequency Schedule

Breed	Frequency of collection (daily)	Duration of trial (days)	Total semen volume (ml)	Total No. of sperms × 106	No. of goats which can be inseminated
Barbari	×1	21	13.8	36077	661
Barbari	×2	21	20.7	33180	763
Barbari	×1	40	20.8	42372	847
Barbari	×2	54	48.0	105484	2109
Jamunapari	×2	54	72.4	159205	3184
Saanen	×1	40	39.9	94680	1893
Saanen	×1	21	12.5	56802	1136
Saanen × Jamunapari	×2	54	74.0	188705	3774

Source : Sahni and Tiwari (1992).

8.3 Anatomy of Male Reproductive Organs

The principal reproductive organs are a pair of testis, the accessory glands and ducts through which the spermatozoa and seminal fluids are discharged at the time of ejaculation.

The testis is covered by a dense connective tissue capsule, the tunica albuginea, from which connective-tissue extensions, the septa, radiate to divide the testis into lobules. Within each lobule are long, coiled ducts called the seminiferous tubules. The seminiferous tubules converge at the apexes of the lobules, where they straighten to form the tubuli recti. The tubuli recti enter the rete testis, a series of irregular anastomosing ducts. Connecting the rete testis and the head of the epididymis, there is series of ducts called the vasa efferentia. Between the seminiferous tubules, there are connective tissues, blood vessels, nerves and the interstitial or leydig cells which secrete the male hormone testosterone.

The testes are formed in the abdominal cavity and in some species they remain there throughout life. In the course of evolutionary development, they gradually came to lie at the posterior end of the body pressing against the body wall. A continuation of this process finally led to an out pouching at this region, forming a sac, the scrotum, outside the body proper.

In addition to serving as a covering for the testicles, the scrotum functions as a thermoregulatory mechanism. This function of the scrotum is shown by actual differences in temperature, the temperature in the scrotum being from 1°C to 8°C lower than that of the abdominal cavity. It has been shown by several workers that insulation or the application of heat to the testicles results in a degeneration of the spermatogenic tissue, the production of abnormal sperm, and temporary sterility. These results are also known to follow periods of intense fever.

The seminiferous tubules voverage at the rete testis, from which there is a series of ducts, the vasa efferentia, which convey the sperm to the epididymis. The epididymis is a continuous, greatly coiled duct varying from a few ft to several 100 ft in length in some species. It is lined with ciliated, tall columnar cells and has some smooth muscle cells in the tubule walls. The epididymis has 3 regions; the head, adjacent to the vasa efferentia; the mid-portion; and the tail, which is continuous with the vas deferens.

The passage of sperm from the tail of the epididymis to the urethra is provided by the vasa deferentia. Each vas deferens consists of a firm tubular structure lined with ciliated columnar epithelial cells and a relatively thick wall composed of smooth-muscle and connective-tissue cells. In the bull, ram, and stallion the wall of the vas deferens is greatly thickened near its junction with the urethra. This area is called the ampulla. It is not present in the boar.

The vas deferens, in combination with the arteries, veins, and nerve supply of each testis, plus smooth-muscle and connective-tissue cells, constitutes the spermatic cord. The vas deferens is a passageway for sperm; the spermatic cord provides the vascular and nervous regulation of the testis.

The seminal vesicles are the largest of the accessory glands of reproduction in the male. They are located in the pelvic cavity at the ends of the vasa deferentia, being an outgrowth of the latter at their urethral end, and each connects with the urethra by means of a duct. The seminal vesicles vary in size; they are several inches in length in the stallion and have a bladder like appearance. In the other farm animals, they are elongated, firm and lobulated, being about 4 to 5 inches long in the bull and the boar and smaller in the ram.

The prostate is compound tubuloalveolar gland made up of a large number of lobules. It is located at the neck of the bladder and tends to surround the urethra. It is not prominent, as are the seminal vesicles, and is easily overlooked.

The cowper' glands, often called the bulbourethral glands, are small, firm, oval glands located on either side of the urethra posterior to the prostate and covered by the urethral muscle in the stallion, bull, and ram. In the boar they are elongated, cigar shaped structures 4 to 5 inches in length.

The remaining portion of the male genitalia consists of the penis. This organ, besides conveying urine to the exterior, has the additional function of conveying spermatozoa into the female genital tract. The penis is made up of muscular and erectile tissue that becomes engorged with blood during the process of erection.

Farm animals exhibit a wide variation in the structure of this organ. In the horse, the end of the penis is rounded and the organ nearly fills the vagina at copulation. In the bull, the diameter is much smaller with the external urethral orifice situated in the urethral papilla at the end and on the left side. The urethral papilla of the bull represents a vestigial Piliform appendage. This is well developed in the ram in the form of a slender appendage which protrudes beyond the glans penis in a twisted manner. The penis of the boar is not provided with an appendage or lateral papilla and the opening to the exterior is situated at the end of the organ and in the centre.

8.4 Female Reproduction

8.4.1 Anatomy of female reproductive organs

In the female, the principal reproductive organs are the paired ovaries and a system of ducts which are concerned with the transport of sperm, the transport of the ovum, the development of the fertilized ovum, and the delivery of the foetus at parturition.

The ovaries are the primary reproductive organs of the female and are homologous to the testes in the male. The ovaries are suspended in the broad ligament of the uterus in the dorsal part of the abdominal or pelvic cavity. In the mare, the ovaries are bean shaped and 2 to 4 inches in total diameter, in the cow being 1 to 1 ½ inches , and in the ewe ½ to 1 inch. The ovaries of the sow are usually larger than those of the cow and are irregular in shape because of the large number of follicles or corpora lutea which occur in all species that produce litters.

Microscopically, the ovary consists of 2 regions, the outer shell, or cortex, and a central portion, the medulla. The ovary is almost completely covered with a single layer of cuboidal epithelial cells, the germinal epithelium. This is a distinct contrast to the testis in which the germinal epithelium is located deep within the seminiferous tubules. All functional ova arise from the germinal epithelium through a series of mitotic and meiotic cell divisions. The ova developed in structural units called follicles. Usually one ovum is produced in each follicle, but thousands of cells are contained in each follicle and contribute to the development of the ovum. The larger follicles are filled with a watery or serous secretion called the follicular fluid, and as these follicles increase in size and protrude above the surface of the ovary they are called Graafian follicles. At certain stages of ovarian activity, the ovary contains one or more large solid oval structures called corpora lutea.

Immediately beneath the germinal epithelium is a layer of dense fibrous connective tissue called the tunica albuginea. Throughout the outer layer of the ovary, the cortex are follicles in all stages of development and at certain time corpora lutea. The medullary (internal) portion of the ovary contains the larger blood vessels as well as the nerve supply and a mixture of connective tissue.

The Fallopian tubes, or oviducts, are modified at the ovarian end into a thin, funnel shaped fimbriated membrane. This portion of each tube is in close association with, and partly surrounds, the ovary but is not attached to it. Because of its funnel shape, it is called the infundibulum or the fimbriated end of the Fallopian tube. This structure is especially desiged to catch the ovum at the time of ovulation. Following ovulation, the ovum rapidly descends into the Fallopian tube proper, where fertilization usually takes place. This portion of the tube is of small total diameter, being about 0.1 inch in the cow. Its wall contains smooth-muscle cells, and the lining epithlium is simple or ciliated columnar epithelium. Both the muscle cells and the cilia participate in the transport of the ovum and the sperm.

In cattle, sheep, goats and swine, the horns of the uterus are well developed, since fetal development occurs here. In swine, the horns are folded, or convoluted, and may be 4 to 5 feet in length. This is an obvious adaptation for multiple fetation. In cattle, sheep and goats, the horns are curved somewhat like the horns of a ram; hence the descriptive name.

In the mare and in woman, the uterine horns are poorly developed, and the body of the uterus is correspondingly larger, since fetal development takes place there. The lining of the entire uterus is called the mucosa. It consists of a row of columnar epithelial cells beneath which are large numbers of uterine glands. These glands produce the secretion which is evident at estrus and participate in the nutrition of the blastocyst prior to implantation. In cattle and sheep, the mucosa contains from 70 to 130 specialized areas for the attachment of the fetal placenta. These structures are called the cotyledons. In the sow and the mare, the placenta is of the diffuse type, is in more or less continuous contact, and there are no cotyledonary areas.

The muscle structure of the uterus are well developed. There is an inner layer of circular and an outer layer of longitudinal smooth muscle. These muscle cells assist by rhythmic contractions in the transport of sperm to the Fallopian tubes, in the support of the fetus during pregnancy and in contracting vigorously at parturition to expel the fetus.

The terminal portion of the uterus is called the cervix neck of womb. This part of reproductive tract had thick walls, is rather difficult to penetrate and is usually filled with a thick, sticky secretion during pregnancy.

From the cervix posteriorly to the urogenital sinus or vestibule, the female genital tract is termed the vagina. In farm animals, it lies horizontly below the rectum and above the bladder. This portion of the tract is relatively simple, both in structure and in function. It consists anatomically of the usual 3 layers of tissue, *viz.* mucosa, muscularis and serosa. The female genital tract terminates posteriorly in the vulva or exterial orifice. With the onset of heat or estrus, these structures become congested and enlarged. The clitoris, the small erectile organ, homologous to the penis in the male, is situated, just inside the portion of the vulva farthest removed from the anus. Stimulation of the clitoris results in a sexual orgasm in the female.

A single attribute, among others, that has sustained the goat as a ruminant species on earth is the relatively high reproductive efficiency in all environments.

To achieve the goal of superior reproductivity, goats have readily adopted, acquired and resorted to limited seasonality, early sexual maturity (both in males and females), vigorous sex drive, high ovulation rate, multiple births and several characteristic features in the oestrous cycle profiles. Obviously, the physiology of reproduction in goats is an elaborate process.

It is, however, admirably organized, sequenced and coordinated (with external and internal stimuli) in such a precision that a series of neurophysiological, endocrinological and behavioural events spontaneously occur in bucks and does repeatedly in cycles.

The elucidation of this complex but novel phenomena of nature would, of course, deserve independent and elaborate presentation.

The reproduction in practice is seemingly a simple but much important means of multiplication of germplasm, which is essential for the production of milk, meat and fibre.The reproduction in does extends beyond production of female gametes (ova) to the provision of site and environment for fertilization, for embryonic and fetal development to full term and finally, for the delivery of the offspring. The nursing of the new-born with nutrition (milk) and care and raising to reasonable independency are further responsibilities of the does.

8.4.2 Breeding season

The pronounced reproductive activities in temperate breeds are generally occurring in the Spring/Autumn season of the year. The length of the breeding season is primarily guided by gene-environment interactions. Most of the Indian breeds exhibit oestrus throughout the year (Kaura, 1952). A few breeds are, however, considered as seasonally polyoestrus with the extended breeding season (Amble *et al*., 1964; Mishra and Biswas, 1966; Singh and Singh, 1974; Singh *et al*., 1978). Jamunapari does come in oestrus after the first rainfall and herbage growth suggesting the possibility of oestrogenic substances in the vegetation at the time (Roy *et al*., 1962). Similarly, in Barbari does, the availability of green phytomass in certain periods of the year and seasonal rhythm modulate the reproductive phenomena (Prasad and Bhattacharyya, 1979a).

8.4.3 Puberty in does

The goat is an exception to other domestic ruminants for the early sexual maturity, both in males and females (Honker, 1952). Puberty in female goats is denoted by the age at which the growing kid shows the first oestrous behaviour, though brief and weak, in life with or without ovulation, The age at which does later become sexually mature, however, depends on the breed, the system of management, level of nutrition and season of birth (Sahni and Roy, 1967). The small and dwarf Indian breeds are known to attain puberty as early as at 150 days in an optimum environment (Ali *et al*., 1973; Prasad, 1974; Mukundan, 1976; Rahman *et al*., 1977). The medium and large breeds attain puberty at a later age (range: 325–550 days). The weight at puberty varies considerably from 9.2 to 18.0 kg in small and dwarf breeds (Ali *et al.,* 1973; Mukundan, 1976; Prasad and Bhattacharyya, 1977; Rahman *et al*., 1977); whereas, it varies from 15.66 to 24.45 kg in larger breeds (Khan *et al*., 1981; Mittal,

1988; Goel and Agrawal, 1989b). There are differences of opinion on the coverage of does at the very first oestrus and on the minimum body weight at which the doe should be bred. In order to gain the maximum reproductive life span, breeding at the 3rd oestrus is generally preferred.

8.4.4 Oestrous cycle profile

Of the 4 classic phases (Pro-oestrus, Oestrus, Met-oestrus and Di-oestrus) of the oestrous cycle, oestrus phase is merely detectable from the external signs and symptoms *e.g.* bleating, restlessness, mounting instinct, frequent micturition, vaginal discharge etc. The onset and cessation of oestrus could be rather precisely detected by the application of teaser bucks. The cyclic events from one oestrus to the next are, however, characterized by (a) compromising attitude of doe towards the buck, (b) transient harbouring of sperms in the fallopian tube, (c) ovulation(s), (d) fertilization of ovum/ova, (e) transportation of embryo(s) to uterus, (f) implantation of foetus(es), (g) gestation, (h) parturition and (i) post-partum oestrus. Failure of fertilization or premature termination of conception promptly leads to the next oestrous (curtailing the otherwise prolonged dioestrus period due to gestation), progressing through the pro–oestrus period (a few hours only). The oestrous cycle length of uncovered does has been studied in several Indian breeds which normally ranges from 17 to 22 days (Ali *et al.*, 1973; Bhattacharyya, 1982; Sharma 1985). Parity does not seem to affect the oestrous cycle length (Prasad and Pandey, 1981). The occurrence of about 10 and 12% of short and long cycles respectively, is an interesting feature in the Barbari breed (Prasad and Bhattacharyya, 1979a). The length of oestrous cycles was significantly shorter during the moderate climatic conditions as compared to severe cool-dry and hot-humid periods (Bhattacharyya, 1982). The duration of pubertal oestrus is usually shorter than the subsequent oestrus in Barbari nannies (Prasad and Bhattacharyya, 1979a). Although the onset and cessation of oestrus can occur at any time in 24 hr, the nulliparous Barbari does show a tendency to exhibit more number of onsets and cessations in the morning hours (Prasad and Bhattacharyya,1979b). The meat breeds generally exhibit pronounced oestrus for longer periods in comparison to the dairy breeds.

8.4.5 Ovulation characteristics

Goats are polyoestrus and seems to behave as intermediary of 'induced' and 'spontaneous' ovulators (Rao *et al.*, 1982). The ovulation rate is an important reproductive trait having high heritability. Information on the ovulation sequence in Indian breeds is scanty. Breed variation exists with regard to ovulation rates (Bhattacharyya and Prasad, 1974; Rao and Bhattacharyya, 1980; Goel and Agrawal, 1991). Ovulation sequence has been studied in detail in Barbari and Black Bengal goats. At least 1 ovulation per nanny in

Barbari and 2 in Black Bengal occur at 32 hr after the onset of oestrus (Prasad *et al.*, 1980; Rao and Bhattacharyya, 1980). The average ovulation rate in nulliparous Barbari nannies was 1.4; 90% of ovulations occurred within 36 hr after the onset of oestrus (Prasad *et al.,* 1980). The ovulation rate in primiparous Black Bengal nannies was 4.0 and 85% of ovulations occurred within 40 hr after the onset of oestrus (Rao and Bhattacharyya, 1980). The right ovary appears to be more active than the left in some breeds (Achuthankutty and Raja, 1971; Goel and Agrawal, 1991). Most of the ovulations seem to occur at 20–40 hr after the onset of oestrus, though breed difference is apparent.

8.4.6 Gestation period

Mean gestation period in goats is 147 days (range: 145–152 days). Several factors are known to affect the gestation period. A positive correlation exists between gestation length and the weight of the kid at birth and weight of the doe at service (Mishra *et al.*, 1979). Parity of the dam affects the gestation period (Sundarsanan and Rajan, 1973). However, sex, season, number of kids and breed do not influence the gestation period (Prasad and Pandey, 1981). In small and dwarf breeds, gestation period ranges from 143 to 147 days; while in larger breeds, it ranges from 146.40 to 156 days (Slater and Bhatia, 1935).

8.4.7 Parturition process

Gestation period for different species of mammals is fairly constant. The various symptoms of approaching parturition are not precise enough to be observed in all cases and, therefore, the appearance of visible labour pains is taken as the onset of parturition in nannies. As the process of parturition advances through dilatation of cervix, series of uterine contractions and relaxation of broad ligaments, the foetus is presented in a position that poses the least difficulty for passing through the pelvic inlet. Normal presentation co-ordinates the natural curvature of the birth canal and the curvature of the foetus (Goel and Agrawal, 1988b). The process of parturition has been studied with time sequence in a number of Indian breeds of goats. The majority of the kiddings occur during the early morning and late evening hours in Jamunapari, Barbari, Surti and Marwari breeds. A number of factors, *viz.* breed, parity, season and litter size, influence the parturition process (Tiwari *et al.*, 1969; Kuriakose, 1981; Bhattacharyya *et al.,* 1983). In small and dwarf breeds, the mean duration of parturition varies considerably from 105.00 to 364.89 min. In large breeds (*e.g.* Jamunapari and Beetal) it varies from 148.58 to 173.54 min (Chawla *et al.*, 1981).

8.4.8 Age at first kidding

Age at first kidding depends on the age of puberty and indicates the post-pubertal reproductive competency in does. This is an important reproductive trait which can be precisely determined. In small and dwarf breeds, it is 422 to 596 days whereas, in large breeds the range is 476 to 753 days. The reproductive performance and breeding efficiency of some Indian goat breeds is given in the Table 8.6.

8.4.9 Involution of uterus

The uterine involution begins with the completion of the 3 stage of parturition, *i.e.* expulsion of foetal membranes. It is considered to be complete with resumption of normal cyclic activity of gonads and re-establishment of the genital tract. The non-gravid and gravid horns of does involute completely by 20 and 40 days postpartum respectively (Shrivastava, 1979). The process of involution is completed earlier in primiparous than pluriparous nannies.

8.4.10 Post-partum interval

The interval between parturition and first post-partum oestrus is an important reproductive trait which signifies the reproductive efficiency in nannies. The earlier a doe returns to oestrus after parturition, the earlier it can be bred resulting in a shorter kidding interval which is desirable for enhancement of life time production. The post-partum interval varies among breeds, lactations, nutritional regime, season and parity (Singh *et al.*, 1986; Prakash and Khan, 1988). In small and dwarf breeds, it varies from 40 to 88.23 days (Ali *et al.*, 1973; Prasad, 1979; Sahni, 1979; Bhattacharyya *et al.*, 1981; Mukundan *et al.*, 1983), while in large breeds, it varies from 129.83 to 192.40 days (Wani *et al.*, 1980a; Chawla *et al.*, 1981; Goel and Agrawal,1989a; Mittal,1991b).

8.4.11 Kidding interval

Kidding interval is defined as the period between two consecutive kiddings and is constituted of the service period (period from parturition to conception) and gestation period. Breed, season and lactation length influence the kidding interval. In small and dwarf breeds, it varies from 219 to 299 days (Ali *et al.*, 1973; Sengar, 1979; Bhattacharyya,1982; Mukundan *et al.*, 1983), while in large breeds it varies from 229 to 359 days (Sengar, 1979; Chawla *et al.*, 1981; Mittal, 1991b).

Table : 8.6 Reproductive Performance and Breeding Efficiency

S.No.	Reproductive traits	Sirohi	Marwari	Kutchi	Jamunapari	Barbari
1	Age at 1st kidding (d)	661.0 ± 16.0	563.1 ± 17.1	777.5 ± 20.4	786.5 ± 23.2	490.7 ± 15.5
2	Tupping %	85.6	87.4	87.7	-	-
3	Breeding efficiency (%) (on the basis of does tupped)	88.2	91.3	83.4	-	-
4	Kidding %	112.5	113.6	110.0	134.8	158.9
5	Prolificacy %					
	Single	87.5	86.4	90.1	68.7	44.8
	Twin	12.5	13.4	10.0	29.8	51.5
	Triplets	0.0	0.2	0.0	2.5	3.7
6.	1st kidding interval (d)	469.2 ± 16.4	375.5 ± 11.1	464.3 ± 20.9	390.2 ± 10.6	290.9 ± 14.1
7	1st service period (d)	235.4 ± 9.9	183.8 ± 7.8	250.6 ± 15.4	243.4 ± 0.1	143.8 ± 15.0

Source: Personal Communications, Bhat (2008)

CHAPTER 9

NUTRITION AND FEEDING OF GOATS

Feed input is an important component of any livestock production system and the same is true for goat also. Its proper management is an important aspect of any goat production enterprise. To maximize production, it is essential to provide adequate/appropriate nutrition. A proper nutrition input is necessary for maximum exploitation of the genetic potential. One of the major factors contributing to low productivity of goats in our country is non availability of adequate nutrition as per the requirement.

The basic physiological needs of animals are for maintenance and production. The production functions include growth, reproduction, lactation and hair production. Nutrient requirement for these physiological functions differ and in their feeding behaviour, individual differences also exist. In the present goat production system in our country, the feeding management practices being followed in organised farms are quite different from those adopted by the farmers in the field.

Adequate nutritional inputs should be provided as per the requirement of different physiological functions. To adopt suitable feed management practices, the goat flock has to be divided into different groups depending upon the physiological functions like growth, pregnancy, and lactation over and above the maintenance.

9.1 Feeding for Maintenance

An animal has to maintain itself before it is able to produce in the form of tissue growth, development of foetus and milk or hair production. Maintenance requirements include the requirements for basal metabolism (BM) plus the normal activity without any production. Energy is the basic need to keep the animals' physiological processes going. The next is nitrogen (protein) which is required to meet the wear and tear of tissues. Mineral and vitamins are also needed to aid metabolic processes and normal tissue functioning.

9.2 Energy Requirement

Energy requirements for maintenance could be met from grazing alone, if a well developed pasture is available. Generally, during the period from January to June in arid and semi-arid areas, the energy availability to the animal is even less than the maintenance requirement thereby necessitating additional supplementation. The TDN content of the pasture, forage is about 50–55% during lush green season and 35–40% during dry months. Assessment, therefore, has to be made at the local level as to whether the pasture available could maintain the animals. If the animals are fed on stalls on legume hay or high quality grass, it is sufficient to maintain the animals. The vegetation resources on the grazing land should be developed in a 3-tier system (grass-shrub-trees). No additional supplementation is required just to maintain the animal if sufficient vegetation is available in the grazing area. The best system of management should be to leave the animals to themselves, in the paddock, although the initial investment is very high to make a paddock.

9.3 Protein Requirement

The protein requirement for maintaining the body weight of the animal is based on the metabolic faecal nitrogen (MFN) and the endogenous urinary nitrogen (EUN) excretion plus the nitrogen requirements for normal growth of their hair, hoof and nitrogen loss from the skin. About 2 mg of nitrogen or 12.5 g of protein are required per Kcal of basal heat production. This value refers to protein actually needed in the replacement of nitrogen. In actual feeding practice, extra amount should be provided to allow for wastage in digestion and metabolism. Provision of twice the amount of nitrogen/protein required per Kcal of basal heat would meet the maintenance requirement of digestible protein.

9.4 Mineral and Vitamin Requirement

If the energy and protein requirements are met through conventional feeds, the mineral and vitamin requirements are also met for maintaining the weights of animals except in specific deficient pasture or feed. During summer months, there is a possibility of Vitamin A deficiency but this has not been demonstrated in clinical form, although there is evidence of low level of Vitamin A in blood during summer.

9.5 Components of Goat Nutrition

Goats are normally reared on browse and pasture forage that other ruminants do not consume. Major part of the feed for goats comes from natural vegetation on common grazing lands, range land and other non-cultivable areas. Unfortunately, no reliable estimates of feed and fodder resources for goats in such areas are available.

Goats generally accept a wide variety of feeds but what is acceptable to one may not be equally acceptable to the other. They prefer to select from the wide variety of feeds and vegetations (preferably leaves) and like fresh fodder, grains, seeds and pellets rather than the wet feeds, silages, chopped greens, soiled forages and hays. Goats have a higher tolerance to the wide variety of otherwise undesirable phytochemical compounds which enable them to consume a wide range of plant species.

Limited information is available on the accessibility of nutrients vis-a-vis requirement under the extensive system of rearing goats. The unmanaged ranges, however, could seldom provide enough forages and top feeds essential to provide the adequate nutrients. Except for the rainy season, the phytomass available in the major grazing tracts of the country lacks in crude protein and energy (Rekib and Patil, 1981).

9.6 Traditional Grazing Practices

About 80–90% of the natural range lands fall under 'Poor' to 'Very poor' classification. The available vegetation, though may yield dry matter up to 2.4 t/ha (Table 9.1), is generally very high in cell wall constituents and silica which lower the dry matter intake and digestibility. It has been reported that the natural grazing lands normally have the capability to produce 1.6 q crude protein and 12,624 Mcal energy from each hectare (Singh,1988; Sharma *et al.,* 1989). However, most of these nutrients cannot be utilized fully as the forage production is mostly restricted to the rainy season and often rested in inaccessible areas, which cannot be profitably collected or stored. Even during the rainy season, more than 80–90% of the available 220 million tonnes of grasses in community grazing lands are utilized by the large ruminants, cattle and buffaloes. The goats have to thrive on the left over vegetation which contains approximately 0.84 to 1.6% DCP and less than 50% TDN (equivalent to less than 2.2 Mcal/kg phytomass) and normally not relished by the other livestock (Bhatia *et al*., 1973; Mali *et al*., 1983). The scarcity of range-land vegetation is thus acute and adversely affects the productivity of goats due to low nutrient intake. The self–generating capability of many of the range areas has diminished to critical levels over the years and the areas of concentrated grazing have patches of bare soil causing erosional hazards.

The goat production under conditions of variable nutrient availability from forages calls for effective agro-silvipasture management to minimize the scarcity during lean seasons. Experiments at CIRG, Makhdoom have shown that the fodder production, under dry land conditions of Yamuna ravines, can be raised on dry matter basis from 2.4 to 8.0 t/ha fodder per annum (Sharma *et al.*, 1989). The forage production and availability period from degraded lands under natural and silvipastoral systems are given in Table 9.1. The low protein level of the grasslands in the tropics can be improved by 2–7% easily through introduction of legumes, *viz.* Siratro, Atylosia and Stylo (Singh, 1987).

Table 9.1 Forage Production and Availability Period from Degraded Lands

Systems	Forage yield (DM t/ha)	Crude protein (q/ha)	Gross energy (Mecal/ha)	Availability
Natural grazing land	2.4	1.6	12,624	July-September
Improved grass land	2.5 – 6.0	3.5	–	July-October
Silvipasture system (2–Tier)	3.5 – 7.5	6.0 – 7.0	24,705	March-December
Silvipasture system (3–Tier)	7.5 – 9.0	10.8 – 12.7	33,800	Full year

Adopted from Singh (1987), Sharma *et al.* (1989)

9.7 Grazing

The goats are alert to the food potential above their heads (due to advantageous setting of eye balls) and can defoliate the smallest branch of tree without damaging the twigs up to the maximum height they reach on hind legs. The highly mobile upper lips of goats enable them to graze as close to the ground as the sheep whilst being more selective than the cattle. Goats can thrive very well in the mixed farming systems with other domestic livestock in the forest, as well as, common grazing land (Table 9.2).

Table 9.2 Grazing Choice of Range Plants by Domestic Livestock, Percent of Intake

Plants	Grazing			
	Horses	Cattle	Sheep	Goats
Grass	90	70	60	20
Weeds	4	20	30	20
Browse	6	10	10	10

Source: Bell (1978)

Of the various nutrients involved in the feeding of goats, energy and protein happen to be more important.

From the data available through metabolic trials on the limited number of animals, it is evident that the protein and energy requirements for maintenance are higher per unit metabolic body weight in goats. Goats have the capability of consuming dry matter to the tune of 5 to 8% of its body weight. As a species, it can utilise lignin and cellulose better than the other ruminants and sustain water deprivation for longer periods. The nitrogen recycling through rumen is also considered better.

The deficiency of major nutrients, energy, protein and dry matter, in the country amounts to 50–60, 50–75 and 80% respectively. Availability of the grass, browse and agro-industrial by-products for goats is approximately to the tune of 40, 9.6 and 48 million tonnes/annum in the country.

9.8 Top Feed Resources

A large variety of tree leaves (top feed) serves as promising feed resource for goats (Table 9.3). It is estimated that the annual production of green leaves for fodder from trees in the country is to the tune of 24 million tonnes (NWDB, 1987). Against the requirement of 1.9 mt of DCP and 17 × 10 million Mcal DE, the tree leaves provide approximately 0.7 Mt DCP and 4 × 10 m Mcal DE to goats. Most of the tree leaves contain 20–40% dry matter, 4–15% DCP and 50–60% TDN depending on the season of harvest but their palatability is 'poor' to 'medium' which limits the energy supply to goats (Saha and Gupta, 1987). The low digestibility of several tree leaves is ascribed to the high lignin and tannin contents.

Many tropical forage plants contain anti-quality compounds for which the palatability is reduced and digestibility decreased, besides metabolic disorders at times. A few of the known anti-quality factors in tropical tree forages are detailed in Table 9.4.

9.9 Cultivated Fodders

In areas where the intensive agriculture is practiced and the land holding is too small, it is desirable to rear goats under stall feeding. However, intensive production of goats either on cultivated fodders or organized pastures is not yet common in India. It has been demonstrated in several studies conducted at CIRG that goats can be reared successfully under stall feeding. Amongst fodder crops, goats prefer leguminous fodders like berseem (*Trifolium alexandrium*), lucerne (*Medicago sativa*), cowpea (*Vigna unguiculata*), guar (*Cyamopsis tetragonoloba*) both as green and hay; they eat less of cereal fodders though

Table 9.3 Nutritive Value of Some Fodder-Tree Leaves, Grasses, Legumes

Name	Nutritive Value (Dry Matter Basis)			
	DCP (%)	TDN (%)	DE (Mcal/kg)	ME (Mcal/kg)
A. Tree leaves				
Acacia catechu (Khair)	4.49	46.33	2.03	1.66
Adina cordifolia (Haldu)	3.50	50.90	2.23	1.83
Aegle marmelos (Acacia)	10.88	56.67	2.43	2.04
Ailanthus excelsa (Ardu)	16.20	63.80	2.80	2.29
Alibizia chinensis (Siras)	4.87	40.23	1.77	1.44
A. lebbeck (Siris)	11.26–17.75	60.48	2.66	2.17
Azadirachta indica (Neem)				
- May	8.38	53.28	2.34	1.92
- Sept.	9.28	42.76	1.88	1.54
Bauhina variegata (Kachnar)	9.16	53.54	2 44.00	2.00
Cordia dichotoma (Lasura)	8.78–10.74	42.45	1.86	1.52
Dichrostachys nutens (Brij babool)	5.93	61.39	2.70	2.21
Ficus glomerata (Gular)	6.69	53.82	2.36	1.93
Ficus lacor - Young	6.27	61.99	2.73	2.24
- Mature	4.04	40.74	1.79	1.47
Ficus religiosa (Pipal)	7.00	38.26	1.68	1.38
Ficus roxburghii (Timlia)	5.74–6.24	54.79	2.43	1.97
Grewin optiva	12.30–12.41	61.35–62.13	2.71	2.22
Leucaena leucocephala	16.37	70.22	3.08	2.52
Madhuca longifolia (Mahua)	6.96	37.04	1.63	1.33
Moringa oleifera (Drumstick)	11.09	61.49	2.71	2.22
Morus alba (Shahtut)	10.68	59.59	2.62	2.15
Ougeinia oojeinensis	4.27	45.47	2.00	1.63
Prosopis cineraria (Khejri)	4.49	40.99	1.80	1.47
Terminalia belerica (Behera)	0.86	54.45	2.40	1.97
Zizyphus mauritiana (Ber)	3.09	30.65	1.35	1.10
Zizyphus nummularia (Jharberi)	5.45	51.10	2.25	1.84
B. Grasses				
Brachiaria mutica	7.5	46.0	2.03	1.66
Cenchrus ciliaris	5.5	57.0	2.49	2.07
Chloris gayana	4.3	56.0	2.44	2.02
Eragrostis curvula	6.5	61.0	2.60	2.18
Panicum maximum	6.6	57.0	2.52	2.10
Paspalum dietratum (Dallis grass)	13.9	62.0	2.65	2.22
Pennisetum clandestinum	6.2	55.0	2.39	1.96
Pennisetum purpureum	9.2	61.0	2.61	2.18
Setaria sphacelata (Moss)	3.4	58.0	2.54	2.12
Urochlo mosambicensis (Sabi grass)	5.8	38.5	1.70	1.40
C. Legumes				
Alhgi pseudalhgi	11.6	49.0	2.14	1.71

Table 9.3 (Contd.)

Name	Nutritive Value (Dry Matter Basis)			
	DCP (%)	TDN (%)	DE (Mcal/kg)	ME (Mcal/kg)
Dolichos lablab (Sem)	16.3	63.0	2.80	2.37
Desmodium spp. (Kudaliya)	14.6	69.0	3.03	2.61
Indigofera endecaphylla (Trailing indigo)	11.5	64.0	2.81	2.39
Medicago sativa (Lucerne)	19.0	63.0	2.67	2.25
Pueraria phaseoloides (Kudzu)	13.0	65.0	2.87	2.45
Stylosanthes spp. (Stylo hay)	9.8	59.0	2.61	2.19
Trifolium repens Always humid climates (White cover or Shaftal)	19.6	74.0	3.07	2.65
Trifolium resupinatum, always humid climates (Persian clover)	13.0	65.0	2.85	2.43

The Local names are given in parenthesis.
Source: Hand Book of Animal Husbandry (3rd Edn., 2002) - ICAR, New Delhi

Table 9.4. Anti-Quality Factors in Forage Supplements

Fodder	Distribution	Toxin	Toxic effect
Acacia leaves	Entire tropical zone	HCN, tannins, fluroacetic acid, oxalates	Laboured respiration, depression, coma and death
Albizia foliage	All over India	Tannins	Depression in dry matter and protein digestibility, impaired kidney and liver functions
Blighia sapida Seeds	Tamil Nadu	Hypoglycin	Hypoglycaemia
Delphinium spp.	Western, central and eastern Himalayas	Alkaloids	Stiffened joints, bloat and death
Indigofera herbage	All over India	Indopicine, 3-nitropropanic acid	Hepatotoxic agent
Lantana foliage	All over India	Lantadene	Hepatotoxic agent, photo-phobia and blindness
Leucaena foliage	Entire tropical zone	Mimosine	Goiterogenic and alopasia
Robinia foliage	Western Himalayas	Robin phytotoxin	
Sambuscus nigra	Himachal	Sambunigrin (Glucoside)	Depressed respiration, coma and death

Source: Hand Book of Animal Husbandry (3rd Edn., 2002) - ICAR, New Delhi

relish oat (*Avena sativa*), maize (*Zea mays*), barley (*Hordeum vulgare*), sorghum (*Sorghum bicolor*), bajra (*Pennisetum typhoides*), etc. Goats are known to relish paragrass (*Brachiaria mutica*), star grass (*Cynodon plectostachyam*), napier grass (*Pennisetum purpureum*) and guinea grass (Panicum maximum) over many improved tropical grasses and legumes. The non-leguminous fodders can be fed with legumes in the ratio of 1:1. Some of the agricultural by-products like gram (*Cicer arietinum*), arhar (*Cajanus cajan*) and bhoosa are very much liked by goats. Nutritive value of promising cultivated fodders and crop residues is given in Table 9.5.

9.10 Nutrient Requirement

The nutrient requirements for goats appear to be practically the same as those of sheep, but in account of their special characteristics mentioned earlier, it is necessary that systematic research work should be carried out to find out their requirements for growth, maintenance, gestation and milk production, so that it may be possible to get maximum amount of milk and meat at a minimum cost. So far the nutrient requirement data of sheep is generally applied for goats, as very little work has been conducted in this direction. On this basis, the energy requirement for goats for maintenance is taken to be about 1.0 kg TDN/day/100 kg live weight, while the digestible crude protein requirements for maintenance and for milk production are 45 to 64 g/100 kg live weight and 70 g per litre of milk, respectively. The calcium and phosphorus requirements of goats are 147 and 72 mg/kg live weight, respectively. The mineral and vitamin needs of goats are equally important. Goats in milk have a very high requirement for sodium chloride, hence salt licks and or mixed mineral licks should be made available. Adequate supply of water for drinking ad lib should also be provided particularly during the dry season.

Most of the Indian workers have worked out the requirements for maintenance only using different methods as per details given below :

9.10.1 Based on EUN and MFN trials

Majumdar (1960a) was probably the first to have attempted observations on the minimum and maintenance requirements of protein employing endogenous urinary nitrogen (EUN) and metabolic faecal nitrogen (MFN) excretion on Jamunapari bucks. In the absence of a completely N-free roughage, he employed wheat bhusa and alkali-treated bagasse in 2 separate experiments. For the adequate supply of energy, vitamins and minerals, a concentrate mixture of sago, sugar and oil was also used in them. The closely agreeing EUN and MFN values of the 2 experiments were pooled to work out the minimum and maintenance protein requirements assuming the biological value of protein as 50 % (Maynard, 1951) in the formula.

Table 9.5. Nutritive value of promising cultivated fodders and crop residues for goats

Name		Nutritive value (Dry matter basis)				Range
		DCP (%)	TDN (%)	DE (Mcal/kg)	ME) (Mcal/kg)	
Avena sativa (Oats)	- Fresh	6.6	62.4	2.7	2.3	U.P., Haryana, Punjab, M.P., H.P., J&K, Bihar, Rajasthan, Maharashtra, A.P. and parts of Tamil Nadu
Bambusa vulgaris (Bamboo)	- Fresh	10.2	48.9	2.1	1.7	Hotter parts of India
Cajanus cajan (Pigeonpea)	- Leaves	19.5	68.0	3.1	2.6	M.P., Bihar, A.P., Maharashtra, U.P. and Karnataka
	- Bhusa	4.7	49.0	2.2	1.7	
Cicer arietinum (Chickpea)	- Seeds	16.3	70.0	3.0	2.7	U.P., Punjab, Haryana, Rajasthan, Bihar and M.P.
	- Bhusa	3.6	47.0	2.0	1.6	M.P.
Cyamopsis tetragonoloba (Guar)	- Fresh	11.3	64.0	2.8	2.3	Haryana, Punjab, Rajasthan, Gujarat, Maharashtra, U.P. and M.P.
Glycine max (Soyabean)	- Fresh	18.0	73.0	3.2	2.8	Northern India, Naga hills and Manipur
	- Meal	46.6	81.0	3.6	3.2	
Helianthus anus (Sunflower)	- Fresh	6.9	55.0	2.4	2.0	Throughout India
Hordeum vulgare (Barley)	- Fresh	8.2	60.5	2.7	2.3	U.P., Rajasthan, Bihar, M.P., Haryana, Punjab, West Bengal, H.P. and J&K
Manihot esculenta (Cassava)	- Fresh	22.1	71.0	3.1	2.7	Kerala, Tamil Nadu and Karnataka
Melilotus indica (Sweet clover)	- Fresh	13.3	63.8	2.8	2.3	Throughout north India

Table 9.5 Contd.

Name		Nutritive value (Dry matter basis)				Range
		DCP (%)	TDN (%)	DE (Mcal/kg)	ME) (Mcal/kg)	
Medicago sativa (Lucerne)	- Fresh	19.9	63.0	2.7	2.2	Always humid climates
	- Hay	13.4	59.0	2.6	2.2	
Pennisetum typhoides (Pearlmillet)	- Fresh	5.8	63.8	2.8	2.4	U.P., Maharashtra, Rajasthan, A.P.,Tamil Nadu and Punjab
Phaseolus aureus (Green gram)	- Fresh	7.7	63.0	2.8	2.3	Throughout India
	- Pods	6.8	57.0	2.5	2.1	
	- Bhusa	3.9	48.6	2.1	1.7	
Saccharum officinarium (Sugarcane)	- Aerial part	2.3	58.0	2.6	2.2	Grown chiefly in U.P, Bihar, Punjab and Maharashtra
	- Bagasse	0.6	51.0	2.2	1.8	
	- Molasses	0.8	92.0	4.0	3.7	
Sorghum bicolor (Sorghum)	- Fresh	2.3	56.0	2.5	2.0	Hot arid and wet humid climate up to an altitude of 1500-1600 m
Triticum aestivum (Wheat)	- Bran	12.5	72.0	3.2	2.6	Punjab, Haryana, U.P., M.P., Maharashtra, Bihar and Rajasthan
	- Bhusa	0.0	40.0	1.8	1.3	
	- Grain	8.9	92.0	4.0	3.6	
Trifolium alexandrium (Berseem)	- Fresh	17.5	70.2	3.1	2.7	Western UP, Punjab, HP, Bihar, Tamil Nadu, Maharashtra & Karnataka
Vigna sinensis (Cowpea)	- Fresh	12.5	62.0	2.7	2.2	Cultivated throughout India
	- Hay	10.2	53.0	2.4	1.9	
	- Seeds	21.1	84.0	3.7	3.3	
Zea mays (Maize)	- Fresh	4.7	68.0	3.0	2.6	UP, Punjab, MP, Bihar, AP and J&K
	- Grain	8.8	89.0	4.0	3.2	
	- Stems	1.2	51.0	2.2	1.8	

English names are given in parentheses.

Source: Hand Book of Animal Husbandry (3rd Edition, 2002) - ICAR, New Delhi

Minimum protein (DP)=EUN×6.25×100/BVg/kg body weight requirement

Experiment	No. of Animals	EUN (g/kg/BW)	MFN (g/100g DM Intake)
Low-N diet	5	0.054	0.40
N-free diet	3	0.050	0.43
Average (0.1285 g/$W^{0.75}$ kg)	-	0.052	0.41

Minimum protein (DP) requirement = $0.052 \times 6.25 \times 2$ g/kg body weight or
= 0.065 kg/100 kg body weight

This when converted to body size of average weight (37.30 kg and body size $W^{0.75}$ kg = 15.09), it worked out to 1.61 g DP/$W^{0.75}$kg.

Maintenance

For calculating the maintenance requirement, the suggestion of Crampton (1961), that MFN losses should also be considered along with EUN, was made use of and the following formula of Smith *et al.* (1965) has been employed:

Maintenance protein requirement = (EUN × 6.25 × 100/BV) + (MFN × 6.25 × 100/BV–1) g/kg BW (DP)

This gave the maintenance requirement of protein as 2.50 g/DP/$W^{0.75}$kg. The energy need for maintenance was also calculated as 141.08 Kcal DE/$W^{0.75}$kg.

The maintenance requirement thus calculated is 0.09% digestible crude protein (DCP) and 0.09% total digestible nutritents (TDN). One interesting point is that for its size, the goat can consume substantially more feed than cattle or sheep, *viz.* 6.5 to 11% of the body weight in dry matter when compared to 2.5 to 3.0% in cattle or sheep. This means that the goat can satisfy its maintenance requirement and can produce milk from forage alone.

Utilizable protein

Further, for calculating the utilizable protein, the value of 'K' in the general formula UP = K $W^{0.75}$kg (where 'W' = weight in kg and K = constant), if taken as 82, it becomes of general applicability to all the farm animals (Mitchell, 1929). The value of K obtained for 8 animals in this case averaged down to 83 as against 69 for sheep (Smuts and Marais, 1939) and 75 for pigs (Smuts, 1935).

9.10.2 Based on metabolism trials and balance studies

Majumdar (1960b) also attempted digestible protein requirements for maintenance from balance studies. He conducted 4 metabolic trials using rape-cake as the sole concentrate and wheat bhusa as the sole roughage on 6 adult Jamunapari bucks. Calcium as $CaCO_3$ and phosphorus as Na_2HPO_4 were also provided as and when necessary to keep the experimental subjects in positive nutritional balance in respect of Ca and P. The protein intake was varied to bring the N balance from an initial negative to a positive one. Of these, the first preliminary trial, which had negative Ca and P balances was rejected. The nutritional balances, as recorded in the remaining 3 trials were then subjected to 4 different methods for working out the requirement using 66% digestibility of the feed protein as under:

Method	DP/100 kg Body Weight (g)	DP/W 0.75 kg (g)
Graphical	122.00	3.87
Statistical	120.00	3.80
Percentage Utilization	122.00	3.86
Harris-Mitchel Formula	111.00	3.52
Mean Value	119.00	3.76

9.10.3 Based on DM/IBN intake

The requirements of energy and protein have also been estimated on the basis of DM/TDN intake values using

1 kg DM = 2.75 Mcal DE with 62.50% DM digestibility
or,
1 kg DM = 2.44 Mcal DE with 55.40% DM digestibility
1 kg TDN = 4.4 Mcal DE

Data of Majumdar (1960a), reported for the DM intake, were also processed by Sengar (1980) for this calculation. Although the experiment was not planned for the purpose, the DM intake was almost of the order of maintenance. In the absence of the DM digestibility, 55.40% digestibility has been used on the pooled DM intake of 761.60 g to arrive at the DE requirement as 163 Kcal $W^{0.75}$ kg. From this the DP requirement has been calculated using the equivalence 1 Mcal = 25 g DP as Protein requirement = 4.075 g DP/$W^{0.75}$kg or = DP/100 kg body weight or = 0.129 kg.

9.10.4 Based on basal calories

Majumbdar (1960b) also calculated the nitrogen requirement from the EUN and MFN values and the N-intake per basal calorie as 12.19 mg (A/B) using the formula:

N requirement = 70 ($W^{0.75}$kg) × A/B

Where, A is total loss of N/basal calories at 0 N-intake and B is BV × TD expressed as decimals.

Since this calculation is based on the use of EUN and MFN and BV and TD, values of the same animals, the requirement work out to 3.52g DP/ $W^{0.75}$kg, which happens to be exactly the same as in case of Harris-Mitchell formula method.

9.10.5 Work at RBS College, Bichpuri (Agra)

Singh and Sengar from 1965 to 1980 conducted a series of experiments on nutrition in the PL480 research projects on goats induding their requirements of energy and protein for different physiological functions employing different methods as under:

9.10.5.1 *Based on metabolism trails and balance studies*

Singh and Sengar (1970) on the basis of N balance data of metabolic experiments involving Barbari and Jamunapari goats in the PL480 Research Project No. A7–AH–18, worked out the maintenance requirement of protein (DP). They employed various methods and used the EUN and MFN values of Majumdar (1960a) in Harris Mitchell formula method. The requirements as grossly arrived at by different methods, on the basis of 26.36 kg body weight and size $W^{0.75}$kg 11.63 kg, remained as:

Method	DP/100 kg Body Weight(g)	DP/$W^{0.75}$ kg (g)
Graphical	109.00	3.45
Statistical	111.00	3.50
Percentage utilization	111.00	3.52
Harris-Mitchell Formula	96.00	3.04
Mean Value	107.00	3.38

9.10.5.2 *Based on DM/TDN Intake*

The requirements of energy and protein were also calculated from DM/TDN intake (Sengar,1980) of the balance data using:

1 kg DM = 2.75 Mcal DE with 62.50% DM digestibility
or
1 kg DM = 2.44 Mcal DE with 55.40% DM digestibility
1 kg TDN = 4.40 Mcal DE

The Dry-matter intake data of the positive balance trials of the PL480 Project No.A7–AH–18 (Singh and Sengar,1970) were also made use of in assessing the energy and protein requirements. The figures arrived at for both energy and protein, respectively, were:

Energy = 138.93 Kcal ME/$W^{0.75}$kg

Protein = 3.40 g DP/$W^{0.75}$kg or = 0.110 kg DP/100 kg body weight

9.10.5.3 *Based on basal calories*

Sengar (1980) worked out the protein requirement for maintenance with the help of basal calories, which worked out to 10.01 mg (A/B) in the formula and the N-intake per basal calories.

N requirement = 70 ($W^{0.75}$ kg) × A/B
This gave the requirement of protein as 3.02 g DP/ $W^{0.75}$ kg

9.10.5.4 *Based on EUN and MFN trials*

Rajpoot *et al.* (1979) reported requirements of protein on EUN and MFN basis. They used 4 large (Jamunapari and Beetal) and 6 small (Barbari and Black Bengal) males on two low-N diets. The minimum and maintenance requirements of protein were calculated as under :

Minimum of DP requirement = EUN g/kg BW × 6.25 × 100/BV g/kg body weight assuming the biological value of feed protein as 50% (Maynard, 1951). On the basis of this assumption the minimum requirement of protein worked out to 55 g/100 kg body weight or 1.42 g/$W^{0.75}$ kg. While it compared well with 55 g/100 kg body weight of Hutchinson and Morris (1936), it was low as compared with the one (65 g/100 kg body weight) reported by Majumdar (1960a). The investigated value of 1.42 g/$W^{0.75}$ kg, while closely agreed with 1.41g/$W^{0.75}$ kg of Devendra (1979a), it compared high with 1.22 g/$W^{0.75}$ kg of Singh and Mudgal (1978).

The protein requirement for maintenance was also calculated after the suggestion of Crampton (1961) that the MFN losses should also be considered along with the EUN. The formula of Smith *et al*. (1965) was, therefore, employed for calculating the same as under:

Maintenance EUN g/kg BW × 6.25 × 100/BV + requirement (DP) = MFN g/100 g DM intake × 6.25 (100/BV–1) g/kg BW.

The values thus arrived at average down to 3.05 g/$W^{0.75}$ or 116 g/100 kg body weight inclusive of the 25% safety margin (NRC, 1971; Banerjee, 1978).

For calculating the utilizable protein requirement, the general formula P = K ($W^{0.75}$ kg), where K is constant was also used. From these studies the value of K for goats was calculated as 0.72 (range from 0.658–0.804). This makes the formula P=0.72 $W^{0.75}$ kg as against P=0.83 $W^{0.75}$ kg of Majumdar (1960a).

9.10.5.5 *Based on energy: protein ratio*

Rajpoot (1981) while persuing a project on the requirements of energy and protein resorted to this method for working out these requirements for maintenance, growth, pregnancy and lactation. Since some of these requirements generally overlap, balance studies have little significance. As such the metabolism trials had their own limitations, specially in case of protein (Singh, 1976). It was, therefore, decided to run short-term feeding trials for determination of energy needs for the different functions precisely and then work out the protein requirements on the basis of almost constant relationship between DE and DP (1kg DE = 25 g DP) as under:

9.10.5.6 *Requirements for maintenance*

Healthy adult non- producing animals (13) were kept on a ration consisting of gram bhusa (dry roughage) and lucerne (green roughage) for 8 weeks. All these animals were individually weighed for 3 consecutive days, caged simultaneously and fed the 2 roughages *ad lib*. Their daily dry mater intake through each component and excretion through faeces individually by each animal, were measured for the last 10 days (Rajpoot *et al*., 1979).

The gross energy values of the total intake and excretion were then determined with the help of Bomb calorimeter. The gain/loss in the body weight during 8 weeks was considered as the storage of net energy in the form of fat and was accounted for from the ingested values using the conversion factor 9.0

Kcal NE/g gain. To convert one form of energy into the another, recommendation (100 GE = 76 DE = 62 ME = 35 NE) of Garren *et al.* (1959) was employed. The maintenance requirement of protein was then calculated on the basis of energy intake (1 Mcal DE = 25 g DP).

The maintenance requirement of energy and protein in this case worked out to 125.00 Kcal NE/$W^{0.75}$kg and 3.13.g DP/$W^{0.75}$kg.

9.10.5.7 *Requirement for milk production/lactation*

Requirement for production of a litre of milk with 3.0% fat is 43 g of DCP and 200 g of starch equivalent (SE), whereas for production of 1 litre of milk with 4.5% of fat, it is 60 g of DCP and 285 g of SE. The nutritional requirement of a goat weighing 50 kg and yielding 2 litres of milk with 4% fat may be met by feeding 400 g of concentrate mixture and 5 kg of berseem or lucerne. The ration should have 12–15% protein content depending on the amount of protein in their hay and in the milk produced. The following concentrate mixture may be used to feed the goat:

- i One part of wheat bran, 2 parts of maize grain and one part of linseed cake, or
- ii 2 parts of maize grain, one part of barley, 2 part of mustard cake and 2 parts of gram husk, or
- iii 1 part of wheat bran, 2 parts of barley grain, one part of groundnut cake, or
- iv 2 parts of gram grain and one part of wheat bran.
- v These mixtures should also contain 2% each of mineral mixture and salt.

The requirements of lactation were determined on the basis of a feeding trial conducted on 19 goats involving 4 Jamunapari, 5 Beetal, 5 Barbari and 5 Black Bengal, which were maintained on a liberal diet right from the advanced pregnancy having concentrate mixture, gram bhoosa and green lucerne. Towards the end of the 2nd month of lactation, they were individually caged for 10 days for recording the intake and outgo of energy (Rajpoot *et al.*, 1979).

Simultaneously, the amount of milk produced by each animal, along with its fat percentage was also recorded. The body weights, as usual, were recorded both before and after the experiment. The total energy requirements of the lactating animals were then worked out as such and then apportioned separately from milk (4% FCM) lactation maintenance and the gain/loss in body weight, if any.

The requirement of protein (DP) for lactation maintenance was then calculated on the basis of energy intake (1 Mcal DE = 25 g DP) as per details given below:

9.10.5.8 *Requirements for growth*

The growth requirements of energy and protein were estimated involving 35 kids of different breeds (2 Jamunapari-A, 11 Beetal-B, 7 Barbari-C and 15 Black Bengal-D), ages and body weights. These were maintained on a liberal ration consisting of a concentrate mixture (barley 25%, maize 20%, groudnut-cake 30%, gram 5% wheat bran 8%, molasses 5%, chalk 2%, salt 2% and mineral mixture 3%) and roughages (gram bhoosa and green Napier) for 8 weeks. They were caged individually during the last 10 days for recording the intake and excretion of energy (Rajpoot *et al*., 1979).

Per day gains in body weight were calculated on the basis of body weights before and after the experiment. From the total energy ingested, the maintenance part was subtracted to arrive at the energy required for the gain.

The protein requirement of these growing animals was then calculated on the basis of energy intake (1 Mcal DE = 25 g DP) as usual.

The consolidated requirement of energy and protein for growth, irrespective of stage of growth, therefore comes to 7.88 Kcal DE and 0.20 g DP per g gain or 232.83 Kcal DE and 5.82 g DP/$W^{0.75}$ gaining about 88.50 per day.

9.10.5.9 *Requirements for pregnancy*

This experiment involved 10 goats of each Jamunapari (A), Beetal (B), Barbari (C) and Black Bengal (D). They were weighed for 3 consecutive days and were kept on a liberal diet from the beginning of the third month of pregnancy and continued up to the kidding stage. Their ration included concentrate mixture, gram bhoosa and green sorghum. They were individually caged during the fifth month of pregnancy for the precise observations on their energy intake and outgo (Rajpoot *et al*., 1979).

Body weights just before and after kidding along with the weight of kid/kids, placenta and placental fluids were also recorded. The gain/loss in the body weight of the animal from the 3rd month of pregnancy to immediately after kidding was then calculated and accounted for from the ingested values as depot fat (9 Kcal/g fat). The requirement of protein (DP) was then deduced from the energy requirement (1Mcal DE = 25 g DP).

The 221.79 Kcal DE and 5.55 g DP/$W^{0.75}$ kg, thus worked out as the total requirement (Maintenance + pregnancy) of goats with a mean body weight of 30.38 kg and gaining pregnancy weight at the rate of 61.75 g/day.

The foetal growth in the last 2 months of pregnancy is rapid and the metabolic rate of the goat rises rapidly. During this period, the content of ration should be increased to the level of production ration. A week before she kids, the doe should be provided with more succulent type of food. For 3 to 4 days after kidding, the level of diet should be lowered and made more fibrous. This is necessary to minimize shock to the goat's udder and the metabolism of sudden flow of milk. After this, feeding should be done at a normal rate.

9.10.5.10 *Feeding of young stock*

Feeding schedule of kids should be such that a weekly growth rate of 0.6 kg is obtained. The kid should be fed 56–112 g of colostrum 4 to 5 times a day depending on its birth weight, for 3 days. From the 4th day onward they may be fed the following ration given in Table 9.6.

Table 9.6 Feeding of Young Stock

Body weight	Milk		Concentrate mixture* per day (g)	Green fodder lucerne or berseem (kg)
	Moring (ml)	Evening (ml)		
2.5	200	200	–	–
1.0	250	250	–	–
3.5	200	200	–	–
4.0	250	250	–	–
5.0	200	200	50	*Ad lib*
6.0	250	250	100	*Ad lib*
7.0	200	200	150	*Ad lib*
8.0	250	250	200	*Ad lib*
9.0	200	200	250	*Ad lib*
10.0	250	250	350	*Ad lib*
15.0	250	250	350	*Ad lib*
20.0	–	–	350	*Ad lib*
25.0	–	–	350	1.5
30.0	–	–	350	2.0
40.0	–	–	350	2.5
50.0	–	–	350	4.0
60.0			350	5.0
70.0			350	5.5

* The composition of concentrate mixture (in parts): grams 20, maize 22, groundnut cake 35, wheat bran 20, mineral mixture 2.5, common salt 0.5.

9.10.5.11 *Mineral mixture*

This should be given as an essential part of ration as minerals contribute to the building of skeleton, physiological functions and production of milk. The more important of these salts are calcium and phosphorus. For maintenance, requirement of calcium and phosphorus is 6.5 and 3.5 g respectively per 50 kg body weight. The mineral mixture may be included in the ration at the rate of 0.2%.

9.10.5.12 *Common salt*

The lumps of salt of fairly good size should be hung up in some suitable place where goats can easily get them or they may be kept in the manger. The provision of salt licks in very important for goats as they secrete a good amount of sodium and chloride ions in milk. The salt often helps to tone up the system and may even have some effect in removing worms from the body. Salt to the extent of 2% may also be mixed with the daily grain ration of goats.

In addition to the above specific requirements, which are primarily of academic interest, an attempt was also made to assess a general purpose requirement of a composite goat flock having different size/body weight animals in different physiological condition. Accordingly, flock of 300 animals involving 84 Jamunapari (A), 72 Beetal (B), 72 Barbari (C), and 72 Black Bengal (D), was separately housed as large breeds (A and B) and small breeds (C and D) with community feeding and living arrangements. The animals were weighed and their DM intake was grossly recorded. Subsequently, the intake of energy in terms of TDN and DE was calculated and the intake of DP was derived with the help of formula.

The relevant information remained as under: 1 Mcal DE = 25g DP

S.No.	Particulars	Breeds		
		Large	Small	Combined
1.	Breeds	A&B	C&D	A,B,C&D
2.	No. of animals	156.00	144.00	300.00
3.	Mean body weight (kg)	25.80	16.65	21.23
4.	Metabolic body size ($W^{0.75}$ kg)	11.45	8.25	9.89
5.	DM intake/animal	703.10	531.10	628.40
6.	DM intake – 100 kg body weight (kg)	2.72	3.19	2.96
7.	TDN intake/animal (taking DM digestibility as 60%)	421.86	318.66	377.04
8.	DE intake/animal (Kcal) (TDN × 4.40)	1856.20	1402.10	1659.00
9.	DE intake/$W^{0.75}$ kg (Kcal)	162.11	169.95	167.75
10.	DP intake/$W^{0.75}$ kg (g)	4.05	4.25	4.20

This intake of energy and protein provided only a general picture and did not pertain to any specific purpose as most of the animals, although empty, were still somewhat growing and some were even pregnant and in lactation. These values, however, provide an indication of a practical situation. Other things being almost the same, the intake of the smaller breeds (Barbari and Black Bengal) remained comparatively higher than that of the larger ones (Jamunapari and Beetal).

9.10.6 Work at NDRI, Karnal

Singh and Mudgal (1978) have also reported the endogenous urinary (EUN) and metabolic faecal nitrogen (MFN) for working out the maintenance requirement of protein in goats. They conducted 2 metabolism trials on almost the same lines as Majumdar (1960a), using castrated Beetal bucks and reported as under:

Trial No.	EUN g/$W^{0.75}$ kg	MFN g/100 g	DM intake
I	N–free diet	0.10027	0.396
II	low–N diet	0.09671	0.445
Average		0.09849	0.4204

Based on these, the DP requirement of maintenance worked out to 106 g/100 kg body weight.

N–free diet and 128g/100 kg bodyweight for low–N diet, with an average of 117g/100 kg body weight. This when converted to metabolic body size of 18.18 kg body weight animals used (8.80 kg), the minimum and maintenance requirements, respectively, come to 1.22 g DP or 1.74 g TP and 2.89 g DP or 4.13 g TP/$W^{0.75}$ kg using their reported digestibility of 70%.

Subsequently, Kurar and Mudgal (1979, 1990) on the basis of balance studies on Beetal goats, reported maintenance requirements of protein and energy, respectively, as 2.12 g DP and 124.27 Kcal/DE $W^{0.75}$ kg.

While concluding it would be desirable to have a bird's eye view of the reported nutrient requirements of Indian goats, based on sound experiments as given in Table 9.7.

As against these, the NRC recommendations for goats in general (NRC, 1981), irrespective of breed, country and climate are as under :

S.No.	Particulars	Requirement/$W^{0.75}$ kg	
		Energy (DE) (Kcal)	Protein (DP)(g)
1.	Maintenance (working)	124.27	2.82
2.	Growth (per-g gain)	8.89	1.95
3.	Pregnancy	217.30	4.79
4.	Lactation (milk production/g kg 4% FCM)	1527.50	57.200

An indepth knowledge of qualitative and quantitative aspects of various nutrients in feed and forage is necessary to formulate the ration and for evolving the strategy on feed resource development specifically for goats. The requirement is markedly influenced by the physiological need and stress of environment. Based on the available information, the daily requirement of nutrients in goats, for maintenance, gestation and lactation can be calculated.

Table 9.7 Summary of Data Collected at IVRI, RFS and NDRI

Particulars	Requirement / $W^{0t.75}$ kg		Reference
	Energy (DE) (Kcal)	Protein (DP) (g)	
I. From IVRI			
A. Based on EUN and MFN			
Maintenance	(141.08)	(250)	Majumdar (1960a)
Minumum –	–	1.61	Majumdar (1960a)
B. Based on balance studied (mean)		3.76	Majumdar (1960b)
1. Graphical method		3.87	Majumdar (1960b)
2. Statistical method		3.80	Majumdar (1960b)
3. Percent utilization method		3.86	Majumdar (1960b)
4. Harris Mitchell formula		352	Majumdar (1960b)
C. Based on DM/TDN intake	163.80	4.08	Majumdar (1960b)
D. Based on basal calories		352	Majumdar (1960b)
II. From RFS College			
A. Based on balance studies (mean)	134.75	(3.38)	Singh & Sengar (1970)
1. Graphical method		3.45	Singh & Sengar (1970)
2. Statistical method		350	Singh & Sengar (1970)
3. Percent utilization method		352	Singh & Sengar (1970)
4. Harris Mitchel formula		3.04	Singh & Sengar (1970)
B. Based on DM/TDN intake	138.93	(3.40)	Sengar (1980)
C. Based on basal calories		3.02	Sengar (1980)
D. Based on EUN and MFN			
Maintenance	–	(3.05)	Rajpoot (1980)
Miniumum	–	1.428	Rajpoot (1980)

Table 9.7 Contd.

Particulars	Requirement / $W^{0t.75}$ kg		Reference
	Energy (DE) (Kcal)	Protein (DP) (g)	
E. Based on energy and protein ratio			Rajpoot *et al.* (1979)
1. Maintenance	125.00	3.13	Rajpoot *et al.* (1979)
2. Growth gaining 88.5 g/day	232.83	5.82	Rajpoot *et al.* (1979)
per gram gain	7.88	(0.20)	Rajpoot *et al.* (1979)
3. Pregnancy gaining 61.75 g/day	221-79	(555)	Rajpoot *et al.* (1979)
4. Lactation as working maintenance	198.00	4.95	Rajpoot *et al.* (1979)
5. Milk production /kg 4% FCM milk	1520.00	45.46	Rajpoot *et al.* (1979)
6. General	167.75	4.20	Rajpoot *et al.* (1979)
III. From NDRI			
A. Based on EUN and MFN			
Maintenance		2.89	Singh & Mudgal (1978)
Minimum		1.22	Singh & Mudgal (1978)
B. Based on balance studies	124.27	2.12	Kurar & Mudgal (1979)

Notes: (1) All requirements are working maintenance unless otherwise indicated
(2) * pertain to minimum maintenace based on alone
(3) Values in parentheses have been used in calculating the NRC recommendations along with those from other countries.

9.10.7 Nutrient intake and digestibility on silvi-pastures

The goats, on the reconstituted multi-layer silvi-pastures, have been found to ingest the herbage of almost constant nutrient concentration throughout the year. The seasonal variability in the grazing preference of the goat in a pasture hardly affects the proximate compostion, intake and digestibility of the ingested pasturage (Bhattacharyya and Sharma, 1993). The inherent ability of goats to improve the quality of feed-intake (by making vigorous selection from the available vegetation) apparently stabilizes the nutrient intake round the year. Moreover, the silvi-pastures (3-tier) are expected to sustain 5 times higher grazing pressure as compared to the natural pastures in the waste lands. A mixed diet is believed to promote the detoxification mechanism in goat' rumen and there is evidence that it is easier for the animal to cope with an identical amount of toxins from different plants rather than from one single source (Brendan and Thomson, 1991).

9.10.8 Water Requirement

Goats are among the most efficient domestic animals in the use of water (next to camel) and possess a low rate of water turn over per unit body weight.

Goats have the ability to conserve body water by reducing losses in the urine and faeces. In arid climate, the water intake through forage is high in goats for their ability or willingness to browse. Consequently, goats are less dependent on the free water source than other domestic species in desert climate.

A safe general recommendation is to provide goats with clean water *ad libidum*. Extremes in water temperature will increase energy requirements. Taste factors will also affect normal water intake (Goatcher and Church, 1970).

Factors affecting the free water intake of goats are the level of lactation, ambient temperature, water content of forage consumed, amount of exercise, and salt and mineral content of the diet. Therefore, the daily range of free water intake may be from negligible amount to several litres. When feeding on dry forages and when water is lacking, the efficiency of reproduction will suffer. Sub-optimal water intake will result initially in the reduced feed intake, followed by reduced performance and gradual starvation.

Goats are often more sensitive and reluctant than other species to drink from foul-tasting water sources. If they are forced to drink poor quality water, the result may be infections or undesirable mineral intake.

9.10.9 Requirement of vitamins and antibiotics

Vitamins A, D and E are required by goats, as the microbes in the rumen synthesize most of other needed vitamins. Vitamin A can be supplied by feeding green forage and yellow maize. One kg of lush green fodder will provide 1500 IU of vitamin A. Vitamin D can be obtained from sunlight. Vitamin E is present in adequate amount in most of the normal rations. Synthetic vitamin A & D may be included in the ration of growing kids. Feeding aureomycin or terramycin increases the growth rate of young kids, reduces incidence of scours and other infection diseases and improving the general appearance of the kids.

9.10.10 Ration formulation

Considerable information is available on the feeding of goats in confinement, which adds up to a higher rearing cost. Since goats are primarily raised on unmanaged pasture and ranges, the problem rests on estimating the actual nutrient intake from the pasture and finding the cost effective ways to meet the supplementary requirements in different seasons. Ensuring of adequate nutrition under these conditions is often based on approximation. Besides the cost of supplementary feeding, it is often believed to be in excess of the actual need. On the contrary, exclusive mixed grazing, on waste land and natural vegetation of arid tropical/subtropical zones, is known to result in nutrient inadequacy and therefore, it demands additional feed supplementation. In formulation of rations, feed ingredients of about similar properties are considered as potential substitutes and exchanges are made within concentrate mixtures in accordance with market price, availability and to some extent feed preference. Avoiding the tedious process of digestibility trials, a nutritionist can estimate the total digestible nutrients of different classes of feeds from regression equations. Unfortunately, no such equations are available for goats but the one available for sheep can suffice.

9.11 Practical Feeding Systems for Goat Production

From a practical standpoint of view of goat production in the tropics, the following 4 major systems of goat management are generally followed:

9.11.1 Subsistence system of production

In this system, a few goats are generally kept by the farmers or weaker sections of the society or by those who like its milk or meat for their domestic needs. The goats may be sent for browsing or stall-fed or both the practices may be followed according to the circumstances and available resources of the goat keepers. However, this system serves a very useful purpose for meeting out the requirements of essential animal products to the needy and poor people.

9.11.2 Extensive system of production

This system is generally practiced in regions where the land is not immediately suitable for agricultural improvement and/or is too difficult or costly to fence. Such type of practice is possible in the ravins and uneven land areas, which are normally not used for agricultural operations and thus can be profitably used for raising of goats, provided some grazing areas, bushes and tree fodders are available. Thus in such areas, rearing of goats involves a minimum amount of labour and expenses because goats are allowed to graze and browse *ad lib* for maximum number of hours in the day and practically no concentrate is offered for their feeding. Under such practices, goats have also been used to assist in bush clearance in Nigeria (FAO, 1966). In this system, some form of housing is provided for the goats and a carrying capacity of about 3 to 112 goats/ha is possible although the same may be much less in semi-arid areas.

9.11.3 Intensive system of production

In this system, goats are completely stall fed and practically do not get a chance for grazing. Thus, this system provided maximum protection from uncontrollable environmental factors and gives complete control over the destructive aspects of the goats feeding habits. Dairy goats and small herds of goats in the urban areas are generally managed in this way.

9.11.4 Semi-intensive system of production

This system covers all degrees of compromises between extensive management and zero grazing, but usually involves controlled grazing of fenced pastures with some supplementary concentrate feeding. In Jamaica (Devendra, 1971), it has been shown that an improved pangola (*Digitaria decumbens*) pastures, the carrying capacity can be 38 to 50 goats/ha. Under Indian conditions, this system appears to be the most economical for goat production particularly when sufficient land is not available for grazing and browsing purposes.

9.11.5 Special characteristics of goat nutrition

In addition to the above mentioned systems of raising goats, it is interesting to note in the feeding of goats that they are specially capable of digesting lignin to the extent of about 16% (Barsaul and Talapatra, 1971) which is not at all digested by other domesticated ruminants. Under certain circumstances or with certain foods in particular those of high fibre content, goats can utilise nutrients much better than many other ruminants. As goats have special feeding habits and on account of their prehensible tongues they are able to graze very short grasses and to browse foilage not normally eaten by other ruminants. Thus, their rustic and hardy qualities enable them to withstand dry environmental conditions much better than cattle and therefore, they are capable for surviving and thriving in environments where vegetation is extermely sparse. Further, the goats are also naturally gifted to consume more dry matter as compared to other ruminants, which help them to obtain most of their nutritional requirement from grazing and browsing by consuming more of the roughages. It has been observed that meat goats generally consume at the rate of 3% of their live weight while the dairy goats are capable of consuming 5 to 7% of their live weight (Mackenzie, 1967). However, the dairy goats in tropics generally consume on an average 4.1 to 5.4% (Devendra, 1970).

9.12 Preparation of Concentrate Mixtures for Feeding to Goats

For feeding to kids creep mixtures containing high amounts (50 to 60%) of maize, jowar or barley with groundnut cake should be prepared so that they may be adequate in energy and protein having about 18 g DCP and 75 g TDN. An example is given below:

Ingredients	Percentage	Ingredients	Percentage
Maize/Jowar/Barley	50	Fishmeal	10
Groundnut cake	20	Mineral mixture	1
Molasses	10	Chalk	1
Wheat bran	7	Salt	1

To the above, 150 g TM-5 or aurofac and 25 g Vitablend or Rovimin should be added per quintal of the creep mixture.

The concentrate mixture for feeding to goats in general should be prepared by using the maximum possible amounts of agro-industrial by-products so that it may be cheap and goat husbandry may become an economical preposition to the goat keepers. Keeping in view the availability of such products according to different regions in the country, the concentrate mixture may be prepared having about 15 g DCP and 65 to 70 g TDN. An example is given below:

Ingredients	Percentage
Wheat bran/Rice bran	45
Groundnut cake/sesame cake	20
Barley/Maize/Jowar	12
Molasses	10
Dal chunies	10
Mineral mixture	1
Salt	1
Chalk	1

Considering the utilization of non-protein nitrogenous substances by ruminants, an attempt on an experimental basis can also be made to include 1% urea in the above conceentrate mixture, thereby reducing the costly cakes and adjusting the ingredients by increasing the cereal grains and molasses. This would help in lowering down the cost of concentrate mixture.

9.13 Rumen Microbes

Rumen, the fermentation vat of nature has been a world of mysteries and a field of curiosities for the scientists. Among so many digestive process that takes place in the rumen, the degradation of cellulose is unique. Firstly, because this compound constitutes the major bulk of daily ration and secondly, because mammalians are devoid of secreting the enzyme, celluloses, the ruminants are thus entrirely dependent on rumen microbes for the release of cellulose enzyme, so vital for the digestion of cellulose part of the feed. The degradation of protein and non-protein nitrogen in the rumen is an exciting evidence of microbial activity. The proteolytic activiy of rumen content has been known for some considerable time, and the conversion of feed protein and non-protein portion to microbial protein has been definitely confirmed.

Goats by nature are wandering animals who prefer to select their own feed through grazing/browsing and can accept a wide variety of plant species. In view of this, the rumen micro-organism of goat may be different than other ruminants which will be interesting.

(i) *History*

It was in 1884 that a scientist Tappenier established identity of rumen microbes.

(ii) *Rumen Flora and Fauna*

Rumen flora and fauna consists of mixed population of bacteria, protozoa, oscillospira and to some extent yeasts and moulds which get fixed in the course of rumen region development.

(iii) *Density of Different Microbes*

Number of different microbes/ml of rumen liquor in different ruminants is given as under:

Type of organisms	Buffalo	Cattle	Sheep	Goat
Total bacterial count × 10^{10} (Direct microscopic)	6.9 – 32.70	5.4 – 31.4	18.0 – 88.0	45.0 – 23.5
Potozoa × 10^4	1.8 – 13.8	0.3 – 19.7	1.4 – 7.8	2.0 – 28.80
Oscillospira × 10^4	4.7 – 16.7	6.1 – 8.52	–	–
Yeasts × 10^3	0.06 – 0.18	08 – 13.0	0.0 – 1.0	–
Moulds × 10^2	–	1.0 – 2.00	0.2 – 1.0	–

(iv) *Components of rumen Micro-organism*

Each group of micro-organisms present in the rumen comprise of the following:

A. Bacteria

The different groups of bacterial present in the rumen can be identified on the basis of their

(i) Biochemical activities
(ii) Oxygen requirement
(iii) Morphology

1. On the basis of biochemical activities

(a) Amylolytic bacteria/starch digesters

They hydrolyse starch in the presence of amylolytic enzyme (amylase) in the rumen. The fermentation products are lactic, formic, succinic and butyric acids. The organisms of this group are *Clostriduim lochhedii* besides *succinogenes, bacteroides, amylophilus, bacteroides, ruminico, vibrio, fibrisolvens, Streptococcus bovis, succinomonas, Amylolytica salomonas, ruminatium.*

(b) *Cellulolytic Bacteria*

This group of bacteria is responsible for the digestion of the cellulose part of the feed. They may be rod or cocii shaped.

Amongst rod shaped cellulolytic bacteria which are in abundant in rumen comprise of: *Bacteroides succinogenes, Butyrivibrio fibrisolvens, Clostridium lochhedii, Clostridium longisoporum, Cillobacterium cellulosolvens* and *Cellulomonas fimi.* Among cocci shaped are *Ruminococcus albus* and *Ruminococcus flavefaciens.*

(c) *Hemi-Cellulose Digesters*

The insoluble carbohydrate of plant material *i.e.* hemi-cellulose is digested in rumen and hydrolysed to xylan. There are 6 species of hemi-cellulose fermenting organisms *i.e. Eubacterium ruminatium, Bacteriodes ruminicola, Bacteroides amylogenes, Butyrivibrio fibrisolvens, Ruminococcus flavefaciens* and *Ruminococcus albus.*

(d) *Lipolytic Bacteria*

This group is associated with the hydrolysis of fat into glycerol and fatty acids. The fermentation products are acetic, propionic, butyric and succinic acids with large quantities of other gases including hydrogen sulphide. Among sugars, glycerol, ribose and fructose are fermented.

The bacterial species concerned with the fermentation of fat are *Veillonella alcolescens* and *Anaerovibrio lipolytica.*

(e) *Protective Bacteria*

They digest feed or plant protein from which microbial protein is synthesised which is ultimately used by the host. They live on other bacteria in absence or shortage of substrates. The organisms included are non-spore forming Proteus supp. *Cornybacterium* and *Micrococcus* and amongst spore formers *Lachnospira multiparus*, *Bacteroides succinogenes*, *Eubacterium ruminanium*, *Aerobactor aerogenes*, *Bacteroides ruminicola*, *Lactobacillus fermenti*, *Streptococcus elsdeni*, *Peptostreptococcus elsdeni*.

(f) *Methanogenic Bacteria*

Though, these bacteria attack acetate, propionate and butyrate in natural state, yet they do not do so in rumen. They occur in very low number. The main organisms responsible for methane production in the rumen is *Methanobacterium ruminantium*. It grows with hydrogen and CO_2 as energy source and converts these to methane and water.

(g) *Ureolytic Bacteria*

These organisms mostly utilize ammonia and urea nitrogen, which come through saliva, from blood and also from non-protein nitrogenous substances. These are then utilized by the host. The organisms concerned with this are *Lactobacillus bifidus*, *Bacteroides amylophilus* and *Proteus mirabillis*.

(h) *Sulphate Utilizing Bacteria*

The rumen micro-organisms need sulphur for growth and utilize from sulphate and also decompose the cellulose too. The organisms involved are *Clostridium nigrificans*, *Lachnospira multiparus*, *Butyrivibrio fibrisolvens* and *Bacteroides* species.

2. On the basis of oxygen requirement

(a) *Facultative Anaerobes*

These organisms are of numerical importance, otherwise they have nothing to do with rumen fermentation. The organisms are:

Coliform Bacillus group, Propionibacteria, Lactobacillus and Streptococci.

b) *Anaerobes*

(i) Spore forming rods *e.g. clostridium*

(ii) Non-spore forming rods:
Lactobacilli, *Ramibacteria*, *Eubacteria*, *Methuno-bacteria*, *Lachnospirae*, *Cillabacteria*, Succinic acid production *Bacteroides*, *Amylophilus*, *Butyrivibrio*, Butyric acid production *Bacteroides*, *Fusobacteria*, *Sucnivibrio*, *Desulfavibrio*, *Selenomonas Berelia* and *Succinimonas*.

(iii) Cocci-*Petostreptococci*, *Rumoncocci*, *Veillonellae*.

3. On the basis of morphology:

Different groups of bacteria on the basis of morphology are:

Cocci	–	50.89%
Roda	–	45.70%
Spirillum	–	0.50%
Vibrio	–	0.40%

B. *Oscillospira*

There is only one species namely *Oscillospira guillermondi* present in the rumen. They are iodophillic in nature.

C. *Yeasts*

The role of yeast is not clear as yet. Probably they may be concerned with the sugar fermentation and cellulose digestion. Nine species of yeasts belonging to 4 genera, *viz. Candida, Trichosporon, Rhodtoirula* and *Saccromyces* have so far been commonly observed in rumen.

D. *Moulds*

Their presence has been reported recently. Their funcion is not known as yet. The moulds that have been identified are *Mucor* Spp. *Rhizopus* and *Aspergillus* species.

E. *Protozoa*

Their population is mainly of ciliates and a few species of flagellates. Among the flagellates, the predominate types are *Callimastix frontalis* and *Monocercomonas ruminantium*. Amongst ciliates, 2 groups *holotrichs* and *entodiniomorphs* are present. These ciliates are not found in nature other than rumen or intenstine of herbivorus animals. Distribution of ciliates in different ruminants is given in Table 9.8.

Table 9.8 Showing the Distribution of Rumen Ciliates in Different Ruminants

Ciliate species	Buffalo	Cattle	Sheep	Goat
HOLOTRICHS				
Butschlia parva	+	–	–	+
Blepharoconus cervicalis	+	–	–	?
Blepharocorys bovis	+	+	–	+
Blepharoprosthium pireum	+	+	–	?
Blepharosphaera intestinalis	+	–	–	?
Blepharozoum zonatum	+	–	–	+
Bundleia posteilita	+	–	–	?
Charon equi	+	+	–	?
Dasytricha ruminantium	+	+	–	+
Didesmis quadrata	+	+	–	?
Holophryoides ovalis	+	–	–	+
Isotricha intestinalis	+	+	+	+
Isotricha prostoma	+	+	+	+
Polymorpha ampulla	+	–	–	?
Prorodonopsis coli	+	–	–	?
ENTODINIOMORPHS				
Caloscolex cuspidatus	+	–	–	?
Cochliatoxum periachtum	+	–	–	?
Cunchaia curvata	+	–	–	?
Cycloposthium dentiferum	+	–	–	?
Diplodinium crista qalli	+	+	–	?
Diplodinium dentatum	+	+	–	?
Diplodinium dentatum	+	+	+	?
Diplodinium psittaceum	+	–	–	?
Diploplastron affine	+	–	+	+
Ditoxium funinucleum	+	–	–	?
Elytroplastron hegneri	+	–	–	?
Entodinium caudatum	+	+	+	+
Entodinium elongatum	+	+	+	+
Entodinium longinucleatum	+	+	–	?
Entodinium rostratum	+	+	+	?
Enoploplastron triloricatum	+	–	+	+
Eodinium lobatum	+	+	–	+
Epidinium caudatum	+	+	+	?
Eremoplastron bovis	+	–	–	?
Eudiplodinium maggii	+	+	–	?
Metadinium medium	+	+	+	?
Ophithotrichum janus	+	–	–	?
Ophryoscolex purkynei	+	+	+	+
Ostracodinium dentatum	+	+	–	?
Polyplastron multivesiculatum	+	+	+	+
Tripalamaria dogieli	+	–	–	–

+ Present; – Absent; ? Not known

Source: Personal Communications, Bhat (2008)

(v) *Functions of Rumen Micro-organisms*

1. Digestion of cellulose, starch, pectin, xylon, pentosans and soluble carbohydrate of the ration.
2. Digestion of proteins and nitrogenous compounds of ration.
3. Synthesis of protein and amino acids from ammonia.
4. Synthesis of vitamins needed for host and also for microbial species.

(vi) *Establishment of Rumen Microbial Population*

The rumen which is very small at birth does not contain any microbe. However, rumen grows fastest followed by the reticulum, the omasum and abomasum and these compartments reach their adult proportions by about 56 days.

The first rumen population of the very young animal is most likely derived from back flow of contents of the abomasum and contain large number of lactic acid bacteria, *streptococci* and *coliform* organisms.

However, rumen microbial population is established in two stages:

Stage I. Establishment of Bacteria

Organisms	Age
1. Lactic Bateria	1–6 weeks
2. Cellulolytic bacteria	After 3 weeks
3. Facultative anaerobes	3–6 weeks
4. Other types of bacteria as found in mature animal	After 6–13 weeks

Stage II. Establishment of Bacteria

Their establishment depends on close contact between young and old animals and also when the rumen pH goes to 6.2. However, the rumen protozoa starts surviving in the rumen, when the animal reaches an average age of 6 weeks. Holotriches establishes the first then entodiniomorphs.

(vii) *Variation in Rumen Micro organisms*

Although, the rumen is a remarkably constant environment compared with the habitats of many micro-organisms, the microbes are subjected to some variation in conditions. The cause of such variations are firstly due to the volume of saliva secreted by the host, which in turn affected the buffering capacity, ionic concentration and surface tension of rumen contents. Secondly, the type and the physical state of feed exert inflluence not only by its gross composition in terms of carbohydrates and protein but also by its content of trace elements and other minor constituents which are needed by the organisms for their optimum growth.

(viii) *Factors Effecting the Rumen Microbial Population*

1. **Feeding:** Both number and type of rumen bacteria and protozoa are affected as a result of type of forage, concentrate mixture, grazing and the size of forage particles.

2. **Salt concentration:** A high level of salt in the ration cause decrease in size and number of protozoa and increases the bacterial count. Higher concentration of salt inhibits the growth of some bacteria, but prevents spore formers for longer period.

3. **Diurnal variation:** Diurnal variation as a result of feed and water, which the animal consumes, effects both number and type of population.

4. **Season:** Seasonal variations are observed in density and the type of ciliate.

5. **pH:** Low pH results into inhibition of protozoal population and protozoa are killed at lower pH. Higher pH results into the activity of methanogenic bacteria. The maximum microbial population is at around 6.4–6.8 pH.

6. **Antibiotics:** When antibiotics and sulphur are fed, rumen bacteria get reduced. In general broad spectrum drugs reduce the all types of microbes.

(ix) *Regeneration of the Rumen Population*

Fluctuations in the number of particular organisms, which appear random, may mean that some organisms tend to die off and are re-introduced by inoculation from outside into the animal. The rumen population shows a remarkable tendency to return to normal through reinoculation or regrowth of residual organisms even after drastic alterations. Experiments have shown even when the rumen population was drastically altered by feeding carbohydrate, changed to normal diet resulted in return to normal of the rumen flora in about 2 days. Smaller bacteria returned to normal numbers within a few days after being decreased in number by starvation of the host for 4 days, and the larger organisms also reappeared within a few days, although the time of reappearance varied in different conditions. Most of the repopulation of organisms takes place due to the growth of residual organisms in the rumen of starved animals, although in some cases, it takes place due to reinoculation from external source.

(x) ***Inter-Relationship***

1. **Between protozoa and bacteria:** Investigations have established that the concentration of bacteria in a rumen from which protozoa have been excluded are greater than in the normal faunated rumen as the feed being the same in 2 cases. Two explanations seem applicable that the protozoa reduces the number of bacteria by feeding them directly, or that the effect is exerted through competition for food. The protozoa using food that would otherwise be available for production of greater concentration of bacteria.
2. **Between protozoa:** There is a mutualistic and antagonistic relationship between the different ciliates of the rumen.
3. **Between protozoa and host:** Host relationship of rumen protozoa is apparently harmless. Most are commensals and few are of definite value to the host. None of them is essential for host as in case of termites. One fifth of the total ruminants' protein requirement is met daily by ciliates.

9.14 Tips for Feeding Goats

In view of the nutrient requirement, feeding habits and behaviour of goats, a few practical tips are given here for organizing the optimum feeding of goats:

1. The energetic grazing behaviour of goats should be exploited by allowing them to acquire most of the needed dry matter through range vegetation and green fodders.
2. A balanced diet for stall-fed goats normally contains dry-matter through concentrates, dry fodder and green fodder in the ratio of l:1:1.
3. Goats relish leguminous cultivated fodders like Lobia, Berseem, Leucerne etc. The non-leguminous fodders can be fed with leguminous fodder in the ratio of 1:1.
4. Feed intake by stall-fed goats is affected by methods of forage harvesting, storage and processing. The ensiling of fresh forages decreases the intake

more than the hay made. Chopping of green forages or hays reduces the intake compared to whole forage, whereas, pelleting of low quality roughages increases the intake.

5. Adult dry goats on browse normally gather enough nutrients from range lands and their surroundings to maintain their body weights. But due to seasonal variations in the nutrient availability of the range lands, goats may exhibit loss of body weights which should be prevented by supplementary feeding of green fodder or concentrate mixture.

6. The goats in later part of gestation, peak of lactation or active growth would essentially require supplementary feeding of concentrate which may be given at the rate of 1 to 2% of their body weights depending on the stage of production.

7. The concentrate mixture should be formulated with alternative cheaper ingredients incorporating agroforest by-products and organic wastes where applicable, avoiding any compromise on the DCP and TDN content. Some of the formulation of economical concentration mixture are given in Table 9.9.

Table 9.9 Alternative Concentrate Mixture

	Item	Parts
(I)	Damaged wheat grain	22
	Deoiled ground nut cake	15
	Wheat bran	30
	Dal Chuni	15
	Deoiled mustard cake	15
	Mineral mixture	2
	Salt	1
(II)	Bajra (Millets)	40
	Deoiled G.N. Cake	25
	Rice Bran	20
	Dal Chini	13
	Mineral mixture	2
	Salt	
(III)	Maize grain or Barley grain	37
	Sun dried poultry excreta	30
	Ground nut cake	18
	Wheat bran	12
	Mineral mixture	2
	Salt	1

Source: Hand Book of Animal Husbandry (3rd Edn., 2002) - ICAR, New Delhi

CHAPTER 10

MANAGEMENT AND HOUSING

Livestock management, as a science, has far less recognition than the other disciplines of animal production. While the importance of skills related to the successful goat husbandry is often enunciated by the specialists, the farmers are inclined heavily on the wisdom gained over generations as a family trade. The worth of rich experience in the management of flocks is seldom transcended by any mode of learning; yet the managers miserably wane in the documentation of their experiences that might have been acquired progressively over the years.

The science and art of goat management, like any livestock management, incorporates the management of breeding behaviour and reproductivity, feed habits and feeding, disease surveillance and prevention, perinatal prudence of does and neo-natal kid care, besides management of personnel, shelter and other infrastructures at the farm complex.

10.1 Systems of Management

10.1.1 Traditional system

Goat raising is mainly in the hands of the weaker sections of the community which either do not possess land or their landholdings are so small that crop cultivation does not provide remunerative employment all the year round. Further, in the major goat rearing areas especially in North Western districts of Rajasthan, grazing and stock watering resources are available only for a few months in a year, compelling shepherds to lead a nomadic life. The system of constant migration is one of the main reasons for the high percentage of illiteracy among these sections of the people. Even head of the family have no chances of education because they are also constantly on the move and are employed by their family for grazing the goat flocks. Due to lack of education, these goat owners are not able to appreciate and adopt improved goat husbandry practices brought to them by extension workers. Goat management has thus remained in a neglected state. Migration and grazing practices have an impact on the present status of goat husbandry in the country.

10.1.2 Extensive system

This system is practiced in large tracts of Deccan Plateau where there are hills and large areas of land unable to be cultivated. Goats are taken out of grazing by women and children in the morning and brought back in the afternoon. This is the cheapest system and practiced over all parts of India where grazing land is available. This is also practiced by nomadic tribes who move for pasture to pasture along with the season with their herds. The nomads keep watch dogs with their herds to protect them and to prevent them trespassing in unwanted areas. Advantages of this system are that it is cheap and provides the production and disposal process simultaneously. The disadvantages are mainly that the animals raised on the system are poor producers besides having poor genetic capabilities, and are exposed to continuous stress.

The diversity in extensive management system from use of goat in control of certain weeds in some areas to the migratory flock system depends on the agro-climatic conditions of the region. In this system, the goats are grazed round the year on natural vegetation with some additional hay or concentrate during lean periods of the year.

Since, goats have a wide grazing spectrum, there is no doubt that in a mixed vegetation complex, the inclusion of goat in grazing system would give for superior utilization of the vegetation than is possible with only sheep or cattle. If a sensible stocking rate is adopted, the inclusion of the goat in common use grazing is not detrimental to its associated grazing or ecosystem.

The diversity in feeding habits of goat in comparison to other species of livestock make this animal more useful in controlling some of the undesirable vegetations. The most common concept of goat influence on vegetation is that of biological bush control. The role of goats in shrub management has been very effective and has been used with less ecological upset than any other bush management measure. In order for goats to exert desirable changes in vegetation, browse species to be changed must first be placed in the reach of goats. The method mostly used for this has been changing in some of the countries.

Goats while grazing on grass or bush, also help in dispersal of seeds and improvement in vegetation. In Australia, goats have long been recognized as being able to control many woody weeds in the vast pasture land. There are reports that in extreme deserts, the grazing of sheep and goats has increased the number of bushes, trees and grass cover (Govt. of India, 1987). Thus, it is unwise to put the heavy burden of all evils to goats for land degradation and desertification.

Various amounts of native grasses and shrubs available in trees cropping systems like coconuts and oilpalm in coastal areas are potentially valuable for utilization by goats and sheep. The approximate yields of herbage in these situations is around 4–6.6 tonnes of dry matter per ha, a good deal of which can be utilized. The potential carrying capacity and increasing offtake of goat meat due to integration can be quite considerable. The yield of fresh fruit bunches of oilpalm increased by 3.52 tonnes of fresh bushes per ha per year after 4 years of introducing grazing in comparison to nongrazed areas (Devendra, 1992).

The management of migratory flocks needs special attention to the extent that over use of the grazing land has to be avoided to control environmental degradation. Other important aspect in management of migratory flocks is that the local goat should not be allowed to mix with the migratory flocks otherwise indiscriminate mixing of breeding results.

10.1.3 Semi intensive system

This system is widely practiced by small and marginal farmers and villagers. The nature and extent of this system depends on the type of crops grown and their suitability to goats. Goats are left to graze/browse on the crop residue when the same has been harvested. The advantages of this system are increased fertility of the land by droppings and urine of these animals, control of wasteful habits, good growth rate, easier management and possible increased crop yields. Goat keepers keep herds of goats which they take to other farmers' fields when a crop has been cut. The farmers pay them in cash or kind for the manure and urine they drop in the field while grazing. A modified form of this system is 'tethering' where the animal is held by a rope about 2–3 meters long, and other end being tied to a tree or post. The animal grazes and browses in the area accessible through the length of the rope. This system is popular with farmers who keep only a few goats. It permits utilization of grass, fodder and bushes in a limited area and keeps a control on the animal and saves labour.

The goats are grazed round the year for a few hours daily and the supplementation in the form of roughage and concentrate is also required round the year. This system of management is suitable for variable type of climate and small to medium flock size. One of the ways to ensure regular supply of limited amount of leaves is to adopt silvipastoral system. Experiments at the CIRG, Makhdoom have shown that fodder production under dry land conditions of Yamuna ravines can be enhanced from 2.4 to 8.0 tonnes per ha per annum on dry matter basis (Sharma *et al.*, 1989). The forage production under silvipastoral system is 7 fold higher than in the traditional land use system which assured availability round the year (Singh, 1988). The fast growing and perennial trees are planted in hedge rows at required spacing and the avenue spaces are utilized for food grain/forage production. The periodical lopping supplements the feed supply round the year. Synthesized reconstituted pasture sustained 5 times higher grazing pressure as compared to natural pasture in the waste land of semi-arid tropics (Sharma, 1990). Most of the present waste land should logically be preserved for growing 3-tier canopy of phytomass for fodder rather than high input agriculture for food grains.

What is needed is to maintain a balance between the number of goats and other livestock and the quantity of feed which is available for the combined population during the lean periods. This will ensure protection to the ecosystem in different agro-climatic zones of the country.

10.1.4 Intensive system

If the agro-climatic conditions are such that the land is mostly under perennial cropping and the grazing land is available during inter-cropping periods, the only choice is to adopt intensive management system. Such areas are suitable for milch breeds. During winter, the roughage and concentrate feeding is done at stalls and substantial grazing is possible during summer after harvesting of the rabi crop. For small flocks, tethering for grazing on roadside communal and arable grazing land can also meet the requirement. Cut- and carry-system of top feeds is another way of maintaining the goats under such agro-climatic conditions.

The pressure of increasing population of human and animals further justify adoption of intensive management system for livestock in general, and goat in particular, to ensure control and degradation of the environment. Some of the 100 million ha marginal and range lands have resulted from poor management strategies in the past. The chronic annual feed deficit even for maintenance have been calculated from 26 to 34% (Devendra, 1992).

10.2 Handling and Care of Animals

10.2.1 Kidding does

The doe should be put in the pen a few hours before parturition. She becomes fussy about 2 or 3 hours before actual kidding. The udder becomes engorged with milk, the belly appears shrunk and the flanks appear rather hollow. The tail head is raised higher than usual as the ligaments of either side relax. There is thick, white, starchy discharge which soon changes to a more opaque substance.

The water-bag appears first. Soon afterwards, within 15 minutes, it breaks and the feet of the kid appear with the head resting on them. Since the doe is fussier and noisier than any other domestic animal during kidding, it is not necessary to care for her unless she is obviously in trouble. In case of undue delay in the appearance of the kid after the bursting of the water-bag, the position of the kid may have to be adjusted. If twins or triplets are to be born there is usually a short period of rest between the appearance of kids.

10.2.2 Kids

The first few days of life of kids are rather vital in the management of kids. With multiple births, weaker kids often require special attention particularly at the time of suckling. It is a good practice to clean the udder with $KMnO_4$ solution each time, prior to suckling. Overfeeding and underfeeding are both deleterious to the health of the kid(s). It becomes essential to detect and isolate the kid suffering from diarrhoea or other ailment and to place the cases under the attending veterinarian. Healthy kids become playful within 72 hours and they require soft bedding and clean grassy paddocks. The kids are very susceptible to cold and special attention is required to monitor the shed temperature with Simple Maximum-Minimum Thermometer and to impose correction of ambient temperature to about 18 °C in winter.

10.2.3 Pregnant does

A temporary increase in milk yield after mating is considered to be an indication of pregnancy, but the first sign that a doe is in-kid is the cessation of the periodical return of oestrus. During the first 3 months of pregnancy, there is little alteration in the shape of the in-kid does. The head of the kid can sometimes be felt from 6 to 8 weeks. An old doe or a young doe which is to give birth to one kid may be very misleading in appearance and show no sign of pregnancy. Six to eight weeks before kidding, young does commence to show udder development, but this is by no means a sure sign of pregnancy as they will frequently show such development and even have milk in the udder when they are not in-kid.

10.2.4 Dry does

No extra care is required for these animals. Regular grazing for 8 to 10 hours on a good pasture is sufficient to maintain their weight/condition.

10.2.5 Fattening stock

Complete separation of kids from their mothers is called weaning. The practices and problems of weaning and care of weaners vary from place to place. The management of weaners play an important part in good goat husbandry because these weaners will be the future breedable animals. The following steps will greatly help in proper care and management of weaners.

(i) Weaning should preferably be done at 90 days.

(ii) Avoid malnutrition, as it will result in stunted growth and susceptibility to worm infestation.

(iii) Provide supplementary feeding and good clean pastures.

(iv) Drench them regularly against various gastro-intestinal parasites, as these are very prone to worm infestation.

(v) Vaccinate them against enterotoxaemia, struck and black-quarter diseases.

(vi) Do not graze weaners in burry and seedy type of pastures which may cause skin irritation to lambs, damage to wool and cause ophthalmic diseases.

(vii) Provide them shelter against vagaries of climate and predation.

(viii) They should have easy access to fresh and clean water and nutritious green pastures.

10.2.6 Male kids

Immediately after birth, the nose of the kid should be cleared of any entangling membranes or mucus to prevent suffocation, and the navel swabbed with tincture of iodine. The kid, if healthy and strong, would stand on its legs and make for its mother's teats. Failure to reach the teats, however, is of no consequence. The kid should be taken away from its mother in birth. Colostrum should be the first food to be given to kids; it clears the stomach and develops immunity in them. The kids should be placed on either pan or bottle-feeding. Use of baby bottles and nipples or baby feeders that lessen air swallowing would be better. Nipples and bottles should be kept thoroughly cleaned. Pan-feeding is fast and efficient. Lukewarm milk (95°F/36°C) should be placed in a shallow pan and the lip of the kid slowly dipped into it. Bottle feeding may cause distension of the abdomen, and pan-feeding distension of the abdomen and scouring. During the first 2 or 3 weeks, kids should be given 0.9 to 1 litre of milk 3 times a day. If the milk is costly, milk substitute can be used when kids are a couple of weeks old. They should be allowed free access to any kid-starter meal. Space for exercise is an important item for the playful kids. Milk can be given twice a day later, but should be discontinued when the kid is 3 to 4 months old. Excessive milk feeding will produce exceptionally nice-looking does, but having improper stomach development. These does will be low milk producers.

10.2.7 Teaser bucks

The number of vasectomised bucks used is 1% of the original number of does *i.e.* for 5,000 does, 50 teasers and even when there are only 1,000 does left in the mob, 50 teasers should still be used. All vasectomised bucks are tested for brucellosis. Two teams of teasers are used and alternated every 3rd day. The time to change teasers is when a large percentage of the does that are presented for insemination have creamy vaginal mucus. This indicates that estrous has terminated and this means that teasers did not detect estrous at the onset, either because there were too many does or they had lost interest.

10.2.8 Breeding bucks

Buck kids, unless from highly pedigreed does or from does with good performance records are rarely worth retaining. They should be castrated shortly after birth or within 2 weeks. Male goats are fertile when quite young and if left with young females are capable of breeding and causing early kidding. Goats for slaughter should be raised on milk for the first 6 weeks. They can be sold or slaughtered when 3 months old for meat, which is considered excellent.

The buck, to be in good condition and well suited for breeding, should be kept on range, and made to cover 3 to 4 km each day. Bucks often become sluggish and slow breeders for lack of adequate exercise, because they are kept confined in small enclosures. For giving exercise, they may be yoked to small carriages used for hauling light loads. A buck is very active during the breeding season. The buck's hooves should be regularly attended to as otherwise foot rot or lameness may develop. Bucks should always be kept separate from the does. They become unduly resistive or excited and waste more energy when kept with does.

10.3 Grazing and Feeding Systems

10.3.1 Rotational grazing

The fodder crops should be included in the grain or commercial crop rotation programme. The practice of keeping the lands fallow for wheat, paddy, gram etc. should be discontinued and fodder crops like sorghum, pearlmillet, cowpea and clusterbean should be introduced in kharif in the rotation. The fodder crops should also be sown with grain or commercial crops in such a way that they do not affect the production of grains. Alongwith sorghum and pearlmillet, the legumes like cowpea, dolichos, clusterbean, clitoria, black gram and green gram, may very easily be grown as companion crops. The legumes will not only provide nutritious fodder for goats without adversely affecting the grain yield but will also improve the soil fertility through nitrogen fixation which in turn will be available to the grain crop. It is essential that the programme and multiplication of seeds and planting materials of recommended fodder crops should be followed.

10.3.2 Silvipasture Grazing

Establishment of silvipasture improves quality of nutrient available ensuring its supply round the year. Experiment conducted at Central Sheep and Wool Research Institute (CSWRI) has indicated that the hoggets are able to achieve 30 kg at yearling age whereas under grazing on natural pastures with some supplementation with grass and cultivated fodder hay it was 25 kg. Similar studies conducted on does in advance stage of gestation and lactation indicated that pasture intake was not adequate for them requiring supplementation to harvest desirable production traits. Another study with weaner kids on 3 and 2-tier silvipasture indicated that kids were able to achieve 22 kg body weight at 6 months of age without concentrate supplementation.

10.4 Supplementary Feeding

10.4.1 Fattening of kids

Post-weaning growth is primarily affected by hereditary factors, plane of nutrition, prevailing meteorological conditions, animal's ability to adapt to the environment and managemental stresses. In agriculturally advanced countries, post-weaning phase of growth is mainly used for fattening and finishing purpose, whereas, in our intensive meat production strategies, the active growth is completed by 5 to 6 months of age depending on the weaning age of kids.

10.4.2 Dry does

No extra care is required for these animals. Regular grazing for 8 to 10 hours on a good pasture is sufficient to maintain their weight/condition.

Flushing is conditioning of does for breeding. If the does are in low plane of nutrition prior to breeding, additional supplementation for about one month has beneficial effect in bringing the does into oestrus. Even without additional supplementation when there is lush green pasture, there is flushing effect. Supplementation of about 250 g of concentrate could bring about flushing of does quite well. If the plane of nutrition of the animal is good prior to breeding, flushing is not at all required.

10.4.3 Pregnant and lactating does

If the does are flushed and good grazing is available, no additional supplementation is necessary during early pregnancy up to about 14 weeks of gestation period. During advanced pregnancy (last 6 weeks) however, extra feeding is essential. During this stage, there is also depression in the intake capacity and feed digestibility. Hence, highly digestible feed need to be fed during this stage.

10.4.4 Bucks

When the bucks are to be maintained they should be only for breeding purpose, otherwise they should be sold for slaughter. In some places in the field, farmers keep large number of bucks in their flocks only for getting the hair clip. Some farmers also have religious sentiments not to slaughter their bucks. This practice should be discouraged. Generally if good grazing is available, bucks do well on grazing alone and no additional supplementation is required. During breeding season, concentrate supplement should be provided. In the absence of concentrate, supplement with good legume hay should be provided in the quantity twice the recommended concentrate allowance. During non-breeding season, if the bucks lose weight, fodder supplement should be provided. About 5 to 6% DCP in the fodder/pasture is adequate to maintain the animal during non-breeding season.

10.5 Breeding Management

Goats are bred either by natural mating or through artificial breeding. Bucks of the indigenous breeds donate good quality semen under proper management conditions throughout the year. However, the bucks of temperate breeds if not protected from high temperature, high humidity and high solar radiation will not produce good quality semen during hot dry and hot humid seasons. Females of tropical breeds cycle throughout the year. Temperate breeds which are affected by hours of day light and breed with declining day length. They come in heat in autumn from August to November, although some may breed up to February.

10.5.1 Natural Breeding (individual, pen and flock mating)

The natural breeding is done by either flock mating, pen mating or hand mating. In flock mating system, breeding bucks are usually turned out in the flock during the mating season at the rate of 2 to 3% of the does all through day and night. It is most widely practised in the flocks of all farmers. Semi flock breeding or pen breeding is done to conserve the energy of bucks and give them rest. In this, the bucks are turned out for service with the flock in the pen during night and confined and stall-fed or grazed separately during the day time. Hand mating is practiced when exotic purebred sires are used, or when it is considered desirable to extend the services of buck over much larger flocks. Goats in heat do not manifest behavioural symptoms. Hence, the teaser bucks are employed for detecting the does in heat. These does are then taken out of the flocks and bred to the designated sire of the flock. In pen and hand mating systems and when teaser bucks are used for heat detection, some dye mixed in grease or simple linseed oil is smeared on the brisket of the buck. This makes it possible to record the date when the does is bred and also to remove them from the breeding flock. The colour of the dye should be changed every 16 to 18 days so that the repeaters can be discovered, if the bred does have not been removed from the flock. This is termed as marking of does by breeding buck and marked does are considered as bred.

10.5.2 Artificial insemination

It is because of this that the possibility of artificially inseminating females exists. From 4 to 8 does may be inseminated from one service of a buck although as many as 30 to 40 have been reported, but the degree of success was not given as 0.1 to 0.2 cc of semen is sufficient for the purpose when there is a heavy concentration of sperm in the semen. Great numbers of sperm are of course needed as they are microscopic in size, and the reproductive tract of the female through which they must migrate to come in contact with the egg, also microscopic, is very large indeed in relation to the size of the reproductive cells. Apparently, most sperm do not survive much more than 18 to 24 hours in the reproduction tract of the doe. Russian research workers have reported impregnating from 300 to 400 does in one season with the semen collected from a single buck. They also reported that 90% of the does so inseminated became pregnant. This is a far greater number than could be bred to one buck using natural methods. From 40 to 60 does is generally considered a reasonable/maximum for a buck in one season.

For the purpose of artifical insemination, artificial vaginas are used for the collection of the semen, although it may be recovered in some quantity from the vagina of a doe. Semen can be preserved for several days with suitable media and low temperature and has been transported long distances and used successfully in experimental tests. The semen is diluted for use and is injected into the vagina or deposited at the base of the cervix through a glass or plastic tube. Two inseminations at about a 12 hour interval, beginning soon after the onset of heat, are often recommended.

Although little use has been made of artificial insemination in goat except in an experimental way, some of the advantages and disadvantages are known. The chief advantages are that the usefulness of a superior sire may be greatly extended, the number of bucks needed for a large flock is reduced, and a flock of considerably greater uniformity should result. Disadvantages include the need for extra equipment and labour, some of which must be skilled in the work, the need for one or more 'teasers' to identify which does are in estrus, and the frequent handling of the does during the breeding season.

10.5.3 Heat detection methods

Visual signs of approaching estrous are, a swelling and redness of the vulva and restlessness or nervousness indicating a desire for company, but the most obvious sign is ridding and in turn being ridden. The breeding occurs only during estrous although the buck is capable of breeding at any time.

Overnight teasing and drafting riddled does off in the morning. This is quite manageable for small lots, but when large numbers have to be inseminated, it takes lot of time as does cannot be presented for AI at proper time during estrous. Under the circumstances, the bucks have to be worked too frequently to provide necessary semen. When large number of does are to be handled, harnessed vasectomised bucks are released with the does at about 5 pm and the mob turned into paddock. They are mustered into the yards by 8.00 am the following morning and marked does drafted off. With few modifications to most goat yards, this can be done at the rate of 4,000/hour.

Marked does are either inseminated to get the labour force acquainted with procedure and a few bucks checked and trained or these are put aside and brought back into the mob at the end of the cycle. Generally, the drafted does are inseminated on the acceptable type of estrous mucus. Obviously, those does which are drafted late in estrous are rejected.

10.5.4 Insemination procedures

Under the insemination technique there are a number of pre-requisites such as equipment for insemination, restraint of does, number of does for insemination and adjustment of inseminations during better part of estrous period, and whether heterospermic inseminations are desirable or not.

(i) Equipments: A speculum, a heat light torch and a syringe with insemination pipette is all the quipment required. There are 3 types of specula in use: (1) the duck bill type, (2) metal barrel, and (3) glass speculum. It is the usual experience that the duck bill type is easy for insertion and gives greater ease and freedom of movement. It can be easily dilated in the vaginal passage and removed in a closed condition after use. It is also easy for setrilization. A glass speculum is also used. A 2 ml glass syringe attached to inseminating pipette with a rubber connection is most convenient for insemination. A fresh insemination pipette should be used for each insemination.

(ii) Restraint of does for insemination: With the help of two attendants, does can be inseminated at the rate of 100/hour, if the operation is streamlined, using fresh neat semen. Various methods of holding does have been used and some methods used in Australia between fifties and sixties are:

(a) Does held upside down in a cradle.
(b) Does placed on a battery of bails on a raised platform.
(c) Does held on a rail, as for marking.
(d) The operator working in a pit and a doe manoeuvred up to a hock bar at the edge of it.

Other methods used in Asia and in South America, include the use of various cradle and jacking devices to lift the hind quarters. These are not easy to operate. The method used in Australia, at present, is for the attendant to saddle the doe facing the tail and with a hand under each flank to throw her hind quarters over a rail 24" above floor level. Attendant keeps the doe's hind legs extended by pressing into her stifles. If the doe's front feet are on the floor, she will not struggle. After insemination the doe is released through a gate in the pen.

A flase floor about 2 feet high is put into the catching pen to enable the inseminator to have the does presented at eye level.

When the doe is placed over the rail, an assistant inserts glass speculum whilst the inseminator loads his syringe. The inseminator then holds the speculum, locates and positions the cervical opening and inseminates the doe. The semen is deposited in the cervix.

A simple wooden crate with sloping platform is considered for restraint. The rear portion is raised to a height of 3 feet so that the inseminator can conveniently inseminate, in standing position.

10.6 Handling and Management of Goat Products

10.6.1 Wool/hair

Skirting is done after shearing. It is removing the objectionable parts such as tags, leg pieces, neck pieces, bellies, locks and stained portions from the body of the fleece.

After skirting, the actual grading of the wool/hair is done by a trained classer. The method of classing in the principal wool-producing countries differs considerably. The grading or classing of fleece is done on visual appraisal of length, fineness (handle or feel). Grading of Indian wools/hair was initiated under FAO/UNDP project based on wool quality, length, colour and vegetable fault (content of vegetable matter, *e.g.* burr, seed). About 90 types were prescribed besides lower line. The Indian Standard Institution (IS2900:1979) evolved a grading system based on micron value. It prescribes almost 120 grades. Thereafter, style grading was introduced in Rajasthan which cuts the grades to 32 only. However, none of the above grading systems is practised in the country to any significant extent. The results led to the following recommendations:

(a) Wool from different agro-climatic regions and breeds may be marketed as such.

(b) At the time of shearing fleece should be properly skirted.

(c) Fleece after skirting may be classed into at the most 2 to 3 grades, like fine, medium and coarse.

(d) Bales from individual lots should be sampled, and these samples be subjected to objective quality assessment through laboratory testing for yield, length, average fibre diameter and vegetable matter content.

10.6.2 Meat

In our country, the word mutton is used in a very broad sense to include goat meat also. However, mutton by definition is the flesh together with the associated tissues such as blood vessels, nerves, glands, fat and bones from carcasses of goat older than 12 months of age. Sometimes, mutton from goat carcasses of 12–24 months of age is called as yearling mutton, whereas that from carcasses older than 24 months of age is called mature mutton. Meat from goat carcasses younger than 12 months of age is called lamb. Flesh from goats should be called chevon (pronounced as “shevon” which means a kid).

Mutton and chevon are substituted for each other once in a while by meat traders in places where there are local preferences and price differential. Although, it is very difficult for a layman to differentiate goat and sheep flesh, the following points might be useful: (a) Lamb is pale pinkish with evenly distributed firm white fat, (b) Mutton is slightly darker than lamb and many a times there is a well develped and thick fat cover. Mutton may also have marbling in it. Both lamb and mutton may have a few wool fibres sticking here and there (c) Chevon is dark red with coarse texture and a sticky subcutaneous connective tissues layer which may have adherent goat hairs.

For the production of wholesome meat with good keeping qualities, sound husbandry methods and cleanliness of the animal during final stages before marketing are necessary. Feed additives and antibiotics should be withdrawn at appropriate period before slaughter. Goat may often become dirty due to diarrhoea or during wet weather. Judicious drying off of wet goat on a clean straw for 12 hours before slaughter is of great value. Detection and segregation of sick goat before transporting them to a slaughter point will avoid transportation and subsequent losses. During transportation from a farm to abattoir, every possible care should be taken to avoid injury, lameness, suffocation or transit fever. Goat should never be lifted by hair during loading and unloading as this will cause bruises leading to carcass trimming losses. Water must be provided to goat(s) during a journey of 36 hours or over. A minimum floor space of 213 cm^2/goat must be ensured during transportation. Goat normally lose up to 3–6 kg of their body weight during transportation depending on the weather and duration of the journey, due to sweating, exhaustion, and excretion of urine and faeces. Stress during transportation and before slaughter affects carcass yields, microbial contents of the tissues, onset of rigor mortis and the keeping quality of flesh and preseved meat. Therefore, goat should be rested at least for 16 hours prior to slaughter with continuous access to water.

10.6.3 Milk

Since goats are small, many goatmen prefer to have the does stand on a milking platform. This also gets the milch animal up off the soiled floor. On the other hand, in India people are used to squatting down when they milk does so there is probably not enough to be gained to warrant the cost of constructing a milking stand. Large commercial dairies in European countries construct a masonary platform 38 to 45 cm above the floor level. The does place their heads in a stanchion and while 10 to 20 does are being milked either by hand or a milking machine, the does are eating concentrate happily. By the time the 10–20 does are milked, they too have finished their grain and are ready to be released for the next group of does. In India, the habit is to milk the does by hand from the left side. This is preferable to milking from the rear.

The secretion of milk takes place day and night but it is slowed up by the pressure of the accumulated milk in the udder. So, milking the animals at regular intervals gives the best results. Feeding should also be done at the same time each day. When possible, it is preferable that the same person manages the does/goats.

Milk is a food and should be treated with utmost cleanliness and sanitation. Keep the hair clipped from around the flank and udder to prevent dirt from collecting, which might fall into the milk. Brush the doe daily or at more frequent intervals. Wash the udder or wipe it with damp cloth before milking. The person doing the milking must have clean hands and clothes. Milking utensils must be clean. Milk the does in a clean environment away from the buck.

10.6.4 Edible and non-edible offal

For effecting an improvenment in quality and to increase the quantity of casings collected for processing, the first essential step is improvement in the conditions prevailing in the slaughter houses. Modernisation of slaughter house and provision of a byproduct wing in them are absolutely essential. However, pending modernisation, an interim measure to improve the quality of the casing immediately would be to make provision for adequate water supply in the existing slaughter houses so that the guts are cleaned properly within the precincts, and with the least possible delay.

In many slaughter houses, guts are not removed soon after the slaughter of animals. This results in deterioration of the quality of guts. As such, it should be made obligatory on butchers to remove guts within prescribed hours. The byproducts wing of each slaughter house should have a processing unit for guts under hygienic conditions. Guts may then be sorted out according to calibre, grade etc. Thus, making it easy for the exporter as well as the processor to do only the final grading. This would fetch a better price for the graded product.

10.6.5 Skins and hides

Hides of all animals, which die, cannot be fully claimed as arrangements for timely flaying of dead animals are not available in many parts of the country. Loss of hides of animals lost in floods, famines and those dying in remote areas has been estimated to range between 4 to 10% in different parts of the country. In the case of skins, the wastage due to non-collection is negligible, the major proportion of skins being recovered from slaughtered stock. The estimated annual loss of skins due to non-collection from dead stock has been reckoned at 1 to 2%.

Though India stands 2nd in the world in the production of hides and 5th in the matter of production of skins, it cannot forge ahead in capitalising on its large production unless stress is laid on quality right from the initial stages of production. As such, improved methods of flaying should be introduced in the slaughter houses and better flaying should be encouraged by payment of premia to good flayers. The All-India Khadi and Village Industries Commission has set up a number of flaying centres in the rural areas through the State Khadi and Village Industries Boards. There is however, an urgent need for establishing more village flaying centres for skilled flaying of fallen animals. Demonstration-cum-training centres should be established in important places for imparting training in curing, tanning and rational utilisation of hides. Facilities need to be suitably strengthened to carry out grading of hides and skins according to Agmark standards. Cold storage facilities for preserving raw hides and skins also need to be provided, wherever possible. Before export, a system of complustory preshipment inspection should be introduced. This measure is of vital importance in ensuring quality.

10.7 Goat Housing

There are no serious problems as far as housing of goats in concerned. It is to provide goats with dry, comfortable, safe and secure place free from worms and against extreme heat and other weather conditions. Dairy goats do not require very costly housing. They require very little extra investments for housing and equipment. A separate building can be constructed at low cost out of bricks and thatched roof. A room of size 3 × 1.5 m is quite sufficient for a doe, but some space of daily exercise is also needed. Hence, a coral adjoining the goat house is sufficient to allow daily exercise. In our country, under village conditions, ordinary goats generally are not provided with any special housing. They are usually kept in 'Katcha' thatched sheds with an open enclosure of thorny bushes. However, they need to be protected against inclement weather, wild animals and predators. Free movement of air in the goat house and thatching over the roof as already indicated is essential, specially during the hot season. Goats are generally susceptible to cold and need to be kept in a comparable and warm place especially during winter months. The shed should have proper drainage. Kids should be housed in a warm place and buck should be kept away from milking goats.

Housing instinct is widely prevalent in the animal kingdom and goats are no exception. Since the species have lived in diverse macro-climates of the Indian subcontinent (Mean Max 30°C, Min 40°C in the Western Himalayan region and Mean Max 49°C Min 20°C in the Trans Gangetic Plains and RH ranging from 20 to 95% in different seasons), they have exhibited remarkable ability to adapt to varying climatic conditions. This, however, does not necessarily mean that goats will not be benefitted from the optimum housing, particularly under intensive system of management.

10.7.1 Orientation of sheds

Sheds with long axis running East-West provide a cooler environment underneath than the one with a North-South orientation. The latter, however, keeps the shed dry and promotes sanitation (because of the sun rays falling inside for a longer period) though this orientation may impose a greater stress on animals, if left underneath during the daytime in hot-arid climate. A North-East to South-West orientation of the shed is expected to provide the maximum benefit from the morning sun and the cooler microenvironment within tile sheds in the hot-arid environment (Kelly *et al.*, 1950). Like the latter orientation, East to West and North-West to South-East oriented sheds provide more opportunity for radiation exchange with cooler North sky at night. One slope of the roof is shaded by the roof line of 'A' shaped sheds in such orientations for a greater part of the day, resulting in lower floor temperature. Thus these orientations favour heat loss from animal body to environment both by radiation and conduction. Paddocks on North side of the shed in hot regions and on South side in temperate climates are recommended for better comfort.

10.7.2 Ventilation

The efficiency of ventilation is greatly affected by the summer and winter directions of the prevailing wind at any place. The air in-lets should, therefore, be provided at the height of the goat breeds to prevent the drought. An attempt to cover sheds from all sides to protect the animals from low air temperatures may result in an increase of humidity up to 90% and ammonia concentration to 20 ppm. Good ventilation in the sheds is very important in hot-dry and warm humid climate to promote the heat loss from animals with increased convection and radiation.

In tropics, the two long sides of the goat shed should not be provided with solid wall, above 1 m from the floor. This open part of the shed walls should preferably be kept uncovered when the ambient climatic condition is warm-humid or hot-humid and should be partially covered when it is hot-dry. Protection from hot-winds in the hot-arid zone needs special consideration. During the winter season in uncertain climates and for most part of the year in temperate zone, the sides may be closed but cross ventilation space is left at the roof height. Panels of thatch or fire proof thatch made from locally available grasses are very cheap and convenient for adjusting the ventilation at will from the two long sides. Thus, the open type shed with adjustment of ventilation should provide the desirable comfort within the shed under variable climatic conditions. Top ventilation in the form of space along the length by providing double 'A' shape roof or chimneys has been, otherwise, found to be efficient outlet in goat sheds.

10.7.3 Roofing material

Roofing material in the goat shed is an important factor which determines the cost of construction and micro-environment within the shed. A wide range of roofing materials is available in different climates, which differ to a great extent in their thermal characteristics. The worth of over 50 materials were tested for use as roofing material in California, ranging from hay or straw, galvanized steel and plywood to several types of plastics. On a summer day, the difference in the radiant heat load, under sheds covered with straw versus galvanized iron or plastics, was of the order of 163 Kcal/h/m^2 of the animals surface.

In a study at Central Institute for Research on Goats (CIRG), Makhdoom, it has been observed that the improved thatch provided the coolest micro-environment during hot part of the year (Singh *et al.*, 1988a). Different roofing materials commonly used in India were tested and they can be graded as, the most to least effective in that order, improved thatch, thatch, asbestos, lite roof with top surface painted white and non-painted lite roof. The life of the improved fireproof thatch is reported to be 8 years as against about 1 to 2 years for untreated thatch (Gupta *et al.*, 1985).

10.7.4 Height and shape of roof

The height at centre in 'A' shaped roof is suggested to vary between 3 to 3.5 m. A height of less than 3 m interferes with the proper ventilation resulting in reduced convective heat loss from animals. In temperate and hot-humid climate, where more height does not provide any additional benefit, a height of 3 m is expected to suffice. The heat loss through radiation from goats to cool sky is curtailed in low roof sheds. The shadow size is not affected by height, since higher the shed the faster the shadow moves. In hot-dry environment, radiant heat load plays a vital role. Within the boundaries of the shed, the radiant load on goats comes from the atmosphere, the roof, the shaded ground and sunny ground. The heat load on goats will be minimum within the shed when the distance from the roof is maximum.

'A' shaped roof is indeed better for hot climatic regions. In the hot weather one side of 'A' shaped roof saves the other half from direct solar radiation by casting its shadow. This helps in cutting down heat gain from the roof of the shelter.

Double roof with both roofs of some or different materials are effective in reducing the heating of shed in hot weather conditions. The additional cost of construction in double roof, however, restricts its use to very high producing animals.

10.7.5 Floor type and space

The surface upon which an animal lies may potentially be a source of thermal and physical discomfort, injury and infectious diseases. The ideal bed needs, therefore, to be hygienic, dry, resilient and reasonably temperature resistant. The relative importance of these 4 criteria differs markedly for different species and classes of farm animals. Deep, clean, dry straw can provide an ideal bed for weaners and growers during the cool period, but a thin layer of straw is likely to be more suitable during warm or hot weather conditions. An inadequate bed of wet, filthy straw fails on all accounts. Perforated or slatted floors are mostly drier and more hygienic than solid floors with limited bedding and thus more suitable for humid and high rainfall areas. The type of floor which provides both comfort and cleanliness with minimal risk of injury should be given preference.

The floor space for different age groups should be slightly more in hot weather conditions in comparison to cool weather. This helps in radiant heat losses and convective removal of heat and water from the animal body. In tropical climates, huddling is disadvantageous for health of the animals and their productivity. The optimum floor space requirement to provide healthy and clean micro-environment have been recommended later on the basis of scattered information.

10.7.6 Shelter surroundings

Surfaces around the shelter are very important in view of the radiation exchange between surrounding surface and the shelter. The temperature of different surfaces varies significantly at the same air temperature. It is well known that the green surface do not heat up as much as other surface like gravel or loose loam. It should be possible to maintain the green vegetation adjacent to the goat shelters except the paddock.

10.7.7 Optimum floor space requirement per goat

The knowledge on airspace requirement per animal is thought to be important as the cleanliness of ambient air, with regard to air-borne wastes from animal body, must be directly proportional to the actual volume of air into which those wastes are dispelled. Conversely, concentration of pollutants is directly proportional to stocking density. Systematic studies to delineate the optimum floorspace requirements of different goat breeds in different environmental conditions have been carried out at the CIRG. Floor space requiurement for goats of various age groups has been summarized in Table 10.1.

Table 10.1 Floor Space Requirement for Goats of Various Age Groups

Sl. No.	Category of goat	Floor space requirement
1.	Adult goats	1.25 to 1.5 m^2
2.	Lactating and pregnant does	2 m^2
3.	Bucks	2 m^2
4.	Kids	
	7 to 90 days	0.5 to 0.6 m^2
	3 to 6 months	0.7 to 0.9 m^2
	6 to 12 months	1 m^2

10.7.8 Kidding pens

Individual spacious pens are essential to house does in late pregnancy. Movable hurdles can also be used for preparing kidding pens. Individual kidding pens are contaminated very quickly, and need frequent cleaning and disinfection. Otherwise, they may constitute an important source of naval infection to kids.

Rearing of young kids needs special attention while planning goat housing and a fly proof shed for kids up to 3 months of age will prove worthy. The pens for kidding should be at the warmest part of the goat-house complex if the kidding is expected in cool weather. The protection of new-born kids from low ambient temperatures is essential to reduce the kid mortality.

10.7.9 Lean to type shed

The cheapest form of building is the "lean to type shed" located against the sides of an existing building. Such a shed for 2 goats should be 1.5 m wide and 3.0 m long. This length provides 0.3 m for the manger and 1.2 m for the goats. The remaining 1.5 m space is sufficient for 2 milking goats with a sub wall between them. The height nearest to the wall should be 2.3 m and on the lower side 1.7 m giving a slope of 0.6 m to the roof which may be thatched or tiled. An open frame window of good size in the lower side and an open framed door should be provided. Arrangement for storing hay or dried feed can be made overhead.

10.7.10 Shelter for buck

The buck should be housed separately. A single stall measuring 2.5 m × 2.0 m with the usual fittings for feed and water could be quite suitable. Only one buck should be kept separately to avoid fighting during breeding season.

10.7.11 Housing for pregnant does and kids

Kids should be provided with separate loose stalls away from adults formals. The walls and doors of these stalls should be about 1.3 m high. A box barred or a log is provided for exercise. One stall measuring 1.8 m^2 can accommodate up to 10 kids. Such loose stalls are also suitable for goats at the time of kidding. All stalls should be provided with an enclosure in which the animal can be let loose during the day. The loose housing system reduces the housing cost and labour.

10.7.12 Quarantine/segregation shed

A small Quarantine/segregation shed of 3.6 m × 5 m is essential when the herd size is large. It should be constructed in the farther corner of the farm and provided with a well fenced yard. It should be divided into 2 or 3 sections. Each stall and the yard should have a separate watering arrangement.

10.7.13 Exercising yard for stall-fed goats

A yard measuring 12 m × 18 m is adequate for 100 to 125 goats. Such a yard should be well fenced with strong woven wires, which should be quite close to bottom. The exercise paddock/yard should be made bigger than the enclosures and should have same shade trees if the stock is to be maintained constantly in confinement. An extra strong woven wire should be used as the goats have the habit of climbing fences and of rubbing the bodies against them. It will be appropriate to have a box of 1 m ×1 m × 60 cm high and stationary steel drum or a log of 30 cm × 2.4 cm size is provided for their exercise.

10.8 Essential Appliances

10.8.1 Feeders

The wastage and contamination of feed with faeces and urine have been the major problems with existing feeders in goat farms. To rectify some of these shortcomings, new feeders were developed at the CIRG based on the basic conception of Blanken *et al.* (1970). The rectangular and hexagonal feeders have been developed with provision of feeding green roughage, straws as well as concentrate. It has been possible to drastically reduce the wastage of all 3 types of feed with the use of these feeders (Singh *et al*., 1988a, 1991). The animals are not able to void faeces or urine in the feeding troughs nor they can keep their legs in the feeders thus avoiding the chance to contamination of feed. The newly developed rectangular as well as hexagonal feeders are effective in saving time, labour and space, and seem to reduce the competitive struggle among animals to reach the feed.

10.8.2 Water troughs

Contaminated water is an important source of infection. The water tanks or troughs should be covered and require regular cleaning. Water troughs, 3 to 4 cm in length per goat, when raised in groups is sufficient. Water tanks with flat valves may be suitable for large-scale intensive goat production. Availability of water in goat house round the clock will be beneficial for attaining optimum productivity under any system of production.

10.8.3 Hay racks

Goats refuse to eat what has dropped down on the ground and are very wastefull. Hay racks are helpful. The bars of hay racks should not be more than 5 cm apart and for feeding there should be a wooden board, fixed about 15 cm below the rack, to catch whatever falls from the rack.

10.8.4 Shed hygiene

The shed management includes constant vigil to general cleanliness, timely repairs and renovations. The sheds would require seasonal spraying to protect the flock from ectoparasites and disinfection particularly prior to kidding operation. The earthern floors, where applicable, are needed to be replaced every 3 months besides disinfection with carbolic acid which takes care of snakes and other reptiles. Annual white washing with lime prior to winter are suggested. Painting of sheds are recommeded prior to monsoon. Application of thatch panels to control the excessive airflow during summer and winter through the sheds in the hot-arid environment is expected to improve the shed micro- climate.

10.9 Grazing Schedule and Supplementary Feeding

Goats normally enjoy grazing and foraging to their choice, except in very intensive system of management, tight scheduling of time and grazing area is obligatory. Depending on the nutrient potential of the grazing area at a particular season, supplementary feeding with concentrate mixture, particularly to lactating/pregnant does and growing kids, are necessary. With extensive grazing in the good pasture and in migratory flocks, such supplementation may not be necessary though grazers prefer to carry ground seeds/cereals and salt for feeding in adversities. A good feeding management is rested on storage of feed and hay and the efficiency of management is soon reflected in the general health of the flock.

10.10 Health Management

Good health management is to ensure timely vaccination according to schedule to all members of the flock. Routine dipping will indeed render the flock free of ectoparasites. The administration of anthelmintic drugs before and after monsoon is always advisable. Trimming of hoofs are absolutely essential for those goats, which do not undertake long walking/grazing on the hard and rocky terrains. Even with the best health care and management, a few animals in the flock may develop illness at any time and a good manager will not fail to detect the unhealthy behaviour of these animals. Isolation of these animals in the sick shed and prompt treatment by a competent veterinarian is obligatory. Routine testing of the flock for Brucellosis and Johne's disease should under no circumstances be deferred to as precipitation of these diseases in the flock will be highly derrogatory in future. Vaccination of pregnant does against FM and colibacillosis and antibiotic administration to control insidious mycoplasmal infection should receive imminent priority. The kids are highly susceptible to FM, colibacillosis and coccidiosis for which preventive vaccination and administration of drugs are necessary.

10.11 Feeding Management

A dairy goat may be considered a miniature cow, so the feeding of a doe is similar to feeding a cow. Feeds suitable for one are suitable for the other. It has been estimated that a goat will consume $^{1}/_{5th}$ to $^{1}/_{8th}$ as much feed as a large cow. To be conservative, it is usually claimed that they consume $^{1}/_{5th}$ as much feed. To get high production from a goat, it is important that she be fed a good ration, much better than the buffalo or cow receives in most instances in the Indian village.

Repeatedly, it has been stressed that well-cured legume hay such as lucerne or berseem are the best and cheapest sources of nutrition for a dairy goat. Cut the grass while it is in the early bloom stage and cure it during sunny weather because rain-soaked hay loses considerable food value. The leaves are the best part of a legume hay. In this connection, it has been observed in Ludhiana district that the goatmen tend to feed the coarse groundnut hay to the goats and the leafy portion is sold or given to other stock. If one is interested in getting milk, it is important that good quality hay be fed and this should include the leafy portion as well.

The bucks should have the same feed as the does but they consume more forage because they are larger. The cultivator must be the judge as to how much concentrate to feed but the answer is to provide enough to keep him in good physical condition. During the mating season, it is necessary to feed more concentrate. An exercise pen is important for a buck because he must have exercise. This helps to keep him in good physical condition for breeding.

CHAPTER 11

GOAT PRODUCTION AND LAND DEGRADATION

11.1 Land Degradation and Goats

Out of 350 million hectares (M ha) of land in India, about 174 M ha could be graded as 'good' but the remaining land is either partially or fully degraded. The classification and distribution of the ecologically unstable lands (often referred to as waste land) have been shown by different authors (Vohra, 1986). Five major factors are generally responsible for the degradation of land mass:

(i) Erosion of the top soil by the action of water or wind. The loss of top soil by water action alone is estimated to be around 12,000 million tonnes per annum.

(ii) Water logging and consequent salinisation and alkalisation mainly due to lack of adequate drainage in canal irrigated areas.

(iii) Floods, which affect an average area of around 9 M ha every year.

(iv) Diversion of good agriculture lands to urban, industrial and other non-agricultural uses. It is estimated that land under such uses will increase from the present level of 18 M ha to around 24 M ha around the year 2010 A.D.

(v) Deterioration of our most productive soil under conditions of multiple cropping with the aid of perennial irrigation and the application of large quantities of inorganic fertilizer.

The grim effects of the land degradation include massive soil erosion, floods, siltation of reservoirs, drought, salinity, changes in the climate and consequent infertility of even cultivable lands. The process of denudation of soil cover is being further aggravated by growing livestock population coupled with unchecked grazing devegetation and illicit felling of trees. The reclamation of these waste lands thus demands an integrated three dimensional farming system for the concurrent sustainable production of agriculture, forestry and animal husbandry.

It is not a mere coincidence that goats are reared in India primarily on waste lands and a significant positive correlation (r = 0.78) exists between goat population and available waste lands in different states. All scientific observations reveal that the contribution of goats to the fragility of the unmanaged waste land is only marginal.

11.2 Desertification and Goats

The lack of understanding of the role of goats in the economy of rural poor, especially in ecologically fragile zones had resulted in not only withdrawal of support for development of the species but it has been blamed for man-made ecological problems like soil erosion, deforestation and ultimate desertification. The goats have been often used as 'scape goat' for the failure of adequate land development and forest management on scientific footing. Contrary to cattle and sheep, goats are mainly browsers and would seldom go for grown up trees. They dislike the leaves of timber trees. Available scientific data prove that goats do not prevent the spread and establishment of grasses which are so essential for soil conservation. The browsing habit of goats on the other hand tends to reclaim saline soils by consuming salt laden leaves of range plants (Shankarnarayan *et al.*, 1985) and contribute fertility to soil by even distribution of essential manure on the land it grazes. Yet goats are prohibited for grazing in some forest land though cattle are allowed to graze, having role in erosion of soil and exploitation of 3–5 times of phytomass. In a controlled study, it has been reported that the intensity of grazing of 2 or 4 goats /ha had no adverse effect on run off and soil loss in normal rainfall years (Prajapati *et al.*, 1988). Despite grazing, the run off ranged between 2.02 and 8.24% and the soil loss between 0.07 and 2.66/ha respectively which were close to 8.50% and run off and 2.40/ha soil loss of the non-grazing decade. As a matter of fact, while sheep and cattle find it difficult to survive with remaining vestige of soil cover, the goats sustain due to their ability to extract livelihood from coarse feeds and shrub vegetation on which no other animal can thrive. In this process, it helps the immediate soil cover to regenerate.

In several behavioural studies conducted at CIRG, goats have been found to defoliate smallest branch of tree without damaging the twigs. The vegetative regeneration was 27.0% more in goat paddocks comprising *Cenchrus ciliaris*, *Dichrostachys nutens* and *Leucaena leucocephala* as compared to sheep paddocks (Sharma and Ogra, 1987). Better performance was recorded in goats than in sheep in all ecosystems for 3 successive years (Harsh and Shankarnarayan, 1982). The regeneration of bushes in Cenchrus pasture dominated by *Ziziphus numrnularia* ranged up to 143%.

Goats, while grazing on grass or bush, also help in dispersal of seeds and improvement in vegetation. In Australia, goats have long been recognised as being able to control many woody weeds in the vast pasture land. There are reports that in extreme deserts, the grazing of sheep and goats has increased the number of bushes, trees and grass cover (Govt. of India 1987). Thus, it is unwise to put the heavy burden of all evils to goats for land degradation and deforestation till serious kitchen-energy crisis is solved for the rural mass and details of industrial need vis-a-vis wood resources, are calculated, projected and objectively managed.

A Task Force to evaluate the impact of sheep and goat rearing in ecologically fragile zone was appointed by the Government of India in 1987 under the Chairmanship of Prof. C.H Hanumantha Rao, Institute of Economic Growth, Delhi University. The Task Force concluded as follows:

1. Sheep and goats did not pose a threat to the ecology as was generally believed. The negative effects, if any, have been vastly exaggerated.
2. Small ruminants made a very important contribution to the survival of the economically weaker section of the society and met the essential requirements of industry, associated with wool, leather and meat, besides earning valuable foreign exchange through exports.
3. The unproductive cattle, intensive agriculture on submarginal land and excessive stocking rate of livestock etc. should be critically examined to improve the situation.
4. Forest departments and Panchayats should play a role for protecting and improving the grazing land resources.
5. In the programme on social forestry and expansion of forest, enough attention has not been given to fodder trees. Fodder trees provide a better autotroph and heterotroph relation under the ecosystem over only timber trees.
6. Small ruminants needed to be given due priority under Central as well as State plans so as to raise per animal production.
7. The existing marketing system for the sheep and goat products entailed considerable exploitation of the rural poor. As a result, economically weaker sections could do very little to improve production or generate further employment in rural areas.

CHAPTER 12

MEAT PRODUCTION

Goat meat is very popular in the tropics, but like other kinds of meat, it is often a luxury. The special importance of goat meat lies in its suitability for domestic consumption; for the nomads and people of the rural areas, the value of goat as a source of meat is far greater than is usually appreciated. As a result, a substantial increase in goat numbers could be quickly achieved, despite the wide spread prejudice in many quarters against goats per se. Such a population increase would be possible because of their fertility and reproductive rates, their efficiency in food conversion, the variety of feeds on which they can subsist, and their relative resistance to disease. Further, they could play a most useful part in helping to destroy bushes and shrubs. Properly managed goats have an important part to play in the utilization of arid areas.

Goat meat is preferred to mutton in Sudan and many parts of Africa, the Middle East, India, Malaysia and Fiji. In Malaysia, the demand for goat meat is very high and it fetches the highest price per unit of any meat sold in the market. Wahid (1965) has reported a definite preference for goat meat in East and West Pakistan even though the price is almost the same as for mutton. In Fiji, the shortage of goats in many areas, and a high prevailing price for goat meat relative to mutton, has put greater emphasis on mutton imports.

The preference for goat meat is probably related to certain special features which make it quite different from that of mutton. Whereas in sheep, the fat is distributed all over the body, in goats visceral concentration is characteristic (Villegas *et al.,* 1938; Williamson and Payne, 1965; Zeuner, 1963b). This in turn affects its succulence and tenderness. Goat meat probably has higher lean meat content. As the meat of males has a very strong taint, it is usual to castrate males that are surplus to breeding requirements.

Whereas, the meat requirements of people living in highly developed countries in the west have been fairly thoroughly investigated, those in the developing countries of the tropics have hardly received any attention. In contrast to these advanced countries, where consumer preferences are well understood, the livestock industries in the tropics are less well aware of market requirements and consumer tastes. Such factors as conformation of the animal, early maturity, tenderness and succulence of the meat, colour, texture and distribution of fat are not adequately understood. While, there are definite preferences for certain meats, of which goat meat is quite important, there is at the moment very little consumer preference for quality meat.

Although, it is recognised that the preference for individual meat differ widely, the actual reasons for the preferences are not clear, and remain speculative. They are probably determined by differences in taste, religion, cost in relation to other meats, proportion of lean to fat and the availability of the meat itself. Exactly how goat meat compares with other meats in consumer preference is not very clear.

Despite this obvious importance of goat meat, very little work has been done to assess the potential of goats in terms of conformation and component parts of the carcass. This also implies that the value of a number of breeds for meat production has not been completely identified. The reasons for this are many and are probably due to a combination of factors. Chief among them is the lack of adequate differentiation of the specific value of indigenous goat for meat or for milk. Often the dual purpose goat serves as a meat animal after its usefulness as a milk animal ceases, the stage at which this is reached being determined by economic and sociological considerations. Although, goat is an important meat animal in India, the information on carcass traits and meat production of various Indian breeds of goats is very scanty.

12.1 Effect of Live Weight on Meat Production

A detailed study of the small East African goat in Uganda has been made by Wilson (1958a, 1958b, 1960). He studied the growth and development of kids from birth to slaughter and evaluated the carcass by slaughter techniques. The nutritional status was found to have considerable effects on live weight gain and external measurements. In kids weighing about 2.2 kg at birth, the greatest linear increase was shown by the measurements of body length and the least increase by the length of the lower hind-leg. Sex differences were apparent for all the external measurements studied when the results were compared on the basis of equal age, male kids having significantly larger measurements than female kids. The effect of the nutritional regimes produced significant differences in all external measurements studied on the basis of equal age and equal weight, with low-plane kids having larger measurements than high-plane kids. A high plane of nutrition has a significant effect on growth rate, high-plane kids reaching 15 kg at approximately 20 weeks while low-plane did not reach this weight until 48 weeks of age. The sex difference in live weight increases markedly after 16 weeks of age. Whereas, the growth of females slowed down to approximately 0.012 kg/head/week, that of males continued @ approximately 0.5 kg/head/ week. The kids showed marked recuperative capacity when changed from a low to a high-plane feeding, indicating the significance of good feeding on weight increase.

Studies on the growth of Kambing Katjang kids, fed to appetite on a high plane of nutrition in Malaysia, indicated a very low rate of weight increase. Kids weaned at 12 weeks of age at an average live weight of 7.5 kg only weighed 10.8 kg at 30 weeks. In comparison, Anglo-Nubian and Jamunapari crossbreds weighed 24.1 kg and 22.4 kg respectively at the same age, just over twice the weight of the indigenous goat (Devendra, 1966b, Pant *et.al.* 1990) indicated that the influence of concentrate supplementation becomes significant only after weaning at about 3 months of age. It is therefore, necessary to ensure adequate supplementary nutrition in terms of energy, protein and dry matter for growing kids (3–9 months) for maximum slaughter weight at an early age, when the surface vegetation and browse are nutritionally inadequate in quality and quantity (Parathasarthy *et al.*, 1983, 1984). Sengar (1975) studied the effect of 3 × 3 planes of nutrition (energy and protein 75, 100 and 125, Morrison, 1956) on Barbari and Jamunapari goats and reported that the lower regime of protein and energy could not sustain the higher growth rate in kids. It is evident that the growth rate in Osmanabadi males is significantly retarded at low level of energy and protein (Gaffar and Blabani, 1986) which ultimately reduces the live weight at slaughter, dressed carcass weight and dressing percentage (Tables 12.1 & 12.2). Kumar and Narang (1961) studied the effect of different planes of nutrition on the performance of Gaddi male kids using concentrate rations and

Table 12.1 Effect of Plane of Nutrition on Daily Growth (g/head) of Kids

Breed	Age	Plane of nutrition[a]			Significance[b]	
		HH	MM	LL	Energy	Protein
Barbari	0 – 6 months	36	33	27	$p<0.05$	NS
	9 – 14 months	42	38	22	$p<0.05$	NS
Jamunapari	0 – 6 months	44	46	44	–	–
	0 – 14 months	39	41	30	$p<0.05$	$p<0.05$

Source: Sengar (1975)

[a] HH, high energy, high protein; MM, medium energy, medium protein; LL, low energy, low protein.

[b] NS, not statistically significant.

Table 12.2 Effect of Plane of Nutrition on Carcass Characteristics of Goats Slaughtered at 24 Months Age.

Characteristics	Plane of nutrition[a]			Significance[b]
	HH	MM	LL	
Barbari				
Live weight at slaughter (kg)	18.7	19.4	11.4	p<0.05
Dressed carcass weight (kg)	9.2	9.7	4.5	p<0.05
Dressing (%)	57.3	58.3	53.7	p<0.05
Edible offals (%)	21.4	20.0	20.6	p<0.05
Jamunapari				
Live weight at slaughter (kg)	20.3	21.0	15.8	p<0.05
Dressed carcass weight (kg)	9.7	9.8	6.1	p<0.05
Dressing (%)[c]	53.2	53.2	55.5	NS
Edible offals (%)[c]	18.9	20.0	22.9	–

Source: Sengar (1975)

[a] HH, high energy, high protein; MM, medium energy, medium protein; LL, low energy, low protein.

[b] NS, not statistically significant.

[c] Expressed as per cent of empty live weight.

Kikuyu grass hay and inferred that kids fed with higher energy with varying levels of protein had significantly higher body weights and heart girth and kids fed with low energy levels had ower body weights.

The studies on the potential response of kids to various feeding regimes emphasise again some of the limiting factors involved in the improvement of the situation. Although, the East African kids were able to show a marked response to improved feeding, they still grew very slowly for meat animals, their responses being limited by an inherently slow rate of growth. Selection of animals with a higher genetic growth potential gives maximum response to improved feeding and in achieving a high efficiency of production.

The possibility of improving the small East African goat by crossing and upgrading it with imported Boer goats was tried over several years at several stations in Tanzania in the 1950s and early 1960s. Although, the crossbreds (mainly ½ and ¾ Boer) had a distinct advantage in growth up to 2 years and the carcass was preferred to that of the indigenous goat, the establishment of the "Boer type" improved breed had to be abandoned because of the very high mortality among the purebred and crossbred Boer goats (Hutchison, 1964). It is worth mentioning that the crossbreds had an advantage in weight-for-age over indigenous goats at 1–1.5 years of approximately 25 to 40 %.

In India, in order to increase the production potential of native goats, they have been crossed with improve native breeds or elite exotic breeds. Crossbreds of Beetal × Black Bengal (BB) and Jamunapari × BB performed better than BB (Table 12.3). However, Beetal × BB was much superior to Jamunapari BB. It had an improvement of 51.6% in slaughter weight, 86.9% in carcass weight and 17.4% in dressing percentage. Similarily crossbreds of Beetal × Barbari performed better than Jamunapari × Barbari or Barbari for slaughter weight, carcass weight and dressing percentage (RBS college Agra, 1990). Crossbreds of Beetal × Sirohi performed better than Sirohi in carcass traits (Arora, 1992). Compared to Jamunapari, the performance of Beetal was better and it could be used for cross breeding to upgrade the meat production of other breeds. It may also be preferred because of better adaptability in most ecologies, confinement and intensive management (Acharya, 1992). The dressing percentage of castrated crossbreds ranged from 44.8 to 46.2 and 42.2 to 44.3 for Alpine × Tellichery and Saanen × Tellichery respectively (Prabhakaran *et al.*, 1979).

The effect of age on carcass composition is marked and is of interest. Female East African goats were killed at birth, at weights of 4.1, 7.3, 11.3 and 13.6 kg and the carcass evaluated (Wilson, 1958b). Since the economic value of the African goat to the goat keeper is due to its meat, edible offal, edible fat and skin, the dissection data were grouped to show how these components

Table 12.3 Carcass Characteristics of Black Bengal, Barbari and Sirohi and their Crosses with Jamunapari and Beetal

Genetic group	n	Slaughter weight (kg)	Hot carcass weight (kg)	Dressing percentage
Black Bengal (BB)	8	7.96 ± 0.46	3.07 ± 0.20	38.61 ± 1.54
Jamunapari × BB	17	11.02 ± 0.45 (38.4)	5.02 ± 0.26 (63.5)	45.20 ± 0.80 (17.1)
Beetal × BB	4	12.07 ± 2.99 (51.6)	5.74 ± 1.77 (86.9)	45.34 ± 3.11 (17.4)
Barbari	9	16.70 ± 1.28	8.30 ± 0.69	49.4 ± 1.02
Beetal × Barbari	2	21.8 ± 7.9 (30.5)	11.1 ± 4.95 (33.7)	50.7 ± 5.01 (2.6)
Jamunapari × Barbari	2	16.8 ± 5.82 (0.59)	7.7 ± 2.45 (-7.2)	45.63 ± 1.40 (-7.6)
Sirohi	4	15.30 ± 1.70	7.60 ± 0.85	44.70 ± 0.25
Beetal x Sirohi	5	19.40 ± 1.03 (26.8)	9.70 ± 0.82 (27.6)	44.50 ± 0.39

Source : Singh and Sengar (1979); Arora (1992)

Figures within parentheses are percentage of improvement over Black Bengal, Sirohi and Barbari.

n = number of observations.

varied with change in live weight. The proportion of meat in the live animal increased from 24.5 to 37.5% between birth and 4.1 kg live weight. The proportion of edible offals fell from 6.4% of the total live weight at birth to 4.8% at 13.6 kg. Edible fat increased from 2.6% at birth to a maximum of 12.5% at 11.3 kg and fell to 8.6% at 15 kg live weight. The proportions of skin fell from 12.4% at birth to 7.2% at 13.6 kg.

The total saleable percentage of the animal (meat + offals + fat + skin) increased from 45.9% at birth to 61.8% at 4.1 kg live weight. Thereafter, as the kid developed, the total saleable percentage fell to 60.8% at 7.3 kg, 57.5% at 11.3 kg and 55.5% at maturity. The maximum saleable portion of the kids is therefore reached at very early stage in the growth and development of the animal. It does not follow, however, that early slaughter before 7.3 kg live weights is desirable, since as Wilson has pointed out, the saleable percentage of the body does not increase in the same way as killing-out percentage or dressed carcass percentage. The decreased total saleable percentage at maturity was due to a decrease in the proportion of edible offals, edible fat, and skin, not of lean meat. In further studies on the effect of plane of nutrition and sex on the carcass composition of kids at 11.3 and 13.6 kg live weight, Wilson showed that although there was a significant treatment effect at 11.3 kg live weight, by the time maturity was reached, the carcasses of the 4 treatments were remarkably uniform with regard to their economic composition. The work done in India indicates that the age at slaughter up to the time of sexual maturity governs the carcass weight. The dressing percentage, carcass weight and lion eye area tend to increase with age (Singh *et al.*, 1991). Bone percent in carcass, however decreased with age and weight. It is further observed (Table 12.4) that the age significantly affects

Table 12.4 Effect of Age on Slaughter Weight and Certain Carcass Traits of Goats

Traits	Age		
	12 months	18 months	24 months
Slaughter weight (kg)	16.64	25.47	34.00
Empty live weight (kg)	14.27	22.69	30.96
Dressed carcass weight (kg)	6.55	11.14	16.08
Filling %	14.54	10.97	9.06
Offals (kg)	1.85	2.51	2.66
Pluck (g)	378.80	519.50	595.00
Omental fat (g)	156.16	509.00	716.50
Live (g)	414.50	529.30	586.20
Testicular weight (g)	99.32	121.40	164.20
Lion %	10.27	12.72	12.60
Leg %	32.22	28.93	24.70

Source: Koul *et al.* (1993)

slaughter weight, empty live weight, dressed carcass weight, offal weight and weight of pluck, omental fat, liver and testicles (Kaul *et al.*, 1993). Non descript male kid's slaughter at about 5 months of age had lower slaughter weight, hot carcass weight, dressing percentage, non carcass fat and higher head and cannons and alimentary canal (empty) than adult animals (Agnihotri and Pal, 1992). Body weight, carcass weight, carcass length, rib eye area increased with advancing age in Osmanabadi male kids (Kamble, 1988). The dressing percentage was 44.5 and 48.1 in Jamunapari kids, slaughtered at 6 and 9 months of age respectively, the yield of leg (30.0–31.6%) was more followed by shoulder (20.8%) in both groups (Table 12.5). Eye muscle was significantly better developed in kids slaughtered at 9 months of age indicating meatiness of carcass (Khan and Sahni,1979). As moisture and ash content declined, fat content significantly increased with age. Tenderness and juiciness decreased after 10 months of age. However, about 80% slaughter goats were between 1–2 years of age. Carcass weight and yield of market goats were not significantly affected due to age up to 1–2 years (Kesava Rao *et al.*, 1988).

From an economic standpoint the total edible and saleable proportions of goats are important indices of their value in any one locality. Between localities, however, they would not be directly comparable because of differences in eating

Table 12.5 Effect of Age on Carcass Traits and Meat Quality of Osmanabadi Male Kids

	Age group (months)				
Parameters	**6-8**	**8-10**	**10-12**	**12-14**	**CD**
Slaughter weight (kg)	13.9	16.8	18.8	20.7	-
Carcass weight (kg)	6.3	8.0	9.2	10.3	0.24
Dressing (%)	45.7	47.8	49.1	49.9	1.24
Carcass length (cm)	53.7	56.7	64.7	66.4	1.30
Rib eye area (cm^2)	10.7	14.0	16.1	19.9	0.32
Lean weight (%)	66.9	69.4	65.4	65.8	-
Lean meat composition					
Moisture (%)	74.1	69.3	67.0	63.3	0.59
Protein (%)	17.4	16.6	16.6	16.3	0.39
Fat (%)	4.5	9.8	12.7	16.0	0.38
Ash (%)	1.2	1.1	1.1	1.0	0.04
Sensory Scores					
Colour (%)	77.7	88.8	88.8	88.8	5.67
Tenderness (%)	88.8	88.8	77.7	77.7	7.27
Juiciness (%)	88.8	88.8	77.7	77.7	6.39
Acceptability (%)	77.7	88.8	88.8	77.7	7.22

Source : Kamble *et al.* (1989)

habits and also in the use of the by-products that constitute the saleable component. This is so in Africa, where Wilson (1958b) has pointed out that the proportion considered edible may differ from what is acceptable in Europe. In the African market, he estimated the total edible and saleable proportions to be 48.3 and 55.5% respectively. It is of interest to note by comparison that in Malaysia, indigenous Kambing, Katjang goats slaughtered at mean live weight of 25 kg, the total edible and saleable proportions were estimated as 61.2 and 81.5% respectively (Devendra, 1966). The higher total saleable proportion for the goat in Malaysia is due to the inclusion of parts of the offal such as the head and bone fractions, which were excluded in Wilson's estimate.

12.2 Dressing Percentage

A number of workers have reported dressing percentages for a number of goat breeds in various parts of the tropics. For crossbred Anglo-Nubian goats, Gatan (1941) recorded a relatively high figure of 51.4%. In Texas, Angora wethers fed on concentrates and alfa-alfa recorded figures of 48–52.6% (Miller *et al*., 1943). The work of Congiu (1954) is of interest because it is one of the very few reports in the literature, which compares dressing percentages in intact males and females and castrated males. He recorded 50.8% for males, 50.0% for females and 52.3% for castrated males. The higher estimated dressing percentage for castrates indicates an advantage of castration additional to its effects in improving flavor and facilitating management. In Tanzania, castration of the Boer local halfbred resulted in increased proportion of the loin and hind-quarters (Hutchinson, 1964). From the Congo, Thienpont and Vandervelden (1961) have reported dressing percentage of 45.6 for local goats, and Henrotte (1961) a higher figure of 50% for similar goats killed at a live weight of about 21 kg. According to Payne (1963), the average dressed carcass weight of goats in Kenya was somewhat higher than for sheep and was about 13 kg. From Wilson's (1958b) data, the calculated dressing percentage for the small East African goat killed at 13.6 kg live weight is 43.5 (total edible minus total edible offal). This figure is probably slightly underestimated as it is based on carcass dissection data, but for practical purposes it allows a comparison.

The Kutchi breed of Gujarat is considered to have a carcass weight which is 45% of live weight (Williamson and Payne, 1965). In West Malayasia, indigenous goats were fed on 3 planes-Guinea grass (*Panicum maximum*) alone ad lib., guinea grass ad lib with limited concentrates, and guinea grass and concentrates fed to appetite and subsequently slaughtered to evaluate the carcasses. The corresponding dressing percentages were 44.3, 47.4 and 51.3 respectively (Devendra, 1966b). These figures indicate that the dressing percentage increased with improvement in the plane of nutrition. A correlation of live weight at slaughter and hot carcass weight gave a highly significant value of 0.92 ($p<0.01$). This result is comparable to estimate of 0.87 for local Sicilian and Barbaresa sheep and Calabrian goats reported by Salerno (1956) and a figure of 0.98 ($p<0.01$) for fattening steers in South Africa (Luitingh, 1963). Work done in India indicated that the increase in slaughter weight resulted in proportionate increase in carcass weight, yield, carcass length and loin eye area. However, there was no significant difference in meat and fat colour between 10–25 kg body weight (Sahoo and Panda, 1987; Singh *et al*., 1991). Traits of commercial importance like proportion of carcass components and boneless lean in carcass significantly increased in heavier weight classes with each age category. Barbari kids had greater proportion of carcass components and lean than Jakhrana kids of similar weight and age (Prasad *et al*., 1992)

The dressing percentages for various breeds of goat in Table 12.6 indicate that, in general, they compare favourably with those for recognised sheep and beef breeds. Apart from a low figure of 35.0% for the Hejaz goat (Epstein, 1946), the average dressing percentage is of the order of 43 to 53%. The evidence in two reports about a better response to castration (Congiu, 1954;

Table 12.6 Summary of Dressing Percentage in Goats

Breed	Location	Live weight (kg)	Dressing (%)	Reference
Anglo-Nubian grade	Los Banos, Philippines	22.1	451.4	Gatan (1941)
Angora	College Station, Texas	33.2	48–52.6	Miller *et al.* (1943)
Local	Merca, Somalia	28.1–42.3	50–52.3	Congiu (1954)
Hejaz	Arabia	19–22	35.0	Epstein (1946)
Short and long haired Varieties	Ruanda Burundi, Congo		45.6	Thienpont & Vandervelden
Local	Lower Congo	20.9	50.0	Henrotte (1961)
East African	Serere, Uganda	13.6	43.5	Wilson (1958b)
Indigenous × Kamori Castrates	Malaya, Tanzania	33.8	48.5	Hutchison (1964)
Crossbred Boer		29.3	44.7–55.4	
Indigenous castrates	Mpwapwa, Tanzania	23.9–30.5	46.5–55.4	Hutchison (1964)
Indigenous × Boer		40.6	52.4	
Cutchi	Bombay, India	36.4–40.9	45.0	Williamson & Payne (1960)
Kambing katjang	Serdang, Malaysia	21.4 25.5 28.6	44.3 47.4 51.3	Devendra (1966b)
Native hair goat	Ankara, Turkey		48–51	Eker (1961)
Soviet Mohair	Central Asian Republic	36–42	42–50	Zelenskii *et al.* (1962); Belov & Polijakova (1962)

Hutchison, 1964), and in another to improved feeding (Devendra, 1966b), justifies the results of further investigations to verify these points. Whether castration should be more widely employed as a routine management procedure, and which of the methods currently available is suitable for this operation, also need to be investigated in relation to the live weight at which goats are slaughtered and equally important, to the dictates of the market.

Dressing percentage (DP) in Indian goat breeds varies from 38–56% based on sex, age, weight and conformation status of animal at the time of slaughter (Table 12.7). It is reported on the basis of hot or chilled carcass weight, live weight or empty body weight with or without kidney and pelvic fat. Small sized black Bengal (BB) goats have lower DP than medium sized Barbari and large sized Beetal goats. The overall live weight, carcass weight and yield were 21.97 kg, 10.85 kg and 48.33% respectively for the market slaughter goats of Jamunapari, Barbari and their Crosses (Kesava Rao *et al.*, 1985, 1988). The DP varied significantly between the 3 subjective grades of conformation, *viz.* poor (46.9), fair (52.7) and good (58.5) of market yearling crossbred castrates (Joshi *et al.,* 1986).

Table 12.7 Effect of Breed on Carcass Yield of Goats

Breed	n	Slaughter weight (kg)	Dressing percentage	Reference
Black Bengal-entire males	10	19.25 ± 1.94	44.68 ± 1.20	Kesava Rao *et al.* (1984)
Barbari-male goats	40	19.18 ± 0.34	42.86	Das *et al.* (1986)
Beetal (NDRI)		16.16	44.70	Misra (1979)
Alpine × Beetal		21.44	49.29	Misra (1979)
Saanen × Beetal		21.50	47.29	Misra (1979)
Market slaughter stock				
Jamunapari – Male	10	17.20 ± 1.04	50.79 ± 0.99	Kesava Rao *et al.* (1985)
– Female	14	18.46 ± 0.87	49.21 ± 1.19	Kesava Rao *et al.* (1985)
Barbari – Male	13	19.57 ± 0.97	17.28 ± 0.72	Kesava Rao *et al.* (1985)
– Female	9	50.41 ± 1.02	48.79 ± 1.36	Kesava Rao *et al.* (1985)
Nondescript – Female	6	17.17 ± 0.75	48.79 ± 1.36	Kesava Rao *et al.* (1985)
Gaddi-bucks		17.70 ± 0.84	38.80 ± 1.26	Kulkarni *et al.* (1992)
Jakhrana	17	23.66 ± 0.99	44.60 ± 0.59	Prasad *et al.* (1992a)

n = Number of Observations

CHAPTER 13

MILK PRODUCTION

Although, not all goats are kept for milk production, goat's milk is consumed in most countries where they are bred, and the value of milk as an important source of nutrients, particularly in areas such as the far East, is widely recognized. An interesting feature about the goat populations of Asia, the largest of which are found in India, Pakistan and Indonesia, is that they exist in countries representing the climatic extremes of the tropics. The large populations of goats in this region (approximately one third of the world population) demonstrate that despite the climatic extremes imposed on them, their production of milk and other products are of importance in the nutritional habits of the people. Furthermore, more goats than cattle or buffaloes can be carried per acre, especially in areas where fodder resources are limited, and they are the only source of milk where cows or buffaloes are not kept.

Average lactation yield of Indian goat breeds ranged between 35.20 and 154.6 kg. Surti and Jamunapari excelled other breeds in milk production followed by Beetal, Kutchi, Jakhrana, Sirohi, Marwari and Barbari. Milk yield of Bengal breed was lowest (35.20 kg). Average lactation length ranged between 111.20 to 243.37 days with highest in Jakhrana (243.37 days) and lowest in Bengal (11.20 days). Average lactation length of milch breeds ranged between 153.08 and 243.27 days.

Alpine × Beetal crosses were superior to Sannen × Beetal crosses and also other crossbreds. Alpine × Beetal crossbreds produced 309.60 kg in a lactation of 240.8 days. Milk yield/day of lactation in crossbreds with Alpine × Beetal inheritance was 1.28 kg as against 1.03 kg of Jamunapari. Superior Jamunapari does are capable of producing 150–200 kg milk in 90 days (Roy, 1991). At CIRG, Makhdoom, the highest record is 362 kg in a lactation period of 225 days (Roy, 1991).

However, these and other factors in favour of goat milk production have not been adequately appreciated. This is because milk production from goats in the tropics is essentially a subsistence enterprise serving the needs of rural people. It is therefore necessary to attempt a realistic appraisal of the overall situation, taking into account the special attributes of goat's milk, assessing the various breeds as regards milk production potential, and considering the efficiency with which it is produced in comparison with other lactating ruminants.

13.1 All India Coordinated Research Project (AICRP) on Goats

The centres for AICRP on Goats for milk production are located at two places, *viz*. (i) National Dairy Research Insitute, Karnal and (ii) Kerala Agricultural University, Mannuthy, Trichur. Another centre at Bikaner (Rajasthan) started working from 1st January 1987 (AICRP on Goats, Terminal Report, 1985).

13.1.1 National Dairy Research Institute, Karnal

The objective of this project at NDRI was to evolve a new breed of milch goat suitable to the agroclimatic conditions of the country and capable of yielding 300 kg milk in a lactation period of 150 days in large size breed such as Beetal. The improver breeds used for this project were Alpine and Saanen. Beetal had lactation yield of 156.9 ± 6.8 kg in lactation length of 186.3 ± 5.7 days. Sannen × Beetal and Alpine × Beetal crosses had lactation yield of 309.6 ± 13.5 and 257.1 ± 6.4 kg in lactation length of 240.8 ± 8.0 and 229.9 ± 3.7 days respectively. Thus, Sannen × Beetal crosses were superior to Alpine × Beetal and registered 97.3% improvement over Beetal.

13.1.2 Kerala Agricultural University, Mannuthy, Trichur

The objective was to develop a new breed involving Malabari which could produce 200 kg milk in 150 days lactation period. Improver breeds were same. Alpine × Malabari crosses were superior to purebred Malabari in growth, survivability, milk yield, fecundity and milk conversion efficiency. At Mannuthy, the average first lactation milk yield of Saanen × Malabari was 174.12 kg in a lactation period of 190 days, which is much superior to purebred Malabari as well as Alpine × Malabari crosses. The crossbreds with 50% exotic inheritance were better than those with higher exotic inheritance. With the increase in generation number, the milk yield also increased to an extent of 24.3% in Alpine × Malabari and 26.2% in Saanen × Malabari from F_2 to F_3 generations. In crossbreds, the milk yield increased with the increase in lactation order up to 5, while in the purebred Malabari, there was a decline after 3rd lactation. Saanen × Malabari crosses had an average milk yield of 211.5 kg at 5th lactation in a lactation period of 200 days registering an improvement of 146.8% over purebred Malabari (Annual Reports CIRG, 1985–91; AICRP on Goats Terminal Report, 1985).

13.1.3 Bikaner (Rajasthan) Centre

The objective of this centre was to improve milk production of Jakhrana goats through selection. Field survey conducted by the centre on the production performance of Jakhrana in its home tract during the period of 1986–87, showed that the average peak yield/day of Jakhrana goats in 1st, 2nd, 3rd and 4th lactations were found to be 2.63, 3.23, 4.23 and 3.8 kg respectively in the breeders flock (Annual Reports CIRG, 1985–91).

AICRP on goats, later on, was transferred into the Network Project on Goat Improvement. Under this Network project goat improvement programme continued at different centres.

PL-480 Project on "Studies on the Combining Ability of Desirable Characters of Important Goat Breeds":

Under this project, major breeds of North-Western Region, Jamunapari, Beetal and Barbari along with Black Bengal from Eastern region were compared for their milk production ability, growth under similar management system and the performances of their crosses. Results of crossbreeding amongst the indigenous breeds indicated Beetal and Barbari among the purebreds and Beetal × Barbari; Barbari × Beetal; Beetal × Black Bengal; Black Bengal × Beetal; Black Bengal × Jamunapari and Jamunapari × Barbari among that breeds as the promising types. The combination of large breeds (Jamunapari × Beetal; Beetal × Jamunapari) did not perform well. Beetal is to be taken as the best amongst the purebreds and Beetal × Barbari amongst the crossbreds (PL-480, Final Technical Report, 1985).

13.1.4 Central Institute for Research on Goats (CIRG)

Elaborate research projects in 1985 were undertaken at this Institute to improve the production performance of Indian goat breeds, *viz.* Jamunapari and Barbari through selection. Comparative performance studies on Sirohi, Marwari and Kutchi breeds of goats for meat and milk production under semi-arid climatic conditions of Rajasthan were also undertaken at Western Regional Research Centre (WRCC) of CIRG at Avikanagar (Rajasthan).

As compared to 1985 milk production, there were 38.6, 39.6 and 42.5% increase in 90, 140 days and total lactation yield in the year 1990 in Jamunapari goats. Year of kidding, parity and dam's weight at kidding had significant influence on 90 days, 140 days and total milk production. Season of kidding was also found to influence 90 days milk yield and lactation length. Barbari goats produced 63.96, 86.89 and 82.18 kg milk in 90 days, 140 days and total lactation. It was noted that milk yield increased over the years for Barbari also.

Lactation performance of Sirohi, Marwari and Kutchi breeds were recorded at WRRC, Avikanagar from 1987 onwards. Sirohi does have a better lactation performance than Marwari and Kutchi does over the years. Overall milk yield was recorded to be 100.9 kg in Sirohi, 95.8 kg in Marwari and 98.4 kg in Kutchi respectively. The lactation length averaged 182.8, 174.0 and 182.0 days in the 3 breeds respectively.

13.1.5 Indo Swiss Goat Development and Fodder Production Project Ajmer (Rajasthan)

The Indo Swiss Goat Development and Fodder Production Project established in 1981, initiated study and development of Sirohi breed in its habitat in the Aravalli Hill Range of Rajasthan.

Least squares estimate of milk yield of Sirohi goats at 180 days was 249 kg. Besides significant effects of differences in parity, litter size, month of kidding and year, a substantial amount of the variation appeared to be contributed by differences between location and herds, which are related to differences in management, availability of fodder resources and to some extent, genetic potential. High and low-producing herds in the same area were found to have differences of over 100 kg milk in 180 days. The total milk yields were considerably higher than earlier results from smaller research station trials and confirmed the importance of Sirohi as a well adapted dual purpose breed with good performance in semi-arid conditions.

Under this project, crossbreeding of Sirohi with Alpine and Toggenburg was undertaken to improve the productivity of the breed. It was observed that increase in milk yield was less than expected and hence, the crossbreeding was stopped.

Encouraged by the findings of better performances under farm conditions, the project now concentrates on selective breeding within the Sirohi breed.

13.2 Special Attributes of Goat's Milk

Goat's milk has certain special attributes, which are of particular significance in human nutrition. The fat globules are small in size, friable and the coagulum is light in comparison to that of milk from the cow and the buffalo. These factors make it easily digestible and explain why goat's milk is commonly prescribed for infants and invalids. The casein of goat's milk is also more easily digested.

Moreover, whereas cow's milk is acid in reaction, goat's milk is distinctly alkaline, thereby making it more useful in cases of hyperacidity. The calcium, phosphorus and chlorine content of goat's milk is also much higher than that of human or cow's milk, but its iron content is low. Walker (1965) has described the therapeutic uses of goat's milk in human medicine.

An outstanding feature about goat's milk in comparison with that of other animals is that the tubercle bacillus is very rare, although it can occur. Brucellosis, on the other hand, is fairly common, and is one of the factors which tend to discourage wider use of goat's milk. However; goats are not known to be any more susceptible than cows or pigs to brucellosis, and can be tested or vaccinated against it. The milk from untested goats should be boiled or pasteurised for human consumption. Contrary to belief, goat's milk no not have a strong odour; which is derived from the buck or, rarely, from over-active scent glands in the doe. The odour is generally due to the presence of capric or caproic acids secreted by the skin glands of the buck and passed on to the female by contact, the milk for instance becoming contaminated by hair falling into the milk bucket. Obviously, this can be minimised by proper separation of milking goats from the bucks and also by the practice of good management and hygiene. Mackenize (1970) refers to a method devised by the Robert Ford Laboratory, Pascagoula, Mississippi, USA for permanently destroying the scent glands. This consists of cauterising the glandular area with a disbudding iron. The main area is immediately behind the horn base and the operation should be done soon after birth, but in hornless males it is possible to do it later. Occasional animals have musk glands in other positions and these can be destroyed in the same way.

13.3 Variation in Yield

The milk yield of Indian goats varies from 0.5 to 5.0 kg. The butter fat content is about 4.5%. A good average grade goat, if fed and kept under proper conditions should yield about 2.25 kg milk/day over a lactation period of 7 to 10 months.

A number of goat breeds in widely separated areas of the tropics are recognised primarily as milk producers. The majority of these breeds, however, do not compare in production with superior, temperate breeds such as the Saanen and the Anglo-Nubian. These and other breeds of dairy goats have in the past been introduced to a number of tropical countries (*e.g.* Kenya, Malaysia, Mauritius and the West Indies) either to be bred pure or for crossing with an indigenous breed to increase milk yields. Maule (1966) reviewed work on the performance of dairy goats in various tropical countries and the following data are taken from this review:

In Israel, where milk recording is widely used, the average lactation yield of Saanen is approximately 500 kg. In 1960, 500 registered goats averaged 1,035 kg with a maximum of 2,293 kg in 227 days and a maximum daily yield of 10.7 kg (Israel: Ministry of Agriculture, 1962). Laor (1969) gives the average lactation yield in 1968 of 322 elite goats (including yearlings) on Jewish farms at 1,006 kg with 183 goats exceeding 1,000 kg. The maximum yield in 1968 was 1,912 kg with a daily average of 6.7 kg.

In Queensland, Australia, production records are available for a number of years, but these do not give breed comparisons (Pegg, 1968). Average annual production rose from 538 kg milk with 4.0% butterfat in 1956–57 to 771 kg with 3.8% in 1960–61; since then, yields have again improved, and the records for 1967–68 showed that 25 goats averaged 886 kg in 226 days with 3.1% butterfat. The highest recorded yield (in 300 days) for a mature goat in Queensland was 2,109 kg milk and 75 kg fat with 3.6% (breed not stated).

In the West Indies, a yield of 1,240 kg was obtained from a Saanen goat over 35 years ago and it is possible that still higher yields have been recorded since, but information on the performance of dairy goats is seldom quoted nowadays. In the British Goat Society Year Book for 1946 the West Indian record 24-hours yield, given by a goat in Barbados under official test, was 9.7 kg, and several yields of over one gallon were obtained at Shows organised by the Trinidad Goat Society in 1945. Recent information gives yields of 818 kg in 250 days and up to7 kg a day.

The recorded yield of 91 milch goats (breed not stated) in South Africa from 1957/58 to 1961/62 was 877 kg with 3.37% butterfat and in 1962/63 1,065 kg with 3.25% butterfat for 91 goats.

The yield quoted, especially for Indian breeds, by different workers often show surprisingly large variations. Wright (1937), in a report on Indian dairying, said that individual Indian goats had given remarkably high yields at the Mission farm, Etah (Uttar Pradesh) and the Cattle Breeding Farm, Hisar and quoted yields of 272–544 kg, with one exceptional yield of 780 kg, probably given by either a Jamunapari or Barbari goat. It is worth mentioning that, except perhaps for the Barbari and Beetal breeds, all the Indian breeds are used for meat production, so that there is often competition between kid rearing and use of the milk for human consumption. This feature can discourage the selection of goats for milk production; nevertheless, dual-purpose animals are more likely to be economic. The high yields of Damascus goats in the subtropics suggest that this breed should be given greater opportunities in tropical experimental breeding.

When permitted to suckle instead of being milked, milk production in the goat appears to be similar to that of sheep in that it is dependent on the number of kids being suckled but independent of the number of kids born. This is evident in the data of Rai and Chorey (1965) for Jamunapari and Barbari goats in India and also for Mubende goats in Uganda (Sacker and Trail, 1966a). Parallel findings in sheep have been extensively studied (Davies, 1958; Alexander and Davies, 1959; Doney and Munro, 1962; Sacker and Trial, 1966b). Rai and Chorey (1965) also reported the effects of the first month's lactation on the live weight of does and found that both for high yielding Jamunapari and Barbari goats, the decline in body weight during the first month after parturition was 15–16%, equivalent to 9 kg. To what extent this reduction in weight is due to lactational stress alone is not clear, since such factors as the nutrition of the does before and after parturition, season of kidding and husbandry practice are all involved. It does emphasize, however, the need for food nutritional management to ensure that weight loss is small. Nevertheless, Mackenzie (1967) points out that loss of weight during lactation is perfectly normal and makes the necessary energy for high yields available without overstraining the digestive system.

Webster and Wilson (1966) state that in the goat, maximum daily milk yield is not reached until between the 8th and 12th week after kidding, giving a flatter lactation curve than that of the cow. This is in agreement with the lactation curve illustrated by Mackenzie (1967) and results in the milk production of the doe being more evenly spread throughout her lactation than is the case with cattle.

13.4 Weaning Practice

Kids are weaned at any age from birth up to 6 months. In meat animals, the tendency is to delay weaning to allow the kid to have maximum benefit of the dam's milk, whereas in milk animals it is usual to wean early so that the milk from the dam can be used for domestic or commercial purposes. In some circumstances, delayed weaning is disadvantageous, since longer the period of milk feeding, the less efficient becomes the conversion of nutrients to milk and milk to meat. This inefficient double conversion can be minimised by a shorter suckling period, depending on such factors as level of husbandry practice, labour input and the availability of milk substitutes for the artificial feeding of kids. The common tendency to delay weaning until about 6 months of age, when the dam's milk supply is declining, is in any case not justified. The attendant disadvantages are twofold. Firstly, the recurrence of oestrus is unnecessarily delayed with consequent prolongation of kidding interval, and secondly, the kid's rumen development is delayed. In both cases, the overall efficiency of production is reduced.

Probably, the most important consideration is the effect of weaning on the growth of kids and on the milk flow of the dams, and the economic benefits of the measures adopted. From a commercial point of view, milk production for the market should not be at the expense of providing adequate milk to support the growth of the kids. Kids are entirely dependent on the milk of the dam or on milk substitutes for their early growth when the rumen has not been fully developed. When milk production for sale is the aim, therefore, adequate measures need to be taken to maintain normal growth. These aspects were examined by Prabhu and Amble (1946) who reported a study of the possibilities of introducing weaning on a large scale as an economic benefit of rearing goats in India. The experiment was initiated by the Indian Council of Agricultural Research and was repeated over a period of 5 years from 1940 to 1945. Twenty does of the same breed (Jamunapari or Barbari), age and condition, which were at different stages of lactation, were divided into two groups of ten each. In one group of does, the kids were weaned at birth, while in the other the kids were left with the dam unitl weaning at 4 months of age. The weights of kids were recorded at monthly intervals until they were one year old. Statistical analysis corrected for sex suggested that suckling does not confer any advantage on the growth rate of the kids. In other words, handfeeding gave as good results as suckling of the kids. An important advantage of this method in addition to more milk being available for sale and human consumption is that accurate lactation records can be maintained to assist in the selection of dams for further breeding.

The analysis by Prabhu and Amble (1946) of the effect of early weaning on the dams' milk flow was not conclusive, but certain features are worth noting. Does whose kids were weaned at birth appeared to give more milk than does which suckled their kids. The increased yields were associated with increased lengths of lactation. To what extent these effects were environmental or genetic is not clear since it was not possible to separate them. If the differences are real, *i.e.* caused by early weaning, then there is an economic advantage that is potentially useful.

Mackenzie (1967) advises that female kids be given four pints (2.3 kg) of milk daily, and males being reared for breeding, five pints (2.8 kg). These amounts seem extravagant in comparison to the eight pints usually allowed for heifer calves in dairy herds, and two pints commonly considered adequate for lambs of early-maturing breeds. Two pints (or 1 kg) would appear adequate for even the larger breeds of tropical goats. Where balanced concentrate rations are available, supported by good quality browse, milk feed can cease at 8 or 10 weeks, but where other available feed is not very satisfactory, milk feeding may continue up to 16 weeks. As with all ration changes for ruminants, milk feeding should not be stopped abruptly, but the quantity should be reduced gradually over 2 or 3 weeks. It may be possible further to reduce the amount of the dam's milk to the kids where cheap substitutes, such as dried skim milk powder, are available.

13.5 Influence of Age

The age of the doe exerts a marked influence on milk production. Maximum yield at a certain age would depend on the interaction between age and all the other factors which influence milk production. These include adaptation, feeding and management, and disease control. It is reasonable to expect that only under conditions, where these factors favour the doe there will be a maximum expression of the genes controlling milk production. Yarkin and Eker (1961) analyzed the performance of 77 Kilis goats over 7 lactation in Turkey and reported that milk production was at a maximum at the 4th after which it fell rapidly. The butterfat percentage only fell after the 5th lactation.

In two herds of Beetal goats originating from different parts of Punjab in India, Amble *et al.* (1964) analysed 800 and 788 lactation records and found that yield declined in the 2nd lactation, then rose to the 5th and declined in later lactations. Milk production in the 5th lactation was however, less than that in the 1st.

In West Malaysia, Anuwar and Devendra (1966) presented lactation data for indigenous goats, Anglo-Nubian indigenous and Wa Anglo-Nubian goats, which indicated that for 36 local goats having 2 to 5 lactation records, milk yield was maximal in the 3rd lactation. For halfbreds, available data for 3 lactations showed that both milk yield and lactation length increased over the 3 lactations. Data on only 2 lactations for the Anglo-Nubian group showed a decline in the 2nd lactation. For various crossbred goats in Puerto Rico, Sanfiorenzo (1962) observed significant differences in milk yields between females of different ages, maximum milk production being obtained from four to six-year-old does.

It is of interest to note that similar trends have been reported for goats in temperate regions by a number of workers. Simmons and Lambert (1937) compared milk production at different ages in America and found that maximum milk production and length of lactation was reached between 4 and 6 years of age. In Germany, Bonnekamp (1939) found that yield rose with age until the 5th lactation, remaining unchanged for the 6th and 7th before declining. Watkin and Knowles (1946) analysed 660 recorded lactations for 300 goats in the United Kingdom and found that about 70% of the goats reached their maximum yields in the 2nd or 3rd lactations and then decreased to the 7th at a rate of 10% per lactation. An interesting feature of this analysis was that lengths of lactation varied enormously, with two thirds of all lactations lasting for 9–12 months. Ronningen (1965) studied the effect of age on milk yield of goats in Norway and noted that the highest milk yields were obtained from goats in their 5th lactation (In India, goat's milk has been found to be richer in fat, protein and total solids than that of Indian and European cows (Table 13.1). The composition of goat's milk shown in Table is considered to be the average composition of goats milk in India (ICAR,

Table 13.1 Average Compositions of Goat's Milk in Comparison with the Milk of Indian and European Cows and Indian Buffaloes

Species	Fat	Protein	Lactose	Ash	SNF	Solids
Indian cows	4.8	2.8	4.6	0.74	8.1	13.5
European cows	3.7	3.4	4.8	0.73	8.9	12.7
Murrah buffaloes (including ash)	6.8	3.9	5.7	–	9.6	–
Goat	4.9	4.3	4.1	0.89	9.3	14.2

Source: Schneider *et al.* (1948), Indian Council of Agricultural Research (1962), Ghosh and Anantakrishnan (1963, 1965).

1962). The figures for Indian cows were the results of extensive surveys reported by Schneider *et al*. (1948) and those for European cows were for all breeds (Davies, 1939). The composition of Murrah buffalo milk has been reported by Ghosh and Anantakrishnan (1963, 1965).

The milk of temperate dairy goats is reported (Mackenzie, 1967) to contain rather less fat (3.8%), total protein (2.9%), ash (0.79%) and SNF (8.68%), which is no doubt associated with their much higher milk yields.

A detailed and extensive review of the composition and characteristics of goat's milk has been made by Parkash and Jenness (1968), who stated that while the range of size of the fat globules is much the same as that of cow's milk, the proportion of smaller globules is greater in goat's milk. Its nutritive value does not differ appreciably from that of cow's milk; the calculated energy content of goat's milk of average composition is 65 kcal per 100 g. Although goat's milk is higher in C8:O Caprylic and C10:O Capric fatty acids than cow's milk, both milk fats are relatively poor sources of dietary essential fatty acids (linoleic, linolenic and arachidonic) and contain approximately 66% by weight of saturated fatty acids.

The same authors also reported that, in view of similarity in amino acid composition of cow's and goat's milk, it is doubtful, contrary to some findings, whether the biological value of goat's milk or its proteins are very different from those of cow's milk. The calcium and phosphorus content of the milk of both species is also similar. However, goat's milk has much lower contents of vitamin B_{12} and vitamin B_6, even so, it contains twice as much of these two vitamins as does human milk. No precursors of vitamin A are found in goat's milk which, unlike cow's milk and also buffalo's milk (Luxminararyana and Dastur, 1968), carries the vitamin intact.

There is some evidence form studies of the effects of goat's milk on the growth rate of rats to indicate that it is better for them that of other lactating ruminants. Das and Guha (1936) made a comparative study of buffalo, goat, cow and human milk as the sole diet of young rats and found that growth declined in that order. Human milk failed entirely to maintain young rats and was only satisfactory when whole vitamin B complex was given. This finding was subsequently confirmed by Basak and Guha (1936) who studied the influence of buffalo, goat and cow milk on the growth rate of rats and found that when fed to appetite, goat's milk produced a significantly greater growth rate than did cow or buffalo milk. No significant differences in weekly live weight gains were found between goat and buffalo milk; both, however, were superior to cow's milk.

Whereas legal standards for cow's milk are recognised, parallel standards for goat's milk do not exist, and in view of the subsistence nature of goat milk production, it is doubtful, if such standards could be enforced. Presumably, the legal standards set for cow's milk, although unnecessarily low for goat's milk would be acceptable. Any use of these could only be applicable where intensive commercial milk production by goats is established.

13.6 Repeatability and Heritability of Milk Production Characters

The genetics of milk production in the goat has received limited attention. This led Mason (1965) to assume that since considerable work has been done in dairy cows and sheep, and since in its main outlines, the situation was the same in both these species, the genetic characters would also be similar in the goat. He considered that the repeatability and heritability estimates for the first lactation were of the order of 0.4–0.6 and 0.2–0.3 respectively.

While, there is no report in the literature where heritability has been determined, four estimates have been reported for repeatability in indigenous and crossbred goats. Sanfiorenzo (1957) analysed 368 records of 139 goats comprising six different types, including two West Indian and four crossbred types and estimated repeatability values of 0.36, 0.42 and 0.09 for milk production per lactation, milk production/day and length of lactation respectively. He attributed the low value of 0.09 for length of lactation to the different management practices to which the goats were subjected. Repeatabilities for milk production characters have also been reported for Beetal and Kambing Katjang goats and for crossbred Anglo-Nubians.

Anuwar and Devendra (1966) also calculated correlations between milk yield and length of lactation for various lactation periods, and found that for local goats these ranged from 0.8 at 1st lactation to 0.9 at the 5th lactation; correlations of 0.7–0.9 were obtained for F1 Anglo-Nubian × Kambing Katjang from the 1st to the 3rd lactation and of 0.7 for two lactation in ¾ Anglo-Nubians. In all cases the correlations were statistically significant. Kartha (1937), on the other hand, reported a lower correlation coefficient of 0.6 for 103 goats and 235 lactations at Hisar in India. The significant feature of the estimates summarised in Table 13.2 and the correlations reported between milk yield and lengths of lactation is that these figures are comparable to those reported for tropical milk cattle (Mahadevan, 1966) and also the milking buffalo (Alim, 1967). Mason's observation (1965) that the genetics of milk production in the goat is similar to that of sheep and cattle would therefore appear to be justified.

Repeatability estimates for milk production ranged from 0.2 to 0.5 (Table 13.2). Four of these estimates are sufficiently high (0.4–0.5) to suggest that there are permanent differences between goats, which could stimulate progress through genetic selection. Two estimates of about 0.4 for milk yield per day tend to confirm this. On the other hand, the values for length of lactation and two estimates for kidding interval are too variable to allow firm conclusions to be drawn from them. These few estimates emphasise, however, that more detailed studies are needed on the various genetic aspects of milk production in the goat so that their implications can be applied in breeding goats for better milking performance.

Table 13.2 Repeatability of Milk Production Characters in Goats

Breed	Location	Milk production	Milk yield per day	Length of lactation	Kidding interval	References
Six types combined data	Puerto Rico	0.36	0.42	0.09	–	Sanfiorenzo (1957)
Beetal	N. India	0.18	–	0.04	0.05	Kumar *et al.* (1962)
Beetal	N. India	0.21	0.35	0.36	0.26	Amble *et al.* (1964)
Kambing Katjang	Malaysia	0.47	–	0.29	–	Anuwar and Devendra (1966)
Anglo-Nubian × K. Katjang (F_1)		0.42	–	0.17	–	–
³/₄ Anglo-Nubian		0.54	–	0.53	–	–

13.7 Efficiency of Milk Production

Although, there appears to be a similarity between goats, cattle, sheep and buffaloes in the genetics of milk production, there is evidence that in terms of live weight, the goat is much more efficient in producing milk than the other 3 species. Table 13.3 illustrates this point where the milk yields of goats, cows and buffaloes in West Malaysia have been expressed as yield per kg live weight. It is to be noted that Anglo-Nubian crossbreds were the most efficient. This feature has also been confirmed by the more general comparison of Webster and Wilson (1966) and Mackenzie (1967). In view of their lower maintenance needs in comparison with cattle and buffaloes, and despite their relatively high metabolic requirements, it is clear that the conversion of nutrients in milk by goats is, in general, a more efficient process. It is thus obvious that the biological advantages of the goat as a milch animal have been greatly underestimated. Current shortages of milk in many countries could be considerably relieved by the better use of milk goats, provided it is accompanied by proper management and disease control. Increasing milk production from goats has already been considered as a temporary substitute for dairy development based on cattle and/or buffaloes in the Far East (FAO, 1967).

Table 13.3 Comparison of the Milk Yields of Goats, Cows and Buffaloes in West Malaysia on a Live Weight Basis

Species	Average milk yield (kg)	Average weight (kg)	Yield per Kg live weight (kg)
Goats			
(i) Local (Kambing Katjang type)	89.5	32.7	2.8
(ii) Anglo-Nubian × K. Katjang (F1)	295.5	41.8	7.1
(iii) ¾ Anglo-Nubian	236.8	55.5	4.3
Cows			
(i) Local Indian dairy cattle (*Bos indicus*)	880.9	363.6	2.4
(ii) Jersey (*Bos taurus*)	1377.3	409.1	3.4
Buffaloes			
Murrah	1814.1	454.5	4.0

Mackenzie (1967) stated that the labour cost per gallon of goat's milk is approximately twice that of cow's milk in commercial herds so that when labour costs are an important factor, the goat is at a disadvantage. However, in many parts of the tropics milk production by goats instead of cows would be practical and also economic. This is particularly true of those areas where the maintenance of cows is difficult or limited because of environmental conditions, and by economic considerations. In such circumstances, the ubiquitous goat with its ability to fend for itself with little management, would be much more economic and useful as a supplier of milk for household consumption. Further research is required on adequate management techniques for dairy goats in the tropics, and on the use of

Mediterranean versus Alpine types in crossbreeding programmes. Critical investigations are required on the factors influencing variations in milk yield between breeds and within lactations, as also on the influence of age on milk production.

13.8 Crossbreeding for Improved Milk Production

Saanen, Alpine and Nubians were imported to India for crossbreeding of native goats. As already mentioned Alpine × Beetal crosses were superior to Saanen × Beetal crosses and also other crossbreds. Alpine × Beetal crossbreds produced 309.60 kg in a lactation of 240.8 days as against Saanen × Beetal crossbreds which produced 257.10 kg in a lactation period of 229.9 days whereas Saanen × Malabari were superior to Alpine × Malabari crossbreds.

There is a good deal of evidence to show that in most cases, introduced breeds have proved very successful in increasing milk yields. This should provide sufficient impetus for further work in this field in those areas where there is room for such improvement.

The potential value of various exotic breeds that can be used for upgrading purposes has already been indicated and attention is also drawn to a review of the performance of these breeds in various parts of the tropics by Maule (1966). Perhaps, the most difficult decision in embarking on a programme of work to improve milk production in indigenous herds is the choice of the breed to be introduced. The actual choice will depend on prevailing local conditions and also on how this environment compares with situations in other countries where the breed has been shown to have adapted and performed well. Of the European breeds, the Saanen and the British Alpine appear to be more promising than the Toggenburg, but both are excelled for tropical use by the Anglo-Nubian. This breed apparently adapts itself to tropical climates better than any other temperate breed. All available information indicates that in every case of its introduction and use, the crossbred progenies have given favourable results; it has the additional advantage of being able to improve meat as well as milk production.

In Trinidad, the Anglo-Nubian is looked upon as a useful dual purpose goat for peasant farmers and gives 1–3 kg milk/day; in Mauritius, where Anglo-Nubians have been kept for some years to improve the size of local goats, yields of 1–1.4 kg a day are usual in a lactation of about 7 months. The Anglo-Nubian has the advantage of a higher butter fat content (4.4–5.6%) in comparison with the Saanen (3.7–4%) or the Toggenburg (3.7–3.8%), but the average milk yield is somewhat lower.

Experience in West Malaysia showed that with one exception, British breeds which were introduced in the late 1940s and early 1950s failed to acclimatise and gave low yields of milk. Of the 4 breeds introduced (British Saanen, British Alpine, Toggenburg and Anglo-Nubian), the Toggenburgs were the least hardy and were soon discarded. The crosses of British Saanen and British Alpine with local goats were superior to the purebred females in milk production, with a maximum daily yield of 3.6 kg by a British Alpine crossbreds. The Anglo-Nubian was by far the best of the imported breeds and was very successful in increasing the size and meatiness and also the milk yield of local goats (Devendra,1962). Anglo-Nubians have also been introduced into East Malaysia, Brazil, Australia (Queensland) and the Philippines.

A breed which has many of the features that have been described in favour of the Anglo-Nubian is the Jamunapari. Both breeds are comparable in milk production, growth rates and mature weights. A factor which may limit the wider use of the Jamunapari, and favour the Anglo-Nubian instead, is the lack of sufficient numbers of Jamunaparis of known genetic background and proven merit.

The potentialities of the Damascus goat and its derivatives (*e.g.* Improved Cyprus) appear to have been neglected in tropical experimentation, but what little information is available, suggests that it may be a most suitable breed for intensive milk production with high-concentrate rations.

CHAPTER 14

UTILIZATION OF BY-PRODUCTS

About 35.0 million goats are yearly slaughtered in authorised slaughter houses in India. In addition, there are illegal slaughters which account for more than 25% of actual slaughters. The goat population has shown increasing trends over the years in spite of the fact that their rate of slaughter is higher than sheep, whose population has almost remained static. Goat contributes Rs 10,800 million from meat, Rs 4500 million from milk, Rs 200 million from hair and Rs 1050 million from manure.

The loss per annum due to non-utilization or under-utilization of various animal by-products from abattoirs in the country is estimated to Rs 1000 million. The reasons for such a situation are lack of modern abattoir facilities; lack of appropriate and economically viable technologies; lack of transfer of technology from laboratory to industry and ignorance of public and meat traders about the benefits of processing animal by-products. When the animals are slaughtered for food, two types of by-products, edible and non edible are produced. This classification is not rigid. Their consumption as food is dictated by a number of factors such as religion, purchasing power, customs and food habits. Sometimes, scientific achievements and technology developments could change non-edible by-products to edible by-products. For instance, horns hoofs are strictly non-edible by-products. But technology has been developed for conversion of keratin protein to keratin hydrolysate, which could be used as flavouring agent in meat soup concentrates.

Following are some of the by-products of goats emanating from abattoirs:

14.1 Hides/Skins

Skins are very valuable by-products used mainly for leather manufacture. It is estimated that 47.3 million pieces of skins are available yearly. Though, India stands second in the world in the production of hides and fifth in the matter of production of skins, it cannot forge ahead in capitalising on its large production unless stress is laid on quality right from the initial stages of production. As such, improved methods of flaying should be introduced in the slaughter houses and better flaying should be encouraged by payment of premia to good flayers.

As against the declining trend in the exports of hides and skins, leather and leather manufactures have been emerging in a significant manner in Indian exports. The average annual exports of this item improved from Rs 63 crores during the triennium ending 1968–69 to Rs 146 crores during triennium ending 1973–74 with the peak having been reached in 1973 when exports amounted to Rs 175 crores.

14.2 Dead Carcass Utilization

Failure to make use of the carcasses of fallen animals is responsible for an enormous wastage of otherwise useful material.

Desiccating plants fully equipped to process animal carcasses should be established on a sub-district, district or regional basis and that incentive payments need to be made to those who bring fallen animals to these plants. Each state should establish a suitable number of carcass utilisation and hide flaying centres with a view to prevent the colossal waste that is occurring at present.

The huge financial loss sustained due to faulty methods of flaying. Curing of hides and no utilisation of the by-products of animals has been observed. This need to be checked by properly utilization by the establishment of modern utilization centres.

14.3 Slaughter By-Products

14.3.1 Bones

A major portion of the bones collected in India is utilised for production of crushed bones and bone grists and a small quantity is used for the manufacture of bone-meal. There were 100 bone crushing mills and about 360 bone digesters in the country at the end of the Fourth Plan. Most of the mills crush bones primarily with the object of exporting crushed bones and bone grits. The bone digesters are working on cottage industry basis set up in various states with the aid of the Khadi and village Industries Commission or the State Government for converting locally available raw bones into bone meal for utilisation as fertiliser. New bone digesters will have optimum use of set up only in areas capable of utilizing the bone, meal, preferably in remote places unconnected by rail or road. Cooperative of bones collectors should be provided with bone digesters on rent. Encouragement should be given by the State Governments for setting up factories for the manufacture of gelatin, glue and Neat's foot oil.

14.3.2 Intestines

Usually casings are made from small intestines. Sometimes, oesophagus, stomach and urinary bladders are also used. Casings made from small intestines are marketed in terms of hanks (= 300 ft). About 8.76 million hanks of goat and sheep casing would be available per annum in the country if all the intestines of slaughter animals are processed. Goat and sheep casings are mainly used for the manufacture of surgical catgut sports gut and in the manufacture of sausages. During 1983–84, the country exported animal casings worth Rs 1.82 crores and surgical catgut worth Rs 23.8 lakhs. Manufacture of casings is on a cottage scale with poor infrastructure and needs considerable improvement. There are three surgical catgut manufacturing plants to meet the entire demand of the country. Cost analysis with returns on processing of small intestines to valuable end products are furnished in Table 14.1.

Table14.1 Analysis Showing Returns from 1000 Feet Small Intestine (Sheep/Goat)

Product/end product	Animal	Quantity	Price (US$)	Value (US$)
Small intestine (Raw)	Cattle	1000 ft.	0.5/100 ft.	5.00
	Sheep/Goat	1000 ft.	0.15/100 ft.	1.50
Sausage casing/1000 ft. of raw intestine	Cattle	1000 ft.	4.8/hank	16.00
	Sheep/Goat	1000 ft.	3.0/hank	10.00
Sport guts (tennis)/ 1000 ft. of intestine	Cattle	33 ft. 6 rolls	7.0/roll	42.00
	Sheep/Goat	33 ft. 3 rolls	7.0/roll	21.00
Surgical suture/1000 ft. of intestine	Cattle	16 doz. tubes	3.0/doz.	48.00
	Sheep/Goat	16 doz. tubes	3.0/doz.	48.00
Intestinal mucosa/1000 ft. of raw intestine	Cattle	15 kg	0.1/kg	1.50
	Sheep/Goat	5 kg	0.1/kg	0.50
Heparin/ 15 kg intestine mucosa (0.03% yield)	Cattle	4.5 g	1.00/g	4.50
	Sheep/Goat	1.0 g	1.00/g	1.00

Source: Personal Communication, Bhat (2008)

14.3.3 Intestinal mucosa

It is obtained while cleaning intestines for making sausage casings and can be used for preparation of heparin, fertilizer and mucosa protein meal as a source of protein in animal feeds.

14.3.4 Rumen contents

The rumen contents (RC) consist of partially digested feed material eaten by the animal before slaughter together with digestive juices and microbial flora. The RC weighs about 10–15% of live weight of the animal. It is either wasted or composted into manure. It causes solid disposal pollution problem in slaughter house. The RC is rich in certain amino acids, crude protein (12–15%), B-complex vitamins and minerals and can be recycled in livestock/poultry feed supplement by ensiling and other techniques. To increase protein content further, RC could be enriched during ensiling. About 4000 metric tonnes of crude protein could be obtained, if RC of both small and large ruminants are collected and processed for animal feed. Animal feeds have been developed by blending wet RC with maize meal (RM meal) and whole blood blended with wheat bran (WBB) meal.

14.3.5 Animal fats

Full and rational utilisation of animals fats which are available in large quantities from fallen and slaughtered animals is highly important. This would not only benefit the livestock producers but would also help in saving foreign exchange worth crores of rupees, which are being spent at present on the import of animals fats. As large, a quantity as 61 million kg of animal fats valued at Rs 8.9 crores were being imported in a year (1972–73).

Slaughter houses which are being modernised should have a by-products plant within their premise or in close proximity so that all available fats from slaughtered stock could be processed. Efforts should also be made for efficient and quick recovery of fats from the dead animals. Since the proportion of fallen stock is very much higher in the case of bovines, a chain of carcass utilisation centres should be established in areas of concentrated bovine population.

14.3.6 Blood

At present blood is being collected only in a few slaughter houses of the country. In developed countries, blood finds several important uses. It is consumed by human beings in the shape of black puddings and blood sausages, and blood albumin is employed as a substitute for eggs in the ice cream manufacture and in bakeries. It is used extensively for industrial purposes, as a fertiliser and is incorporated as an important constituent for the manufacture of stock feeds. In India, only blood collected from sheep, goats and pigs is utilised for human consumption. It is consumed by the weaker sections of the society after frying it with spies or boiling it with rice. Occasionally it is mixed with wheat flour in the preparation of chapaties.

Albumin fraction obtained from blood is also used as human food and as repalcement of egg albumin. Except for methionine, blood is rich in other essential amino acids like lysine, leucine, phenylalanine and valine. Blood contains 20% solids, out of which 80% is crude protein. All processes in vogue for blood meal production require heat for drying. In the conventional batch process of dry rendering technique, it takes more than 8 hours for drying at temperature above 100°C. It significantly affects the nutritive value of blood meal. A simple method of preparing blood meal is by passing steam directly into blood, allow it to dry and pulverise into a meal. The cost analysis with return by processing blood into various valuable end product is given in Table 14.2.

Collection of blood from all the slaughtered animals is highly important. When incorporated in the livestock feeds, it would provide a valuable source of

Table 14.2 Analysis Showing Returns from 1000 kg of Animal Blood (Sheep/Goat)

	Product/end product	Uses	Return (kg)	Price (US$)	Value (US$)
	Liquid blood				
1.	Blood meal/100 kg liquid blood	Feed supplement manure	200.00	0.40	80.00
2.	Spray/flash dried/100 kg liquid Blood	Leather finishing plywood adhesive	150.00	1.00	90.00
3.	Frozen plasma/100 kg liquid	In sausages & liver paste	550.00	0.50	275.00
4.	Spray dried plasma/1000 kg	In sausages & liver paste	55.00	5.00	275.00
5.	Spray dried haemoglobin/ Feed haemoglobin/1000 kg blood	Feed for minks	40.00	1.00	40.00
6.	Hydrolysed RBC/1000 kg blood	In soups, sausages	40.00	3.00	120.00

Source: Personal Communication, Bhat (2008)

animal protein and as a fertilizer it would enrich the soil. Uncollected blood in a slaughter house becomes a serious sanitary problem. It quickly clots, choking drains, septic tanks etc. and rapidly decomposes serving as an ideal medium for bacterial growth. Blood collection on efficient lines will be possible only in modern slaughter houses, as collection has to be done speedily only and without dilution with water. Otherwise, processing would be prolonged making moisture removal highly expensive.

14.3.7 Horns and hoofs

Horns and hoofs constitute a very small portion of animal by-products but because these are rich in keratin and have considerable value as fertilizer after conversion into meal, these have much economic value. The horn core is particularly rich in ossein which is used in developed countries for manufacture of gelatin. The export trade in buffalo horns and antlers fetched Rs 6.2 lakhs in a year 1972–73. It is desirable that wastage in the collection of horns and hoofs is reduced as much as possible and export trade in these items is increased. Further, horns and hoofs left in the country should be processed for the manufacture of gelatine and the unutilised portion converted into meal for use as fertilizers.

14.3.8 Meat unfit for human consumption

The annual production of useless meat, *i.e.* meat condemned for human consumption and meat which remains adhered to the bones and other tissues is estimated to be of the order of 23,000 tonnes. Since useless meat is excellent source of nitrogen in poultry feed and fertilizer for tea and coffee plantations, such meat should be converted into meat meal and should not be wasted as is being done at present.

14.3.9 Hair

Ordinary goat hair is used in manufacture of ropes, rugs and gunny bags as it does not fit into textile manufacture. Mohair and Pashmina is used in manufacture of knit wears and shawls in north temperate regions. Mohair is blended with wool, polyester and acrylic fibres for producing hosiery goods and shawls.

14.3.10 Glands and organs

A number of biochemicals and pharmaceuticals of importance could be prepared from the glands and organs of slaughtered animals. Bile juice which is wasted can be collected from gall bladder and used as raw material for preparation of bile acid and salts. Bile can be concentrated and used in the manufacture of detergents.

14.4 Goats Manure and Composting

Goats dropping improve the fertility of soil considerably and penning of goat in harvested fields (goat folding) brings in additional income to the flock owners.

CHAPTER 15

GOAT RECORDS

It is essential to maintain the necessary records at an organised goat farm to know about inputs and outputs. This will help in working out the economy of goat production per unit of area.

The recording system is required to be simple, accurate, and capable of collecting the required information and finally kept up to date. The data collected should be used for proper analysis. In goats, growth rate, feed conversion, wool/hair quality, dressing percentage are highly heritable traits. Prolificacy, milking capacity and wool hair weight are traits of low heritability. More profits can be obtained if more attention is paid to characters of high heritability. Having decided breeding programme, a farmer must then choose the parent stock.

Good records should identify the offspring with the parents. They should also give date of birth, sex and final disposal of individual. This information together with desirable confirmation is the basis upon which selection for flock improvement should be made.

There should be a system for ear marking, tattooing and or tagging. In case of short ear breeds, tattooing on inside of thighs or tails be practiced. A combination of tattooing and tagging is desirable.

Institution of selection would require identification of animals, recording of their pedigree and their performance. In small flock situation with essentially uneducated farmers, it is difficult to convince them of the need for such recording. On-station recording is more convenient, but the population size is generally small to provide sufficient intensity of selection and even accuracy of selection, especially where progeny performance is the criterion. Further, it does not allow production of required number of superior males. The subjective assessment in the farmers' flocks done by the farmers themselves does result in some improvement. Objective assessment both in terms of quality and quantity would be more desirable in determining genetic merit especially of the males to be used as breeders. In addition to the production, reproduction and survival are also important traits and will need recording and due consideration in selection. This could be done through selection among sire families.

The recording of performance of pedigreed progenies of a large number of sires is extremely important when selecting for milk, the milk having medium heritability and being limited to the female sex. On-farm recording of dairy goats is being practiced in Europe and America where large commercial dairy herds are maintained under intensive management. Once a month AM-PM milk recording is sufficient to determine lactation yield. The recording may start after first fortnight. In case of goats which are suckled, the milk recording may be done by separating the kids on the days of recording or the recording be started after the weaning especially in dual purpose goats (milk and meat).

In case of meat production, recording of body weight at weaning and around 6 months of age may be sufficient. Where there are possibilities of studying individual feedlot performance, the fortnightly body weight gain along with fortnightly feed consumption, efficiency of feed conversion, carcass yield and quality could also be recorded and involved in selection of breeders specially males. For improving prolificacy in meat breeds, selection of males from larger litters and on their body weight at 6 months can help in improving mothering ability and reproductive performance in addition to body weight gains.

In case of improving fibre production, recording of first 6 monthly or preferably annual cashmere production and assessment of fleece quality of at least all available males would be desirable.

15.1 Various Records

15.1.1 Livestock inventory

This register gives information about the individual identification, percentage and date and reason for sale or culling.

15.1.2 Wool/hair weight and quality

Wool/ hair evaluation should be limited to weight of each clip, staple length at a particular site, average fibre diameter, medullation percentage, visual assessment of fineness and making careful examination of undesirable characters such as black fibres or coarse fibres.

15.1.3 Growth records

Body weights at birth 3, 6, 9 and 12 months of age are important. Birth weight is related with vigour, survivability and growth rate. Weaning weight is used to measure total kid production of an ewe. Six month weight is correlated with market and yearling weight and is highly related to adult weight and doe productivity. Body weights of each kid are required to be correlated to age, sex, type of birth, season, and age of dam.

15.1.4 Prolificacy

Twin producing does are more profitable than single producing doe excepting in areas where grazing conditions are poor.

15.1.5 Health

Information on causes of death and incidence of diseases helps in planning prophylactic health programme.

In addition to above data, the following records should be kept properly.

- Livestock account record
- Breeding record
- Kidding record
- Shearing/wool/hair yield record
- Mortality record
- Sale of animals and wool/hair record
- Purchase of livestock, medicines and equipments records.

15.2 Guidelines for Recording of Data

All recording of data should be done by the Farm manager or Technical officers available to him.

15.2.1 Body weights

The body weight should be recorded at birth and after every 7 days up to 28 days, every 15 days up to 268 days and thereafter every 30 days up to 358 days. The earlier weights (10 kg) should be recorded closer to 25 g and weights up to weaning (20 kg) closer to 50 g and later weights closer to 100 g.

15.2.2 Body measurements

The body measurements, *viz*. heart girth, length, and height at withers closer to 1 cm should be recorded at birth, weaning, 6, 9 months and 1 year with the help of a metallic tape marked in cm and after placing the kid on a flat even surface with upright standing position.

15.2.3 Wool/hair yield and quality

(i) Immediately after birth, fleece/hair colour and the extent of body covered with the colour should be recorded.

(ii) The record of six monthly greasy fleece/hair weight of all animals should be made closer to 25 g. The exact age at shearing should also be recorded.

(iii) Samples of fleece from approximate 1" × 1" area (in case density is not to be determined, otherwise exact area should be marked and complete sample taken) from mid-side should be clipped with a sharp scissor as closer to the skin as possible and used for recording clean yield percent, average staple length, average fiber diameter, medullation percentage and density.

15.2.4 Culling and disposal

(i) All kids which are stray (of unknown parentage), highly deviating from the characters of a genetic group in colour and quality of wool/hair and the kids which are extremely poor in growth (less than 2 standard deviation below the weaning weight) would be culled at weaning.

(ii) All male kids should be ranked on the basis of 6 months body weight and those below the mean should be culled.

(iii) Out of the rest 50%, the requisite number in each genetic group should be retained on the basis of wool/hair weight and hair quality (average fibre diameter). While retaining, efforts should be made to have largest numbers of sires represented.

(iv) All does not breeding/kidding in two consecutive seasons should be culled (this may not be strictly followed in exotics if otherwise the condition of the animal is good).

(v) All does beyond the age of 7 years or those that become gummers earlier should be culled.

(vi) All bucks that have been used for 2 consecutive breeding seasons and have been mated to at least 10 does of each genetic group should be removed.

(vii) All does showing extremely poor mothering ability should be culled.

(viii) The culled/extra animals should be disposed off to the state Governments for breeding purpose if suitable or through the contractor on the basis of their live weight. The information about the extra males suitable for breeding should be circulatd among all the development agencies.

(ix) The disposal of culled animals, through the contractor should be the responsibility of Farm manager. Farm manager will obtain the necessary sanction for disposal and writing of these animals, giving age, live weight and reason for culling.

In India, it is not customary for farmers to keep any records in goat keeping. Breeding, lactation and feeding records are essential for proper management and production of goats. Since goats are mainly kept by rural poor who are illiterate, it becomes difficult for them to keep records. This problem is insurmountable till such time they get adult literacy. However, it is possible to devise simple-record keeping for people who have minimum education and can do record keeping of simple events like breeding cycle, pregnancy, kidding, feeding, and lactation.

CHAPTER 16

GOAT DISEASES

Disease is a major impediment against a successful goatary. Identification and prevention of diseases in time are essential features of a health, security programme. Infectious diseases like goat pox, infection of lentiviruses, pasteurellosis, clostridiosis, colibacillosis, coccidiosis, chlamydiosis, mycoplasmosis and parasitic infestations are responsible for heavy mortality in goats. Chronic wasting diseases (paratuberculosis, corynebacterium infections, brucellosis and parasitic infestations) are not immediate killers but they account for serious economics losses. The diagnosis of goat diseases, based only on clinical symptoms is difficult and inaccurate. However, the important clinical symptoms help to detect the sick goat at the earlier. The symptoms, treatment and prophylaxis of the important goat diseases are described below:

16.1 Abortions and Embryonic Losses

There are many disease factors that are responsible for causing abortions and embryonic losses in goats. *Chlamydia*, *Leptospira*, *Mycoplasma* and *B. melitensis* infections are the main causes responsible for abortion in late pregnancy (Table 16.1.) Pathak (1968) and Kulshresta *et al.* (1978) reported 1–1.6% losses as a result of *B. melitensis*. The high incidence of *Chlamydia* infection (20–30%) associated with abortion has been reported in goats (Krishna and Rajya, 1985). Mycoplasma (*M. agalactiae*, *M. Oculi*, *M. arginini*, serogroup 11) has been associated with inflammatory conditions of female genital tract and abortion in goats (Dhanda *et al.*, 1959; Singh *et al.*, 1975).

Pathak and Singh (1984) and Tiwana *et al.* (1984) reported that the prognosis of *B. melitensis* infection being poor, elimination is the only solution. Antibiotics can be used in the preliminary stages of *Chlamydia* and *Mycoplasma* infection in goats.

Table 16.1 Causes of Abortion and Embryonic Loss in Indian Goats

Disease	Cause	Incidence of infection (%)	Pregnancy Loss (%)	Source
Early embryonicloss	–	–	23 – 29	Achutankutty & Raja (1971)
Late abortion	*B. melitemsis*	2.15	1 – 1.6	Pathak (1968) Kulshresta *et al.* (1978)
	Chlamydia	20 – 30	7.9	Jain (1977) Krishna & Rajya (1985) Singh (1973)
	Mycoplasma	17 – 18.1	–	Tiwana *et al.* (1984) Singh *et al.* (1974–75) Pathak and Singh (1984)
	Leptospira	–	–	Upadhye *et al.* (1980)
Still birth	*Chlamydia*	–	3 – 8.25	Jain (1977) Krishna and Rajya (1985)

16.2 Neonatal Kid Mortality

Still birth and neonatal mortality present problems, particularly under intensive management conditions. Neonatal losses can be classified into four groups: still births as a result of *Chlamydia* infection; new born (0–10 days old kid) mortality due to foot and mouth disease, young mortality (1–4 weeks old) as a result of *Escherichia coli* infection (colisepticemia, colitoxemia, colibacillosis) and coccidial infections of kids (13 months old) (Sharma and Dutt, 1968; Bhatia and Pande, 1969). The incidence of *E. coli* infection in kids up to 30 days old was 39.3%; mortality, 5.4–19.0% (Vihan *et al.*, 1988). The mortality in the preweaning stage varied from 8 to 72% (Table 16.2).

Table 16.2 Neonatal Mortality (%) as a Result of Various Diseases

Disease	Incidence (%)	Mortality (%)	Source
Colibacillosis + Colisepticemia	39.3 – 60.0	5.4 – 19.0	Vihan *et al.* (1986) Vihan *et al.* (1987) Vihan and Singh (1988)
Chlamydia	–	62.8	Krishna and Rajya (1985)
Coccidiosis	12.0 – 32.3	5.1 – 12.3	Vihan *et al.* (1987)
Foot-and-mouth	–	3 – 16	McVicar & Sutmoller (1972) Vihan *et al.* (1987)

Source: Bhat (1988)

Neonatal mortality can be prevented by improving the level of nutrition in advanced stages of pregnancy (last 6 weeks), ensuring hygienic condition in the kidding sheds, providing proper bedding and ensuring early feeding of colostrum. The monoclonal antibodies derived from the caprine species are desirable for therapeutic purposes for intestinal infections, particularly in the treatment of neonatal kid diarrhoea caused by *Escherichia*.

16.3 Mortality in Kids

Mortality in kids is one of the most important factors for the success of any goat enterprise in the country. The success of goat farming is measured in terms of either the meat and milk produced or of the number of kids successfully reared in a given season. Whatever the measure used, it is based upon the natural ability of the goat to grow and breed. Growth and reproduction however, are two life processes dependent upon health for their full expression. By and large, morbidity and mortality rates are higher in new born animals than any other age groups and the prevention of losses among them assumes special importance. There are number of factors and diseases which effect the extent of mortality in kids.

The kid mortality during different stages (Table 16.3) indicates that 50% of the death inflicted in the first month and about 25% in the first week of birth, another peak in death during 3rd and 5th month when their milk is completely withdrawn from the diets (Singh and Sengar, 1978). The kid mortalities are varying from 10 to 74% in crossbreds as well as in native breeds of animals.

The kid mortality during different calendar months indicate that it is distributed throughout the year. However, February to April and again from October to December has been treated as the peak season of kid mortality. The kids suffer most mortalities during winter months (Singh and Sengar, 1978), rainy season being the worst in Mukteswar when 50% kids were lost. The best season from survival point of view were the cooler months of the year.

The birth weight of kids was an important factor, lighter kids had a significantly higher mortalities than the heavier ones (Mittal, 1976; Khan *et al*., 1978).

An incidence of mortality in kids up to 16 weeks of age groups is recorded as 44.23% (0–1 weeks), 32.05% (1–4 week) and 24.35% (4–16 weeks). Highest mortality (43.59%) in kids were observed during winter and onset of rainy season (Paliwal *et al*.,1978).

Besides these certain factors, there are many common diseases, *viz.* pneumonia, diarrohea, septicaemia (colibacillosis), contagious ecthyma, goat

Table 16.3 Kid Mortality at Different Ages (Percentage)

Breed	0-7 days	8-15 days	16 days to 1 months	1-3 months	3-6 months	6-9 months	9-12 months
Jamunapari	12.4	17.52	5.11	6.20	8.99	7.54	0.98
Beetal	30.15	12.97	5.96	9.18	7.22	3.50	2.33
Barbari	2.375	11.87	5.42	7.91	11.15	2.50	1.80
Black Bengal	23.53	22.50	7.50	9.33	6.32	1.42	0.17

Kid mortality during different calendar months

	Jan	Feb	Mar	April	May	June	July	August	Sept	Oct	Nov.	Dec.
Jamunapari	5.11	2.94	11.63	16.75	10.22	7.33	4.34	13.17	10.22	8.05	5.11	5.11
Beetal	3.22	7.00	13.53	9.74	17.81	1.61	1.61	9.74	6.01	1.66	20.54	7.57
Barbari	3.26	13.95	18.19	14.93	7.57	0.00	5.27	7.57	6.46	9.72	4.30	8.75
Black Bengal	4.78	15.10	10.50	9.81	7.28	5.29	2.76	4.65	6.54	11.69	12.60	8.93
At Makhdoom Jamunapari	4.16	1.04	1.04	0.0	1.04	1.04	0.00	8.33	8.33	4.16	6.25	2.08

Overall 3.75

Source : Singh and Sengar (1978)

pox, enterrotoxaemia, arthritis (joint ill) and coccidiosis which are reponsible for higher mortalities in kids.

The new born kids are particularly susceptible to infection because it is lacking certain protective mechanism such as antibodies due to failure to pass the placental barrier certain protective minerals and vitamins and poorly developed reticulo-endothelial system. Therefore, incapable of developing an active immunity or dealing satisfactorily with invading organisms.

(a) *Pneumonia*

Multifold etiological factors are responsible for pneumonia. Virus and mycoplasma are considered to be the initiating factors and subsequently complicated by secondary bacterial involvement such as *Pasteurella multocida*. It may also be due to ordinary cold or living in unfavorable conditions.

Clinically, it is manifested by an increase in respiratory rate, cough, abnormal breath sounds on ausculation and in most bacterial pneumonias by evidence of toxaemia and develops a high temperature which may be proceeded by a chill, loss of appetite, uneasiness and dullness, etc. The death of the kids occurs in pneumonia due to interference with gaseous exchange between the alveolar air and the blood. In bacterial pneumonias, as the added effect of toxins produced by the bacterial and necrotic tissue. Cyanosis is most likely to develop at this stage and be less pronounced when hepatization is complete and blood flow through the part ceases. Haematological observations usually reveal a leucocytosis and a shift to the left in bacterial pneumonias.

Postmortem findings

Gross lesions are usually observed in the anterior and dependent part of the lobes. Purulent exudates in the bronchioles and lobular congestion or hepatization were generally observed.

Treatment

With adequate treatment in the early stages, bacterial pneumonia usually responds quickly and completely with antibiotics but viral pneumonia may not respond at all or may relapse after an initial response. The final outcome depends on the susceptibility of the causative agent to the treatment available and the severity of the lesions when treatment is undertaken. The diseased animal should be removed immediately to some dry, warm place where there is plenty of fresh air and no strong drought. Turpentine oil with camphor should be applied on the chest. Nursing is very essential. Plenty of fresh clean water and some warm milk should be given.

(b) *Contagious ecthyma*

It is a contagious and highly infectious viral disease of kids characterised by the development of pustular and scabby lesions on the muzzle and lips.

The death may occur due to the extension of lesions into the respiratory tract. The morbidity due to this disease may be as high as 50% but the mortality rate may reach up to 5% if kids are badly cared.

It is caused by dermatotropic ungulate pox virus, composed of at least six immunologic strains.

Clinical findings

Recovered kids from this disease are solidly immuned for 2 to 3 years, but no antibodies appear to be passed in the colostrum. The lesions develop as pepules and then pustule stages which are not generally observed than as thick tenacious scabs covering a raised area of ulceration, granulation and inflammation. The lesions may appear as discrete, thick seals 0.5 cm in diameter or be packed close together as a continuous plague. The effected kids suffer a severe setback because of restricted suckling and grazing. Sometimes it may lead to a severe gastro-enteritis and broncho-pneumonia.

Treatment and control

When the disease is encountered for the first time, attempts should be made to check spread by the immediate isolation and treatment of affected kids with antiseptic ointment or astringents lotions.

In the early stage of an outbreak, the affected kids should be isolated and the remaining vaccinated. The vaccine is prepared from a suspension of seals in glycerol saline painted on to a small area of scarified skin inside the thigh or by pricking the ear with an icepick dipped in vaccine. Vaccination is completely effective for at least 2 years. The vaccine does not cure the affected cases.

(c) *Naval ill (Joint ill)*

The usual pattern of development in neonatal infection is a septicaemia, with a severe systemic reaction or septicaemia with few or no systemic signs followed by localization in various organs. If the portal of entry is the naval, local inflammation occurs, the so called "Naval ills".

It is especially found in kids born under crowded and not too clean conditions. Under such conditions, the bacteria lying in pens or sheds soon gain entrance into the blood stream through the low wound of the naval. From the local infection at the naval, extension may occur to the liver or via the urathus to the bladder or systematically to produce septicemia. In blood borne infection, localization is most common in the joints producing suppurative or non-suppurative artheritis.

Causative agents

Most common bacteria are *E. coli. Staphylococcus* and *Mycoplasma*. Lal Krishna *et al*. (1978) reported 4 cases of joint ill in kids from the bacterial infection of *Staphylococcus aureus* and *E. coli.*

Clinical findings

Inflammation of the joints cause pain, hot and swollen with consequent lameness. In many of the infections, there will also be an accompanying omphalophlebitis and evidence of lesions in other organs e.g. liver, endocardium and meninges. The joints most commonly involved are the hock, stifle and knee. The prognosis in case of acute suppurative artheritis is never good. Death occurs in about 7 to 10 days.

Lal Krishna *et al.* (1978) reported on postmortem that bilateral consolidation of apical, cardiac and anterio-ventral portion of diaphragmatic lobes of the lungs with multiple pin-head sized yellowish white focci which on cutting revealed suppurative mass.

Treatment

Control of an infection after it settled in the joints is difficult because of the poor blood supply to the part and the low levels of antibacterial drugs which develop in synovial fluid after their parentral administration, however, if this condition can be quickly and accurately diagnosed and the responsible germs identified, it may be possible to treat it successfully by the use of one of the more recent antibiotics. If the causative bacteria are not identified, a combination of penicillin and streptomycin I/M should be given. The parentral or intra-articular administration of cortisone or hydrocortisone or the oral or parentral administration of phenylbutazone is also recommended.

(d) *Goat pox*

It is caused by ungulate poxvirus highly contagious which is antigenically distinct from sheep poxvirus. All ages of goats are susceptible but a malignant form with a general distribution of lesions occurs in kids. It is most serious, causes death in 50% of affected animals. Major losses may occur in each new crop of kids. Goat pox in sheep is more severe than sheep pox and lesions occur on the lips and oral mucosa, the teats and udder. The goat poxvirus affords solid protection in sheep against both goat and sheep poxes but sheep pox vaccine does not protect goats against goat pox. Goat pox vaccine should be routinely used at 3 months of age.

Clinical findings

In kids, the malignant form is more common type. There is marked depression and prostration, high fever and discharges from the eye and nose. Affected kids may die during this stage before typical pox lesions develop. Skin lesions appear on skin and on the buccal, respiratory, digestive and urogenital tract mucosa.

Treatment and control

No specific treatment is adviced. Quarantine of infected premises, vaccination by tissue vaccine absorbed on aluminum hydroxide has been followed by no ill effects and by the establishment of immunity in 2 weeks which persist for 9 to 12 months.

A killed vaccine with adjuvant provides protection for 5 months (Sharma and Dhanda, 1970). Vaccination in the face of an outbreak is unlikely to prevent deaths during the subsequent weeks.

(e) *Colibacillosis*

By far the commonest disease entity of new born kids caused by *E. coli*. It may occur in two forms in kids, one most common septicaemic, other one enteric form. This is occurring when the new born animals are kept in groups in close confinement. The development of disease is rapid and considerable expense is involved in treatment and control of the disease.

The organism is very common in the environment and most adults develop an immunity to it.

Toxins which was produced by the serotypes of *E. coli* causes damage to vascular endothelium leading to transudation from vessels, specially into serious cavities in septicaemic colibacillosis, other one causes the enteric form of the disease. It shows that gastric hypotonicity and hypomotility are primary response in the enteric from and that the resulting gastric distention precedes the appearance of diarrhoea. The important abnormality of function is the out pouring of fluid and sodium and bicarbonate ions leading to fatal dehydration.

When the disease is confined to the gut, the condition responds reasonably well to treatment and death is due to dehydration and electrolytic imbalance. Animals which recover from the septicaemia may develop lesions due to localization in other organs. Artheritis is a common sequel in kids. Meningitis and pneumonia may also occur.

Clinical findings

Collibacillosis is always septicaemic and peracute. Two age groups appear to be susceptible, one to two days of age and 3 to 8 weeks old. Peracute cases are found dead without premonitory signs. Affected animals are depressed and weak. Anorexia is complete, there is marked increase in heart beat and although the temperature may be high initially, it falls rapidly to subnormal levels when diarrhoea and sometimes dysentery appear. Acute cases show collapse and occasionally signs of acute meningitis manifested by a stiff gait in the early stages followed by recumbency with hyperaesthesia and tetanic convulsion. Chronic cases are usually manifested by arthritis.

Necropsy findings

Submucosal petechial haemorrhage fibrinous exudate in the joints and serous cavities. Organism may be cultured from the gut, the mesentric lymphnodes, spleen, heart, blood and cerebrospinal fluid.

Treatment

Antibiotics including streptomycin, tetracyclines, neomycin and chloramphenicol, nitrofurazone are highly effective against *E. coli*.

Streptomycin should be administered at the rate of 20 mg/lb body weight daily. The tetracyclin, neomycin, and chloramphanicol at 5 mg/lb body weight daily for 3 days. It is standard practice to administer oral preparation of these drugs in conjunction with the parenteral injection. When treatments are administered orally, they should be given at least twice daily and continued for 4 days.

Appropriate supportive treatment will also be given such as fluid therapy, 100 ml. dextrose saline or belamyl.

Antihistaminic may be used with enteric-toxaemic collibacillosis although the importance of vitamins particularly vitamin 'A' in the protection of mucosal against infection is well established.

(f) *Enterotoxemia: (Over eating disease, pulpy kidney)*

Enterotoxemia is an acute infectious but noncontagious, highly fatal characterized by sudden death, convulsions, caused by absorption of a toxin produced by the bacterium *Clostridium perfringens* type C & D. The toxin is produced in the intestinal tract, the condition is therefore an auto-intoxication. The major toxin of type D is innocous epsilon prototoxin which is converted to lethal epsilon toxin by trypsin. It is characterised by sudden onset of a profuse haemorrhagic enterits, diarrhoea, convulsions and sudden death.

The factors influencing susceptibility is over-eating or enteritis from some other cause, causing stasis of the intestinal tract. Stasis allows for absorption of greater than normal amounts of toxin, with the rapid production of symptoms. The seasonal incidence is higher during spring and early summer months, specially when rainfall or irrigation has provided sufficient moisture for fast growth of nutritous plants. The course of illness in kids affected with entrotoxaemia is short and may develop during the night, consequently, commonly find the dead kids during morning. Affected animals separate from the flock and during warm weather, seek the shade. Respirations are very rapid and may include oral breathing, saliva drips from the mouth and the body temperature elevates. The animal may manifest abdominal pain. The morbidity rate averages 5 to 10% but most affected kids die.

Postmortem findings at necropsy reveal a large amount of undigested feed in the abomasum, distension of pericardial sac with clear fluid and heamorrhages on the heart and thymus are characteristic findings.

Prevention

The incidence can be prevented or reduced by vaccination and management. An effective multi-component vaccine is available and should be administered to pregnant goats or at 3 months age of kids. Protection extends from birth to 5 weeks of age.

(g) *Coccidiosis*

It is a contagious enteritis of young kids caused by infection with Eimeria sp. It is characterised by haemorrhage, diarrhoea, anaemia depression and emaciation.

Coccidiosis occurs where animals are housed or confined in small areas, The morbidity rate is usually low. The pathogenic coccida of goats have in most instances not been identified, although *Eimeria arlongi* and *Eimeria faurel* are known to occur.

The source of infection is the faeces of clinically affected or carrier animals and infection is acquired by ingestion of contaminated feed and water or by licking the hair coat contaminated with infected faeces.

Clinical Symptoms

A mild fever may occur in the early stages but in most clinical cases, the temperature is normal or subnormal.

The first sign of illness is usually the sudden onset of severe diarrhoea with foul smelling, fluid faeces containing mucus and blood. Severe straining is characteristics, severe dehydration emaciation and complete anorexia are usual.

Kids generally die of haemorrhagic anaemia some days before oocysts appear in the faeces. The period during which oocysts are discharged in significant number is quite short (3 to 6 days). A count over 5000 oocysts are discharged and is considered to be significant.

Postmortem

Congestion, catarrhal enteritis and thickening of mucosa of the caecum, colon, rectum and ileum are the characteristics gross changes.

Treatment and control

Coccidiosis is a self limiting disease and signs subside spontaneously in survivors when the multiplication stages of the parasite has passed. Sulphadimidine, sulphabromezathin or phathalylsulphathiazole are the drugs of choice. The treatment should be continued for at least 14 days.

In an outbreak, the clinically affected animals should be isolated and treated with one of the more efficient drugs. The control of coocidiosis depends largely upon hygiene and avoidance of over crowding. As far as possible, the young kids should be separated form adults which provide source of infection.

(h) *Diarrhoea or scour*

Too much dampness or more of fresh green leaves in wet condition or too sudden change in feeding or excess of concentrate, chill and warm are responsible for this disease. It stands to reason that the cause must be searched and removed. The trouble is closely related to indigestion.

In order to get rid of the initiating factor in the intensive, 4 oz. of linseed oil and 2 oz. of caster oil should be given. Strenacin and Gastina tablets are found quite beneficial for treating diarrhoea among kids.

Scouring in kids is generally the direct result of artificially milk feeding when milk is given either too hot, too cold or is given more at a time than the kids can digest or the bottles are not well cleaned. The quantity of milk for a day or two should be reduced. For the prevention of diarrhoea at early stages, it is recommended to give preventive doses of chloromycetin palmitate syrup and furoxone continuously for 3 days to develop resistance against digestive disorders.

16.4 Chronic diseases in adult goats

Gastrointestinal infections (coccidiosis, mycobacterium paratuberculosis) and gastrointestinal parasites (*Haemonchus* sp., tapeworm and immature *amphistomes*) are common in goats, maintained in confinement and when assembled from a wide variety of sources. These diseases of the digestive tract retard the growth, weaken the physic and stunt the animal (Mukherjee, 1963; Shastry and Ahluwalia, 1972; Singh, 1987; Vihan *et al*., 1986). The internal and external parasites generally render the infected goats weak and anemic because of their constant feeding on host tissues. The absorption of essential nutrients from the reticulo-rumen and intestine is greatly interfered with, causing malnutrition. The determination of worm load and the use of suitable drugs in an infected goat flock is the simplest way to control parasitic infestations. (Table 16.4).

The prognosis of chronic infections like M. paratuberculosis is poor and elimination is the only solution.

Amongst various types of parasites, three categories are recognised:

(1) Helminths: *nematodes* (round worms), *cestodes* (tape worms) and *trematodes* (flukes).

(2) Protozoa: unicellular parasites, affecitng digestive tract (*coccidia, amoeba*, etc.), blood (*trypansomes*, *piroplasm* etc.) and organs (*leishmania*, *trichomonas*).

(3) Arthropods: insects (blood sucking flies, blow flies, mosquitoes, lice, bugs, fleas) and *arachnids* (ticks and mange mites).

Table 16.4 Preventive Measures for Parasitic Infestation

Disease	Curative agent	Drug and dosage	Remarks
Fascioliasis	*Fasciola gigantica*	Zanil, 1ml/3kg BW	100% effective
	Fasciola hepatica	Trodex, single dose injection 1 ml/40 lb BW	90–100% effective
	Dicrocoellum dendriticum	Hexchlorophane, 15 mg/kg BW in alcoholic solution	Effective and non-toxic
Immature	*Paramphistomum cervi*	Neguvon, 50–70 mg/kg BW	94% effective
Amphistomiasis	*Cotylophoron cotylophorom*	Hexachlorethane, 1mg/2kg BW	50% effective
Nematodiasis	*Haemonchuscontortus*, *Ostergia* spp.	Nilverm, 15 mg/kg BW	Safe and effective
	Trichostrongylus colubriformis, *Cooperia* spp.	Thiabendazole, 50 mg/kg BW	2 or 3 times/year
	Strongyloides, *Bunostomum*, *Trignocephalum*	Fenbendazole, 50 mg/kg BW	90–95% effective
Monieziosis	*Moniezia expansa*, *Moniezia bendeni*	Yomesan, 75 mg/kg BW	
	Avitellina spp., *Stilesia* spp.	Lead arsinate, 0.5–1 g in gelatin capsule	
Mange	*Sarcoptes scabiel*, *Psoroptes* spp.	0.4% dipping with Cythion or Melathion	2 or 3 times/year

BW = body weight; 1 lb = 0.454 kg

The first two categories are commonly grouped as endo-parasites while the last one as ecto-parasites. In nature, multiple infections are common. In some pathogenic forms, the symptoms are typical while in most of the cases, there are almost identical syndromes and there is usually resemblance and/or overlapping. Secondary pathogens of bacterial or viral origin may also complicate the disease condition. Many pathogenic species of helminths and protozoa cause sufficient damage the hosts' tissue before their eggs/larvae/cysts are passed out. A clinician, charged with the treatment and control work need to be fully conversant with the pathogenic species and the nature of their development inside the host.

For an effective treatment and suitable control of the parasitic diseases, the veterinarians have to be familiar with the various diagnostic aids including collection, preservation of parasites and parasitic infested tissues and their dispatch for specific identification .

16.5 Diagnostic Methods

16.5.1 Diagnosis by clinical symptoms

In helminthiases, only a few infection have pathognomonic symptoms and/or lesions *e.g.* nasal granuloma, coenuriasis (gid disease), parafilariasis (Summer Bleeding). In most of the gastrointestinal parasites, there are identical syndromes such as diarrhoea dysentery, oedema of the dependent parts, particularly in the submaxillary space (bottle jaw), unthriftiness, anorexia, anaemia, hidehound condition etc. Most of protozoan disease show typical symptoms *viz.* piroplasmosis, trypanosomiasis, leishmaniasis and trichomoniasis Similarly nasal bot in sheep, warble in cattle, cutaneous myiasis and mange among ectoparasites, are easily diagnosed.

16.5.2 Diagnosis of faecal examination

Examination of faeces is usually done to detect the eggs and larvae of helminths, oocysts and other protozoan stages, parasiting mainly alimentary canal and organs *viz.* liver and lungs. In acute cases, there may be no stage in the faeces inspite of the actual worm infestation particularly in their early stage of invasion. Hence, a negative finding in the faeces is no surety of freedom from internal parasites. To be sure, the faeces from the suspected case should be examined repeatedly. Mild purgative is also resorted to in doubtful cases because, the worms of their portions may be expelled and identified.

The examination is done by two methods:

(a) Direct examination (b) By concentration method

The following procedure should be adopted:

Collect the faecal sample preferably directly form the rectum or collect by a neat spatula, a small quantity from the top of freshly voided mass which is not touching the earth or any other material as the soil is inhabited by a number of free-living nematodes and other organisms. If collection is to be done from more than one animal/ bird, the spatula should be washed every time. The examination of faeces should be done while fresh or after fixation in equal quantity of 10% formalin. If sporulation of coccidian oocysts has to be studied, a part of fresh sample should be kept in 2.5% potassium dichromate solution. After collection, the faeces/pellets should be broken and a uniform mass be prepared. If it is hard, a small quantity of clean water, preferably saline or tap water may be mixed to form homogenous mass. If the animal is suspected for blood fluke infection, a small portion of the fresh faeces should immediately be fixed in 10.5% formalin before examination as the fully embryonated of blood flukes hatch out at once when in contact with water. The homogeneous faecal mass should first be strained to remove coarse material from the faeces. The examination of faeces is usually conducted first by direct smear preparation and subsequently by concentration. The concentration is done by floatation and centrifugation.

(a) Direct examination

In heavy infections, the parasitic stages are usually observed in direct smear preparations. This is done by putting a small quantity of faeces with a toothpick, clean broom stick, match stick on a clean slide and mix it with a drop of water to have an uniform mass. Spread it over a small area and cover with a clean coverglass. The smear should be as thick as through which light can easily pass. Examine it under a microscope. Move the slide in a zig-zag manner from one end to the other so that the complete field under coverslip can be examined.

(b) Concentration method

This method may be adopted when there is light infection and there is large faecal mass containing much coarse particles including undigested pieces of straw. As most of the helminthic eggs are lighter than the specific gravity of either sheather's sugar solution or saturated sodium chloride solution or zinc sulphate solution, the advantage is taken to get these eggs floated. This is done by centrifugation or floatation. The former is more accurate.

About 10–15 ml of uniform suspension of faeces (1 part of faeces mixed with 5 parts of water and strained through a sieve of about 1 mm mesh) is taken in to clean grease-free 15 ml centrifuge tube and fill it up to its rim with required quantity of water. Take another 15 ml centrifuge tube completely filled up to its rim. Put both the tubes in front of each other and centrifuge at about 100 rpm for 2 minutes. Remove the supernatant fluid, if it is clear otherwise repeat the centrifugation. The deposit at the bottom should again be mixed with water and centrifuge. This process should be repeated till all colouring matter in the faeces is removed. Afterward, mix the sediment with sufficient volume of either of the solutions mentioned above and fill the tube up to the rim. Put a clean round coverslip over it in a way that no air bubble is left between coverslip and the fluid. Centrifuge it for two minutes. Gently pick up the coverslip and put on a clean slide and examine under the microscope. In case no facility is available for centrifugation, the tubes containing faecal mass may be dissolved in either of the solutions and leave it in vertical position for 30–40 minutes. Except for trematode and most of the cestode eggs, eggs of nematodes and oocysts will float on the surface of the liquid.

For examination of the trematode eggs, it is also necessary to examine small portions of sediment directly. In case of lungworms, the first stage larvae are passed out in the droppings. The larvae, when in large numbers, can be seen in direct smear preparations. A few pellets may, however, be placed in a petridish with a small quantity of lukeworm water. Shortly, the larvae will come out in the water. If the faecal sample is too soft, it should be spread out uniformly in a petridish and make small hole dips by glass rod. The larvae will migrate into such depression containing water.

The eggs of various helminths are identified on the basis of specific characters in respect of size, shape, colour and contents. There is, however, difficulty in differentiation of different bursate nematode eggs.

For more detailed examination, eggs counting and faeces culture are done.

16.5.3 Diagnosis by blood smear examination

(a) Thin blood smears:

Absolutely clean grease-free slides should be used for preparation of blood smears. A drop of blood from a fresh puncture of the ear vein or in any part of the body may be placed in the centre of one side of the slide. The edge of another slide be touched with blood drop and place it at an angle of less than 45 and spread it gently over the slide. After drying in the air, the smear is ready for staining thin blood smears. Staining of blood smears should be done in accordance with the instructions available for different stains, *e.g.* Leishman, Giemsa, Wright's or Field's stain. If stains are not available, the air-dried blood smears should immediately be fixed in methyl alcohol or rectified spirit before despatch to a laboratory for diagonsis.

(b) Thick blood smears

This is used for quick examination of blood parasites *e.g.* trypanosomes and mirocrofilariae of different filarial nematodes. Thick blood smear is not useful in avian or camel blood because of their nucleated erythrocytes. It is first dehaemoglobinised before staining it in distilled water for 15 minutes until colour has disappeared. It should be fixed in methyl alcohol before staining.

16.5.4 Immunodiagonostic tests

In these tests, the diagnosis is based upon the presence of antibodies in the host specific to particular parasites. Such antibodies can be detected by serological tests like precipitaion test, agglutination test, complement fixation test in diagnosis of Trichomonensis, Trypanosomiasis, schistosomiasis, or by allergic tests like Casoni skin test diagnosis of hydatid disease. Antigens used in these tests are derived from adult parasites or their stages.

16.6 Slow-growing Viruses

Slow-growing viruses like Visna/Maedi and Jaagsiekete occur in Indian goats, causing death by invasion on the respiratory system. A virus belonging to the enterovirus group is responsible for the caprine arthritis encephalitis syndrome (CAE). The disease is revealed with progressive lameness due to arthritis and hind quarter paralysis due to encephalities caused by viral aggression. This disease has been reported in a few Asian countries (Singh, 1988), particularly in Thailand. The prognosis of this disease is poor and elimination is the only remedy.

16.7 Contamination of Goat Meat

Edible organs like the liver, lungs, heart, and brain are frequently damaged because of the invasion and migration of parasites. Fascioliasis in goats is an important disease that reduces the value and extent of goat meat (Arora and Iyer, 1966; Gill *et al.*, 1983). Other parasites include hydatid cysts and coenuriosis, which directly damage the meat and indirectly affect the goat production. The prognosis of fascioliasis is good if effective drugs are given at the right time. The prognosis of both hydatidosis and coenuriosis are bad; elimination is the only answer. Improved serodiagnostic tests are needed for detection of hydatidosis in goats.

16.8 Zoonotic Disease Factors

The evidence of zoonotic diseases occurring in goats is fairly common in India (Pathak, 1968; Rao, 1971; Charan *et al.*, 1973; Jain, 1977). The goat diseases that can infect man include tuberculosis, brucellosis, chlamydiosis, and cysticercosis/taeniasis. To improve the quality of chevon, proper carcass inspection in the slaughter houses is necessary.

16.9 Problems and Constraints for Preventive Health

Several new vaccines and other preventive measures have become available. The implementation of adequate preventive measures in a country like India requires an augmented infrastructural facility at village level. There is paucity of skilled veterinary assistants (labour) in India and most of the problems of goat farmers remain unattended and unsolved, perhaps due to an overemphasis on the health of cattle and indifference to goat diseases.

16.10 Diagnostic Facilities and Immuno Biologicals

The disease prevention and efficient health control require adequate laboratory facilities and diagnostic antisera and antigens. There is also a need to develop an early diagnostic test to detect subclinical cases of M. paratuberculosis in goats. Disease monitoring and surveillance system have not been adequately geared to cover the goat diseases.

16.11 Vaccination Schedule

In intensive and semi-extensive systems of goat rearing, it becomes essential to rigidly follow the preventive vaccination schedule (Table 16.5) for flock-health security. While raising goats under extensive systems, particularly associated with migration, the vaccination schedule would require period adjustment to avoid vaccination en route.

Table 16.5 Vaccination Schedule for Ggoats

Name of disease	Nature and name of vaccine	Dose	Route	Duration of Immunity	Time of vaccination
Haemorrhagic Septicaemia	Formalized culture of *P. septica*	2 ml.	I/M	1 year	June/July
Enterotoxaemia	Formalized culture of *Cl. Welchii*	2 ml (1st dose) 2 ml (booster dose) at interval of 10 days	S/c S/c	6 months	On pregnant stock one month before kidding & in kids at 3 months of age
Clostridial infections	Multicomponent clostridial vaccine (*Cl.chauvoei/Cl. septicum*)	2 ml 1st dose 2 ml (booster dose) at interval of 14 days	S/c	1 Year	Adult stock June/July
Black Quarter	Bacterial suspension of of killed *Cl. chauvoei*	2 ml 1st dose 2 ml (booster dose) at interval of 14 days	S/c	1 Year	June/July
Anthrax	Anthrax spore vaccine	1 ml	S/c	1 year	June/July
FMD	Polyvalent (O,A,C. & Asia 1)	5 ml	S/c	6 months	Oct./Nov. and April/May
Rinderpest	Tissue culture Rinderpest	1 ml	S/c	5 years	June/July
Lungworm	Irradiated *D. filaria* (infective larval) vaccine	1000 larval Ist & IInd dose after 10 days	Orally Orally	1 year	

Source : Singh (1985)

CHAPTER 17

ECONOMICS OF GOAT REARING

Goat is a multi-functional animal and plays a significant role in the economy and nutrition of landless, small and marginal farmers in the country. Goat rearing is an enterprise which has been practiced by a large section of population in rural areas. Goats can efficiently survive on available shrubs and trees in adverse harsh environment in low fertility lands where no other crop can be grown. In pastoral and agricultural subsistence societies in India, goats are kept as a source of additional income and as an insurance against disaster. Goats are also used in ceremonial feastings and for the payment of social dues. In addition to this, goat has religious and ritualistic importance in many societies.

17.1 Scope for Goat Rearing and its National Importance

The goat contributes a lot to the stability of small holder farming system by providing financial resource and organic manure. A large population of small farmers and landless lobourers in Asia depend on goat for their sustenance. In pastoral and agriculture subsistence societies, goats are kept as a source of investment and as an insurance against disaster. They are also used in ceremonial feastings and for the payment of social dues. Goat manure is prized as garden manure in many countries. Not much work on the socio-economics of goat has been done; however, there are some reports available indicating benefits of goats vis-à-vis other livestock species.

Goat make a valuable contribution to the livelihood of economically weaker sections of the society. Amongst the livestock owners goat rearers are the poorest of the lot.

Realizing the importance of goat in the agrarian economy of the country, various developmental activities have been taken up by Government of India.

The Central Government had established Central Institute for Research on Goats at Makhdoom, Farah, Mathura District, Uttar Pradesh. Financial assistance is available for goat production from NABARD through Nationalized Banks. Subsidies are also available for weaker sections.

17.2 Economics of Goat Rearing

A number of studies on the economics of goats to the keepers have been conducted. The studies conducted in Himachal Pradesh revealed that labour was the main component of cost in feeding and rearing of goats in the stationary production system and it was much higher than in the migratory system. The feed cost proportion was relatively low (Raut and Nadkarni, 1974). A peasant family in Andhra Pradesh earned Rs.140/doe allowing for capital loan interest on it and expenses incurred (Sriram Murthy, 1977). Work done at Central Arid Zone Research Institute (CAZRI), Jodhpur indicated that under good nutrition goat rearing may provide Rs 250/goat a year to a small farmer besides 2 quintals of manure and benefits of clearing abnoxious weeds and thorny bushes (Ghosh and Khan, 1980).

A study on economics of goat production starting with a unit of 5 animals to unit of 100 animals with different sections of farmers was undertaken by Central Institute for Research on Goat (CIRG). Net income of Rs 212/goat a year from 10 heads and Rs 252/goat a year from 50 heads has been reported. No income difference with 50 or 100 animal units was observed (Sharma *et al*., 1982; Saini *et al*., 1986). Singh and Ram (1987) reported economic contribution of goats based on a study of 226 goat keeping households in Punjab. The gross income per head was highest for the small flocks because these were managed better. The net income was higher for the small flocks in the plains than that for the flocks in hilly region. Milk contributed the highest (66.6–84.4%) income followed by the sale of animals (13.4–30.3%) and manure (1.0–3.4%).

A survey on socio-economics of goats from Rajasthan reported by Ahuja and Rathore (1987) indicated that the number of goats increased 3 times between 1951 and 1983. The goats accounted for 28 to 31% of the value of livestock assets and 16 to 19% of gross receipt from crops and livestock. This further indicated that investments for small goat units are modest and there is quick pay off due to fast multiplication. It also indicated that risk in goat farming is much lower when compared to crop production.

Acharya and Patnayak (1974) worked out the relative economics of rearing cattle, sheep and goats in the hot arid regions of Rajasthan, while primarily maintained on free range grazing or browsing with some supplementary feeding to cattle during lactation, the goats were 130% more economical than cattle and 123% more economical to sheep. Raut and Nadkarni (1974) reported that income derived from goats under both stationary and migratory systems of management was substantially higher than from sheep. Swain *et al.* (1983) found indigenous goats 2.5 times more economical than sheep on free range grazing under semi arid conditions of Rajasthan. Sharma (1987) reported that goats provide significantly more meat and milk per unit live weight per year than cattle, sheep and camel. A field study conducted at Tilonia (Rajasthan) on relative economics of 1 buffalo unit and 5 goat units for 4 years revealed that profit per annum was Rs 1755 from buffalo unit and Rs 1945 from the goat unit (Table 17.1). Arora and Swain (1989) reported economics based on productivity of 30 sheep and 30 goats with one breeding male from each of the two species over 3 years that a family can earn on an average Rs 690.49/year after paying 10% interest on the loan incurred on the non-recurring assets at the beginning, repaying 10% loan and meeting other expenditures on the maintenance and upkeep of animals. However, when the net profit was opportioned to sheep and goat, there were loss of Rs 132.96 in the case of sheep and a profit of Rs 823.45 per year from goats. The difference between assets and liabilities at the end of 3 years in sheep was Rs 429 and Rs 2885 in goats, which can provide an additional income to the family of Rs 1104.66 per year.

17.3 Financial Assistance Available from Banks/NABARD for Goat Rearing

NABARD is an apex institution for all matters relating to policy, planning and operation in the field of agricultural credit. It serves as refinancing agency for the institutions providing investment and production credit for agriculture and rural development. It promotes development through a well organised Technical Services Department at the head office and Technical Cells at each of the Regional Offices.

Loan from banks with refinance facility from NABARD is available for starting Goat farming. For obtaining bank loan, the farmers should apply to the nearest branch of a Commercial or Co-operative or Regional Rural Bank in their area in the prescribed application form which is available in the branches of financing bank. The Technical Officer attached to or the Manager of the bank can also help/give guidance to the farmers in preparing the project report to obtain bank loan.

For goat rearing schemes with very large outlays, detailed reports will have to be prepared. The items of finance would include costs of assets like development of land, construction of sheds, purchase of equipments, purchase of breeding stock, rearing cost of animals till it generates income etc. The cost of land is not considered for loan. However, if land is purchased for setting up a goat farm, its cost can be treated as party's margin as per the norms.

17.4 Scheme Formulation

The scheme can be prepared by a beneficiary after consulting local technical persons of State animal husbandry department, commercial farmers etc. If possible the beneficiaries should also visit progressive goat rearers and government/military/agricultural university farms in the vicinity and discuss the profitability of goat rearing. A good practical training and experience in goat rearing will be highly desirable. Nearness of the Goat farm to a veterinary aid centre and breeding centre should be ensured.

The scheme should include information about land, livestock markets, availability of water, feed, fodders, veterinary aid, breeding facilities, marketing aspects, training facilities, experience of the farmer and the type of assistance available from State Government.

The scheme should also include information on number and types of animals to be purchased, their breeds, production performance, cost and other relevant input and output costs with their description. Based on this, the total cost of the project, margin money to be provided by the beneficiary, requirement of the bank loan, estimated annual expenditure, income, profit and loss statement, repayment period etc, can be worked out and included in the scheme.

17.5 Requirements of a Good Project

A format developed for formulation of goat rearing schemes is available at the financing institutions. The scheme so formulated should be submitted to the nearest branch of bank. The bank's officers can assist in preparation of the scheme or filling in the prescribed application form. The bank will then examine the scheme for its technical feasibility and economic viability.

A) ***Technical Feasibility – This would briefly include***

i) Nearness of the selected area to veterinary dispensary, goat breeding centre, marketing outlets for fattened kids/meat and the financing bank's branch.

ii) Availability of good quality animals in nearby livestock markets. The distribution of goat breeds in India are given in Table 3.2 and 21.1.

iii) Availability of training facilities.

iv) Availability of good grazing ground/lands.

v) Availability of green/dry fodder, concentrate feed, medicines etc.

vi) Availability of veterinary aid/breeding centers and marketing facilities near the same area.

B) *Economic Viability – This would briefly include*

i) Unit cost of animals
ii) Input cost for feeds and fodders, veterinary aid, insurance charges, etc.
iii) Output costs *i.e.* sale price of live animals, manure/penning charges, etc.
iv) The average unit cost (indicative only) of goat rearing units is assumed for calculating project cost.
v) Income-expenditure statement and annual gross surplus.
vi) Cash flow analysis.
vii) Repayment schedule (*i.e.* repayment of principal loan amount and interest).

Other documents such as loan application forms, security aspects, margin money requirements etc. are also examined. A field visit to the scheme area is undertaken for conducting a techno-economic feasibility study for appraisal of the same. The model economics of goat rearing unit of 50+2 under semi intensive system is given in Table 17.1 to Table 17.11.

Table 17.1 Relative Economics of Sheep and Goat Production under Free Range Grazing System Production

	Sheep	Goat
No. of animals at the beginning of the project		
(a) Females	30	30
(b) Male	1	1
No. of animals added from outside to maintain the flock strength during 3 years	13	Nil
No. of adult animals culled during 3 years and replaced by young ones born in the flock	Nil	16
Average annual adult mortality (%)	31.25	7.14
Average young mortality (%) (both sex)		
(a) 0–90 days	16.39	12.19
(b) 0–180 days	56.66	28.16
(c) only females 0–1 year	63.15	35.29
Lambing/kidding percentage	82.48	84.19
Milk yield (kg)/ lactation	–	53.34
Lactation length (days)	–	152
Wool yield per clip (kg)		
(a) Adult	0.553	–
(b) Young at 1st shearing	0.427	–
Weight (kg) at market age of males (6 months)	14.55	14.95

Table 17.1 Contd.

	Sheep	Goat
Fluid cash (Rs) available/year after paying 10% of the loan + 10% interest on the loan incurred at the beginning + maintenance and uptake of animals	(-) 132.96	823.45 (Net Rs 690.40)
Difference between assets and liabilities at the end of 3 years	429.00	Rs.2885.00 (Rs 1104.66 per year)
Interest + 10% payment of loan during 3 years both for sheep and goat sum to Rs 4980.60		

Source: Personal Communication, Bhat (2008)

Table 17.2 Economics of Goat Farming-Techno-Economic Parameters for 50+2 Units

	No. of Bucks	2
	No. of Does	50
A.	**Production Traits**	
i	Age at Maturity (Months)	10-12
ii	Kidding interval (Months)	8
iii	Kidding percentage	85
iv	Twinning percentage	60
v	No. of kiddings per year	1.5
vi	Sex ratio	1:1
vii	Mortality (%) Adults	5
	Kids	15
viii	Saleable age of kids (months)	8-9
ix	Culling of does (% per year) from second year onwards	20
B.	**Expenditure norms**	
i	Space requirement (st.per head)	
	Buck	20
	Doe	10
	Kids	4
ii	Cost of construction (Rs.per sft)	35
iii	Cost of equipment (Rs.per adult animal)	10
iv	a) Cost of green fodder cultivation (Rs./acre/season)	1,500
	b) No. of acres	2
v	Concentrate feed :	
	Adult does (one month before breeding and one month after kidding *i.e.* per kidding)	6.75 kg/month
	Bucks (two months per breeding season)	7.5 kg/month
	Kids (for 30 days)	3.75 kg/kid
vi	Cost of conc. Feed (Rs./kg)	5

Table 17.2 Contd.

	vii	Labour (No.)	1
		Labour wages (Rs.per month)	1250
	viii	Insurance (as percentage of the cost of breeding stock)	4
	ix	Veterinary aid (Rs./adult/year)	20
	x	Water, electricity and other misc. expenses (Rs./adult)	10
C.	**Income norms**		
	i	Sale price of Bucklings (Rs./kid)	1000
	ii	Sale price of Doelings (Rs./kid)	900
	iii	Sale of culled does (Rs./doe)	1200
	iv	Sale price of culled Buck (Rs./buck)	1500
	v	Sale value of male/female kids (Rs./kid)	600
	vi	Income from manure is not assumed as itis used on the own farm	
	vii	Sale of gunny bags (Rs./bag) (13.3 bags/tonne)	10
D.	**Repayment norms**		
	i	Repayment period (years)	6
	ii	Grace Period (years)	1
	iii	Interest rate(%)	12

Source: Personal Communication, Bhat (2008)

Table 17.3 Economics of Goat Farming – Flock Projection Chart

S.No.	Years	1	2		3	4		5	6	
	Kidding Number	**I**	**II**	**III**	**IV**	**V**	**VI**	**VII**	**VIII**	**IX**
	Particulars									
A	Opening Stock – Adult Bucks	2	2	2	2	2	2	2	2	2
	Adult Does Purchase Stock	50	48	40	45	38	42	45	38	42
	Replaced Stock	0	0	10	5	12	8	5	12	8
B	Births during the Year									
	Male Kids	34	33	34	34	34	34	34	34	34
	Female Kids	34	33	34	34	34	34	34	34	34
C	Mortality									
	Adult Does (5%)	2	3	0	2	3	0	2	3	0
	Male Kids (15%)	5	5	5	5	5	5	5	5	5
	Female Kids (15%)	5	5	5	5	5	5	5	5	5
D	Sales during the Year									
	Adult Bucks	0	0	0	2	0	0	2	0	0
	Adult Does	0	5	5	10	5	5	10	5	5
	Bucklings	0	29	26	29	29	27	29	29	27
	Doeings	0	19	23	17	21	24	17	21	24

Table 17.3 Contd.

S.No.	Years	1	2		3		4	5		6
	Kidding Number	I	II	III	IV	V	VI	VII	VIII	IX
E	Closing Stock at the end of Year									
	Adult Bucks	2	0	2	2	2	2	2	2	2
	Adult Does	48	40	45	38	42	45	38	42	45
	Buckling Transfer	0	0	2	0	0	2	0	0	2
	Doelings Transfer	0	10	5	12	8	5	12	8	5
	Male Kids	29	28	29	29	29	29	29	29	29
	Female Kids	29	28	29	29	29	29	29	29	29

Note: (i) Kidding (%) – 85, (ii) Number of kiddings per year 1.5 (iii) Twinning Rate (%) – 60
Source: Personal Communication, Bhat (2008)

Table 17.4 Economics of Goat Farming – Cash Flow Statement

S.No.	Particulars	I	II	III	IV	V	VI
I.	**Costs**						
1.	Capital Cost*	111340					
	Recurring Cost						
a)	Green Fodder Cultivation Cost	9,000	9,000	9,000	9,000	9,000	9,000
b)	Feed Cost						
	Bucks	225	225	225	225	225	225
	Does	5063	5063	5063	5063	5063	5063
	Kids	1200	2400	1200	2400	1200	2400
c)	Medicines/Vet. Charges	1040	1040	1040	1040	1040	1040
d)	Insurance	3376	3376	3376	3376	3376	3376
e)	Misc. (Water/Electricity charges)	520	520	520	520	520	520
f)	Labour wages	15000	15000	15000	15000	15000	15000
	Total Recurring Expenses	35424	36624	35424	36624	35424	36626
	Total Costs	146764	36624	35424	36624	35424	35546
II.	**Benefits**						
a)	Sale of Animals						
	Sale of Adult Buck	0	0	3000	0	3000	0
	Sale of Adult Doe	0	12000	12000	12000	12000	12000
	Sale of Bucklings**	0	25500	25000	56000	29000	56000
	Sale of Doelings**	0	39900	17000	45000	17000	45000
b)	Sale of Gunny Bags	174	208	174	208	174	208
	Total Income	174	107108	61174	113208	58174	113208
c)	Value of Closing Stock						104100
	Scrap Value of Shed & Equipment						11576

Table 17.4 Economics of Goat Farming – Cash Flow Statement

S.No.	Particulars	I	II	III	IV	V	VI
	Total Benefits	174	77608	57174	82208	61174	228884
	New Benefits	146590	42062	21750	46662	25750	193338
	NPV Cost at 15%	231093					
	NPV Benefits at 15%	272797					
	NPV at 15%		41703				
	BCR at 15% DF		1.81:1				
		IRR	25%				

* Excludes the capitalized cost for fodder cultivation for one season, supplementary feed, insurance, veterinary aid, labour wages and misc. expenses.

** Bucklings and Doelings cost is assumed as Rs.1,000.00 – Rs.950.00

Source: Personal Communication, Bhat (2008)

Table 17.5 Economics of Goat Farming – Repayment Schedule

	Bank Loan : Rs126,449					Interest Rate %: 12%		
Years	**Income**	**Expenses**	**Gross surplus**	**Surplus**	**Loan**	**Interest**	**Repayment Principal**	**Net Surplus**
I	174	0	174	126449	15174	0	030	174
II	107108	36624	70484	141622	16995	16994	30622	22868
III	61174	35224	25750	111000	13320	13320	1000	11430
IV	113208	36624	76584	110000	13200	13200	40000	23384
V	78174	35424	42750	70000	8400	8400	20000	14350
VI	354286	36624	317664	50000	6000	6000	50000	261664

Expenses during first year are other than capitalized amount.

Source: Personal Communication, Bhat (2008)

Table 17.6 Economics of Goat Farming – Income to the Farmer

S No.	Particulars	Year1	Year 2	Year 3	Year 4	Year 5	Year 6
1.	Net Income	122,048	199,331	145,206	237,036	156,213	234,683
2.	Monthly Income	10,171	16,611	12,100	19,753	13,017	19,557
3.	Assets with Farmer	174,480	170,480	166,000	162,000	158,000	154,000

Source: Personal Communication, Bhat (2008)

Table 17.7 Economics of Goat Farming – Cost of the Project

S.No.	Particulars	Total Cost (Rs.)
1.	Goat Stall: at 10 ft/does, 20 sft/buck and 4 sft/kids (maximum kid strength assumed 150)	60,000.00
2.	Goats: does at Rs.2,100 and bucks at Rs.2,400 each	109,800.00
3.	Equipments: Feeder and water trays, chop cutter and grinder, cutters, mixers	4,680.00
4.	Fodder: – Fodder cultivation charges at Rs.1,700.00 per acre for 2.5 acre	6,250.00
5.	Feed: Concentrated feed	8,400.00
6.	Insurance Charges: @ 3% of animal cost	3,294.00
7.	Misc. charges Rs.15/- per animal	2,600.00
	Total	**195,024.00**

Source: Personal Communication, Bhat (2008)

Table 17.8 Economics of Goat Farming – Means of Finance

S. No.	Particulars	Total Cost (Rs.)
1.	Farmer's contribution (10%)	19,500.00
2.	Government Subsidy (33.33%)	65,000.00
3.	Loan from the Financial Institutions (56.67%)	110,500.00
	Total	**195,000.00**

Source: Personal Communication, Bhat (2008)

Table 17.9 Economics of Goat Farming – Annual Turnover

Particulars	Year 1	Year 2	Year 3	Year 4	Year 5	Year 6
Annual Sale of:						
Kids	66.00	154.00	85.00	170.00	88.00	171.00
Culled buck	1.00	1.00	2.00	1.00	2.00	1.00
Culled Does	6.00	14.00	36.00	21.00	39.00	19.00
Dead Animal	19.00	12.00	24.00	14.00	29.00	13.00
Manure (Adults)	4.00	1.00	2.00	2.00	3.00	1.00
Manure (Kids)	15.00	11.00	22.00	12.00	24.00	12.00
Milk(1 lit/doe/day)	7,500.00	7,500.00	7,500.00	7,500.00	7,500.00	7,500.00
Selling Price of:						
Kids	1,500.00	1,500.00	1,500.00	1,500.00	1,500.00	1,500.00
Culled buck	2,400.00	2,400.00	2,400.00	2,400.00	2,400.00	2,400.00
Culled Does	2,100.00	2,100.00	2,100.00	2,100.00	2,100.00	2,100.00
Dead Animal	100.00	100.00	100.00	100.00	100.00	100.00

Table 17.9 Contd.

Particulars	Year 1	Year 2	Year 3	Year 4	Year 5	Year 6
Manure (Adults)	3.00	3.00	3.00	3.00	3.00	3.00
Manure (Kids)	2.00	2.00	2.00	2.00	2.00	2.00
Milk(Rate per lit)	7.00	7.00	7.00	7.00	7.00	7.00
Annual Turnover:						
Kids	99,000.00	231,000.00	127,500.00	255,000.00	132,000.00	256,600.00
Culled buck	2,400.00	2,400.00	4,800.00	2,400.00	4,800.00	2,400.00
Culled Does	12,600.00	29,400.00	75,600.00	44,100.00	81,900.00	39,900.00
Dead Animal	1,900.00	1,200.00	2,400.00	1,400.00	2,900.00	1,300.00
Manure (Adults)	12.00	3.00	6.00	6.00	9.00	3.00
Manure (Kids)	30.00	22.00	44.00	24.00	48.00	24.00
Milk	52,500.00	52,500.00	52,500.00	52,500.00	52,500.00	52,500.00
Total Turn Over	**168,442.00**	**264,025.00**	**210,350.00**	**302,930.00**	**221,657.00**	**300,127.00**

Source: Personal Communication, Bhat (2008)

Table 17.10 Economics of Goat Farming Annual Expenditure

Particulars	Year 1	Year 2	Year 3	Year 4	Year 5	Year 6
Fodder cultivation charges Rs.4,000/- per acre for 2.50 acre	10,000.00	10,000.00	10,000.00	10,000.00	10,000.00	10,000.00
Cost of Concentrate: 200 gms per day per doe for 5 months before kidding @ Rs.10 per kg.	7,500.00	9,750.00	10,200.00	10,950.00	10,500.00	10,500.00
Cost of Concentrate: 250 gms per day per buck for 12 months for breeding buck @ Rs.10 per kg.	900.00	1,350.00	1,350.00	1,350.00	1,350.00	1,350.00
Cost of Veterinary Expenses at Rs.100 per animal per annum	5,200.00	5,200.00	5,200.00	5,200.00	5,200.00	5,200.00
Insurance Charges @ 3% of animal cost	3,294.00	3,294.00	3,294.00	3,294.00	3,294.00	3,294.00
Total Expenditure	**26,894.00**	**29,594.00**	**30,044.00**	**30,794.00**	**30,344.00**	**30,344.00**

Source: Personal Communication, Bhat (2008)

Table 17.11 Economics of Goat Farming – Interest Calculation & Repayment Schedule

Particulars	Year 1	Year 2	Year 3	Year 4	Year 5	Year 6
Term Loan						
Opening Balance	0.00	175,500	140,400	105,300	70,200	35,100
Availed during the Year	175,500	0.0	0.0	0.0	0.0	0.0
Repayment during the Year	0.0	35,100	35,100	35,100	35,100	35,100
Closing Balance	175,500	140,400	105,300	70,200	35,100	0.0
Rate on Interest	0.10	0.10	0.10	0.10	0.10	0.10
Interest	8,775	15,795	12,285	8,775	5,265	1,755
Total Interest	**8,775**	**15,795**	**12,285**	**8,775**	**5,265**	**1,755**

Source: Personal Communication, Bhat (2008)

The analysis made in Table 17.2 and subsequent tables indicate that a unit of 50+2 provides a monthly income of Rs 10,000/- to Rs 19,000/- in addition to creation of assets of more than Rs.1,50,000/- with the farmer. Thus goat unit alone is highly financially viable.

17.6 Sanction of Bank Loan and its Disbursement

After ensuring technical feasibility and financial viability, the scheme is sanctioned by the Bank. The loan is disbursed in stages against creation of specific assets, purchase of equipments and animals. The end use of the loan is verified and constant follow-up is done by the bank.

17.7 Unit Cost

Each Regional Office of NABARD has constituted a State Level Unit Cost Committee under the chairmanship of RO-in-charge and with the members from developmental agencies, commercial banks and co-operative banks to review the unit cost of various investments once in six months. The same is circulated among the banks for their guidance.

17.8 Margin Money

NABARD has defined farmers into three different categories and where subsidy is not available the minimum down payment as shown below is collected from the beneficiaries.

S.No.	Category of Farmer	Beneficiary's contribution
a)	Small farmer	5%
b)	Medium farmers	10%
c)	Large farmers	15%

17.8.1 Interest Rate for ultimate borrowers

Banks are free to decide the rate of interest within the overall RBI guidelines. However, for working out the financing viability and bankability of the model project we have assumed the rate of interest as 12% per annum.

17.8.2 Security

Security will be as per NABARD/RBI guidelines issued from time to time.

17.8.3 Repayment Period of Loan

Repayment period depends upon the gross surplus in the scheme. The loans will be repaid in suitable half yearly/annual installments usually within a period of about 5–6 years with a grace period of one year.

17.8.4 Insurance

The animals may be insured annually or on long term master policy, where ever it is applicable. The present rate of insurance premium for non IRDP schemes is 4% per annum.

17.9 Package of Common Management Practices Recommended for Goat rearing

Modern and well established scientific principles, practices and skills should be used to obtain maximum economic benefits from goat rearing. Some of the recommended practices are given here under:

I. *Housing management*

i) Construct shed on dry and properly raised ground.

ii) Avoid water-logging, marshy areas.

iii) In low lying and heavy rainfall areas, the floors should be preferably elevated.

iv) In temperate Himalayan region the floor may be made of wood.

v) The shed should be 10 ft. high and should have good ventilation.

vi) Bucks should be housed in individual pens.

vii) Does can be housed in groups upto 60 per pen.

viii) Provide proper shade and cool drinking water in summer.

ix) Dispose of dung and urine properly.

x) Give adequate space for the animals. The housing space required for goats of various age groups is given in Table 10.1.

xi) Avoid over stocking or crowding

II. *Selection of breeding stock and it's management*

i) Immediately after release of the loan, purchase the stock from a reliable breeders or from nearest livestock market.

ii) Animals in good health and having good physical features must be purchased in consultation with Veterinarian/ Bank's technical officer.

iii) Purchase animals which are ready to breed and in prime stage of production.

iv) Identify the newly purchased animals by suitable identification mark.

v) Vaccinate the newly purchased animals against the diseases

vi) Keep the newly purchased animals under observation for about 15 days and then mix with the general flock.

vii) Unproductive animals should be culled promptly and should be replaced by the newly purchased animals or farm born one

viii) Animals are to be bred at the interval of 8–9 months for maximum productivity.

ix) Cull the old animals at the age of 6 years and above.

x) Avoid the kidding during peak periods of summer and winter.

III. *Feeding management*

i) Ensure Bushes/shrubs for browsing of animals

ii) As an alternative to above, supply of cultivated fodder from own farm or from surrounding farms may be ensured.

iii) Offer roughages *ad lib*.

iv) As a thumb rule $^{2}/_{3}$rd of the energy requirements should be met through roughages. Half of the roughages should be leguminous green fodders and rest half should be grasses/tender tree leaves.

v) In the absence of good quality green fodders, concentrates must be considered to replace them.

vi) Kids should be fed colostrum upto 5 days of age. Later on they can be put on kid starter rations.

vii) Green leguminous fodders should be offered adlib to kids from 15 days onwards.

viii) Provide salt and water to kids at all times

ix) Additional concentrates should be given to bucks and does during breeding season.

x) Care should be taken to meet the nutrient requirements as recommended Table 9.6.

IV. *Protection against diseases*

i) Be on the alert for signs of illness such as reduced feed intake, fever, abnormal discharge or unusual behaviour.

ii) Consult the nearest veterinary aid centre for help if illness is suspected.

iii) Protect the animals against common diseases.

iv) In case of outbreak of contagious diseases, immediately segregate the sick animals from healthy one and take necessary disease control measures.

v) Deworm the animals regularly.

vi) Examine the faeces of adult animals to detect eggs of internal parasites and treat the animals with suitable drugs.

vii) Provide clean and uncontaminated feed and water for minimizing the health disorders.

viii) Strictly follow the recommended vaccine schedule.

V. *Breeding care*

i) It should be planned to obtain 3 kiddings in 2 years period by adopting optimal management conditions.

ii) For every 25 does one buck should be provided in one breeding season.

iii) Breed the animals 12 hours after the onset of the first symptoms of heat for maximum conception.

iv) Unbreedable animals must be examined thoroughly as directed by veterinary doctor for prompt elimination of causes for anoestrum or cull them if necessary.

VI. *Care during pregnancy*

In advanced stage of pregnancy the does must be transferred to either kidding pens or separately earmarked space for kidding with in the main shed after thoroughly disinfecting it. After kidding, the does should be provided with warm bran mash for two days.

VII. *Care of kids*

i) Take care of new born kids by providing guard rails.

ii) Treat/disinfect the naval cord with tincture of iodine as soon as it is cut with a sharp knife.

iii) Protect the kids from extreme weather conditions, particularly during the first two months.

iv) Dehorn the kids during first two weeks of age

v) Male kids should be castrated for better quality meat production.

vi) Vaccinate the kids as per the recommended schedule

vii) Wean the kids at the age of 8 weeks

viii) Proper selection of kids on the basis of initial body weight and weaning weight should be initiated by maintaining appropriate records for replacing the culled adult stock as breeders.

ix) Additional feed requirements of lactating does must be ensured for proper nursing of all the kids born.

VIII. *Marketing*

The marketable products of goat farming include the fattened kids, manure, and culled animals. Marketing avenues for the above products are slaughter houses and individual meat consuming customers and agriculture farms. Therefore availability of either slaughtering facilities or traders who will purchase live animals should be ensured to convert the fatteners into wholesome meat and meat products. Further, demand for manure from nearby agriculture farms must also be ensured.

CHAPTER 18

MARKETING OF GOATS AND THEIR PRODUCTS

The population of goats in India has shown increasing trend from 47.20 million in 1951 to 115.279 million in 1992 indicating an increase of 144%. The cattle and sheep population during this period showed an increase of only 31.7% and reduction of 15% respectively. Goat population in India accounts for 19.8% of the population of 581.317 million in the world in 1992. The world goat population in 2003 increased to 771.5 million and there was proportionately further increase in the country (124.36 million) as per FAO (2006).

With the economic developments, the house-wives living in big towns and cities have been looking for meat cuts that require less preparation for cooking and to better retailing services in general. Thus, the meat marketing system is enlarged to provide wider range of meat and meat products including that of goat meat and in particular the range of processing facilities including canning, sausage making, smoking, precooking and other related enterprises.

Marketing of live goats and goat meat in India is in a primitive way though the goat population has been increasing. Goats are not raised for meat purpose and thus the dressed weight of carcasses is less and there is lack of marbelling in meat, this leads to receipt of less price from vendors and low per capita availability of meat. The transport facilities for live goats and their meat are inadequate for domestic consumption as well as for export. Same is the situation for preservation and storage of meat. There are many problems in marketing of goat commencing from procurement of goats to sale of meat, including exploitation by middle-men and lack of market infrastructure, movement, sale at slaughter houses, enforcement of quality control on meat, absence of grading of goats and meat.

18.1 Marketing Agencies

Producer can sell his goat and hair through a number of different agencies. These include private enterprenuers and cooperative associations.

18.1.1 Private Enterprises

These can be divided into local dealers, state dealers, central market dealers, commission merchants, brokers and manufacturers.

18.1.2 Cooperative Associations

These may also be divided into local, state wide and nation wide. There are now agencies which correspond to all of these designations. All of them may perform a similar service, *i.e.* assist in getting the hair/wool from the producer to the manufacturer, but the objective may be different and the character of the services vary widely.

There are no organised markets in India. The village "beoparies" or agents of big merchants or of mills buy hair/wool from the goat farmers mostly on per goat or per fleece basis, against money advanced to the goat farmers.

18.2 Marketing of Live Goats and Goat Meat

18.2.1 Marketing of Live Goats

Goats from villages are taken to the nearby livestock markets by the goat farmers/keepers or the small itineraries (traders). The goat farmers sell their goats to the local traders or in the market whenever they are in need of money without taking adequate measures for obtaining the optimum price for their animals. Normally, they do not prepare animals for the slaughter purposes by feeding extra energy in the form of concentrates. In the absence of any coordiation among the goat farmers, there is no cooperative system of marketing of their goats. They are also illiterate and are not well versed with the market information and they squarely depend upon the brokers/buyers in the local market.

The local traders, who buy goats from goat farmers/keepers also look forward to brokers as they are also not fully conversant with the prices prevailing in the major markets and the price at which they can sell the goats in local markets. The goats, which are assembled in the local markets are taken to the local slaughter house and some of them are sent to the big town/city (terminal) markets. Some of the big traders take the animals to those markets and at times the wholesalers also collect animals from small markets to market them in big markets directly or through commission agents. In the small markets the mediators are called "brokers or middle men" and in the big markets they are called as "commission agents". Though, these people are supposed to help the producers/local traders/big traders to market the goats in the respective markets and charge the fee from them. Most of the mediators collect commission from both the sellers and the buyers.

So far there has been no undertaking or contractual arrangement between the goat farmers and the butchers or the management of the big meat processing units for purchase of goats on regular basis that too laying down number of animals required, the type of animals and the size of the animals, the delivery and payment schedules etc.

The livestock markets are mostly weekly markets and they are owned privately or trust or by local bodies (panchayat/municipality/corporation). Most of the livestock markets where goats are being marketed do not have the facilities even for shelter. The Bureau of Indian Standards (the then ISI) has laid down specifications for regulated market yards for cattle under 1788–1961 and in spite of it, no tangible achievement has been made to adopt the minimum requirements for marketing of goats. Some of the agricultural marketing departments of states have also brought marketing of livestock under Agricultural Produce Markets Act (APMA) notified in different years by various states. They are unable still to bring all the livestock markets under their fold as some of them are controlled by local bodies or fair committees or private individuals. In the absence of any regulation being complied with and in the absence of certain State Governments bringing livestock under the above said Act, the marketing of goats is still in a primitive way and it is yet to receive adequate attention.

The goats are assembled in city markets and they are normally grouped in lots of 10 or 20. Sometimes they are also sold as truck load numbering around 200.

Though, there is no system of licensing or legalising the merchants, brokers, workers for transshipping and stall dealers in small towns, but all the functionaries in the markets particularly those maintained by local authorities or the State Governments governed by the APMA are licensed and annual fee for the same is collected. Most of the goats which are brought to livestock markets are sold in the same market and few of them are taken back or moved to the neighbouring markets.

In most of the markets, the transaction takes place after examination of the animals by the buyers, though the brokers/commission agents help them in negotiating prices. In spite of the sellers and buyers being mostly illiterate, it is heartening to note that the meat yield could be arrived, at the stage of live goat itself based on the examination of animals for its musculature, fat distribution etc. The prices are settled by one of the following ways in the markets:

(i) Direct or private treaty where buyers announce their offers publically or through secret whispers to sellers,

(ii) Negotiations through brokers directly,

(iii) Negotiations under the cover or hatha system, and

(iv) Auction.

Though the auction or the bid system of purchases is ideal, most of the transactions take place under the cover or hatha system and most of it through the brokers. It is difficult for even an educated person to know from outside what happens under the cover or hatha system. In some cases advance payment is made to the small traders/goat farmers and in the regulated markets, it is only formalising the deal that takes place. Goats are also sold in some of the annual fairs in the country.

In several markets, the real number of animals transacted is not known as the collection of fee in the markets is by auction and the successful bidder do not maintain proper accounts with a fear that profits will be known to all the concerned including the bidder during the next year. This coupled with the frequency of the markets has created lot of problems for improving the conditions of the markets in the country. In the present system of marketing which is in very primitive way, all the functionaries in the marketing have more than their due share in the buyer's rupee resulting transfer of very small sum to the producers. This has resulted in the inadequate attention to improve the conditions of goat meat producers in the country.

18.3 Marketing of Goat Meat

Goat meat is marketed in villages by slaughtering one or two animals once in a week or a group of people joining together and slaughter the animals and sharing the cost of the meat so obtained. There is not much cost on meat due to slaughter of animals in villages. Though, the total produce is marketed there itself, the main problem that could arise in marketing of skin is on account of spoilage of skin due to inadequate preservation. In small towns goat meat is marketed directly to the consumers through unhygienic meat shops. The condition of the room is bad to worse. The equipment used will be rusted and/or not sharp. The facility for hanging carcass or meat is not appealing. Since the time gap between the slaughter and the sale is very short, the deterioration in quality of meat is less. In big towns and cities, most of the meat is also sold on the same day and the quantity of the meat, retained for subsequent day or later to be marketed in the same palce, will be less then 10%. Most of the problems that arise in spoilage will be restricted to 10% of meat, which is sold on next day or on subsequent days. The sale of meat in supermarkets has not picked up significantly so far. The major problems involved in marketing of meat are:

1. Lack of education at the level of butchers.
2. Absence of hygiene or poor maintenance in slaughter houses and lack of transport and storage facilities.
3. Most of the meat shops are also not ideally suited though meat is highly perishable.
4. Cost of transport and refrigeration.
5. Ignorance of consumers about the quality of the meat produced and marketed.

Meat is produced in the slaughter house. However, the meat animals are produced in distant rural places and are required to be brought to place of slaughter and sold to butcher. The animals are generally purchased by middle men and transported on foot or by trucks to the nearest slaughter house where they are sold in lots. Meat produced is also required to be transported to the selling centres and cold storages as the case may be. Thus, the transportation of livestock from their producing centres to the slaughter house and transportation of meat from slaughter house to the market places forms an important aspect of meat trade. Unless, specific requisite care is taken during transportation, there is likely to be a great loss, depreciation, contamination and deterioration endangering the health of the consumer. Further, this involves considerable economic losses due to mortality and shrinkage in live weight. While transporting the live animals, care should be taken to eliminate all factors which may contribute to deterioration of the condition of animals' health and bodily condition which ultimately influence the quality of meat. There is thus a need for providing market yards for sale of live animals in major goat rearing areas. The market yard should have facilities for feeding, watering and holding animals for a few days. Meat when produced is very vulnerable and very much prone to get affected with extrinsic factors like temperature, contamination, dirt and microbes, etc.

Studies conducted on contamination indicate the following losses:

Transport and storage	50% or over
Dirt and skins of animals	33% (approx.)
Pollution in abattoir atmosphere	5
Visceral content in normal conditions	3
Miscellaneous utensil personal etc.	3
Halving, quartering and packing carcasses	2

It is therefore, essential that utmost care be taken during transportation to safeguard meat from contamination. Meat carcasses must be transported on hooks in specially built vans with metal bodies and with ventilation arrangements. Whenever possible, transportation of meat by a refrigerated van is the ideal form of transport. Meat when being sold in butcher shop is in the naked form and a major quantity of meat sold thus exposed to atmosphere, temperature, dust, infection, etc. As such, this kind of marketing is not ideal. However, the trend of sale and purchase especially in urban city complexes are fast changing.

Processed meat in small and large packs and in semi-prepared conditions in proper packaging is being marketed. This kind of marketing has many advantages. The consumer demand for smaller packs can be prepared in the processing unit itself, where high standards of hygienic conditions can easily be maintained, and thereby completely eliminating contamination during transportation. Besides, in smaller packages, as per the needs of consumer demand, much larger quantity can be transported from processing units to greater consumer areas, economising on the transport cost. At sale outlets, these smaller packs are preserved in a proper chiller cooler for days or weeks together. This is a great breakthrough in marketing of meat as compared to old traditional system of sale at the butcher's shop. Goat farmers, cooperatives can set up market yards and facilities for slaughter, meat packaging and utilization of slaughter house by-products in the major goat rearing areas with a view to have organised meat marketing. Goat and their products can also be marketed abroad more profitably by these cooperatives. This will also generate employment opportunities to the needy persons.

A few or all goat breeders should join together to form an export group and select an experienced person who knows the products and the foreign markets. Each goat breeder should contribute a reasonable amount and make a central fund for meeting expenses of the export group. A committee of goat breeders may supervise the affairs. Price must be fixed by this committee according to the quality of the hair/wool, skin and buck, etc. A price list and illustrated catalogue should be published and the cost of the same be met from the central fund of the cooperative export group. Success of such a group will be excellent, if pursued persistently, honestly and without any rift or discontent among the participating members. Such cooperative societies can provide promising results in eliminating middlemen's profit and provide more returns to the actual producers.

This approach will be healthy and effective only when the participating members work with a sense of responsibility, honesty, fair trading and sacrifice for the prosperity of industry leaving aside individual self-interest or greed when circumstances demand.

18.4 Export of Live Goats and Goat Meat

18.4.1 Export of live goats

Live goats were exported from India to various Middle East countries since 70's. The government of India used to lay down the limits for export of goats. Export of goats to neighbouring deficient countries like Pakistan and Bangladesh takes place as unrecorded trade across the frontiers. The number of goats exported through this route to the neighboring countries is not recorded. There has been certain restrictions for export of goats like quarantine and certificate of health being issued by a veterinarian under the export policy. It is evident that export of goats is not economical compared to the export of goat meat.

18.4.2 Export of goat meat

Export of meat, as it is to Middle East countries, has started in a small way around the year 1973. There has been rapid increase in the export of meat while the export of goat meat is restricted on account of or by way of ceiling laid down by the Government of India. This is in order to control the domestic prices of mutton and goat meat. While there were few problems in export of live goats like sourcing animals from disease free zone, goats weighing more than 35 kg, quarantine, health of goats. The problems in export of meat have been tremendous commencing from poor hygienic conditions prevailing in several slaughter houses, inadequate inspection, meat processing units not conforming to the requirements laid down, lack of education and conception at the level of workers handling meat, inadequate exposure of the workers on the hygienic conditions, processing and handling, packing of meat. All these are relevant as the importers have been complaining of receipt of poor quality meat from India. Though it is a normal phenomenon for shifting of exports from one country to the other due to the economic interest, but drastic reduction in the exports, on account of poor quality is certainly going to hamper the prospects of export to other countries. In order to maintain the uniform level of exports, regular and prompt supplies and consistent quality play a prime role. Therefore, the progress depends upon capable, efficient enterprises and on resourceful and experienced men to run the meat industry. The Govt. of India had laid down rules for enforcement of quality on meat meant for export by the Directorate of Marketing and inspection. Export inspection agencies and State Directorate of Animal Husbandry have been recognized as agencies to carry out preshipment inspection on meat prior to export. Still the rules as anticipated are not being implemented by the State Government, which are carrying out the inspections presently. AGMARK Specification for Mutton (Sheep) and Chevon (Goat-Meat) is given in Table 18.1

AGMARK Specification for Mutton (Sheep) and Chevon (Goat-Meat)

General

Mutton and chevon shall be obtained from healthy animals and slaughter in slaughtered houses. The animals shall be subjected to ante-mortem inspections by Veterinarians designated and/or of recognized local agency in respective states by the Agricultural Marketing Adviser. It shall not be treated with coloures, dyes, additives, preservatives and chemicals.

Special

Types : Mutton shall be of the following types:–

I. Type A : Fresh, chilled carcasses
II. Type B : Fresh, frozen carcasses
III. Type C : (Bone-in) fresh, chilled
IV. Type D : (Bone-in) fresh, frozen
V. Tyep E : Deboned (Boneless) fresh, chilled
VI. Type F : Deboned (Boneless) fresh frozen

Table 18.1 Mutton Shall be of the Following Grade and Standards

Commercial name	Grade designation	Special characteristics	Bacteriological standards
Mutton	Standard grade	The dressed carcasses or meat shall have/be	The sample per lot drawn and tested shall fulfil the following requirements
		Boneless cuts to be entirely free from bone pieces, wood dust, metal pieces or other undersirable matter	Total plate count not exceeding 10 microganisms per gram
Chevon	Standard Grade	Free from slime, discoloration, malodour and structural alterations	*Escherichia coli* not exceeding 100 per gram
		Lean	
		Firmness in consistency *i.e.* will not pit on pressure	

CHAPTER 19

GOAT HAIR PROCESSING AND TECHNOLOGY

The hair/fibres harvested from goats can be classified into two major types *i.e.* Mohair and Pashmina (cashmere). Mohair primarily produced from Angora goats is larger in quantity as compared to Pashmina. The major world centres for production of mohair are South Africa and the USA. There has been rapid rise in Angora goat numbers and mohair production in Australia and New Zealand. The major uses of mohair are hand knitting yarn followed by mens' and women's clothing and woven furnishings. Major centres for processing mohair have traditionally been in Europe (U.K., Belgium, France, West Germany, Spain and Italy). Lately India and Republics, which constituted the formerly USSR, have become major importers of mohair.

Major centres of production of cashmere (Pashmina) are China, Mangolia, Iran and Afganisthan. Smaller quantities are also produced in the Republic of former USSR, India, Turkey and Pakistan. Countries like Australia, New Zealand, have taken up the production of cashmere. China continues to be the major supplier with more than 50% of world cashmere production.

The method of harvesting cashmere and the interaction of this with the problem of cold stress will be of paramount importance if the industry expands outside traditional centres of production. Some of the related areas are the photoperiod-induced seasonality of cashmere production. Animals may respond to seasonal climatic changes and the period of down production may not conflict with the high nutrient demands of parturition and lactation. Seasonal shedding permits the harvesting by combing in traditional centres of production. The seasonality can be altered through the use of melatonin.

The work done on quality of mohair in India (AICRP on goats) indicate that the quality traits of mohair in Himachal Pradesh (HP) and Texas Angora revealed that the mohair from Texas Angora was more greasy, densely populated and comparatively with more medulated fibres. It was observed that the purebred female kids from two strains with respect to fibre quality did not differ significantly. However, the crossbreds obtained by using bucks from these two sources produced different results. The ½ Angoras produced by H.P. Angora bucks were significantly inferior in staple length, fibre diameter and medullation to that of crossbreds obtained by using Texas bucks.

The ¾ Angora animals produced by using either purely HP Angora or Texas Angora bucks significantly differed in staple length, fibre diameter and medullation. The progeny of Texas Angora males was superior over HP Angora bred animals in all the fibre characters studied. Although, number of observations were limited in the case of $^7/_8$ Angora, the trend was much similar to that of ¾ Angora.

19.1 Skin Follicular Characters

The quality and quantity of mohair is principally governed by the number and type of follicles present in the skin. In view of crossing non-woolly goats with wool (mohair) producing goats, the study of skin follicles, particularly primary and secondary follicles assumes a significant role in judging the production potential and in serving as a tool for early selection in goats. The overall performance of skin follicular characters in Angora, local and crossbreds indicated that the number of primary follicles/mm^2 is practically same, while the secondary follicles/mm^2 go on increasing with the increase in Angora inheritance. This has reflected in increasing s/p ratio from 0.96 in local to 6.98 in $^7/_8$ Angora. The secondary follicles and s/p ratio in the progeny sired by Texas Angora particularly in ½ Angora was boosted to a great extent.

Comparison of follicular characteristics of HP Angora and Texas Angora, revealed that the two strains of Angora were not significantly different for s/p ratio. The sex differences in both the strains, however, were found significant for secondary follicles/mm^2, total follicles/mm^2, s/p ratio and percent primaries. The number of primary follicles/mm^2 were not affected by sex of the animal.

19.2 Goat Shearing

A trial on Angora was conducted to ascertain the effect of animal shearing. It was observed that shearing once a year is not comfortable for the animals under the climatic conditions of Rahuri (Maharashtra). Also, there was loss of mohair due to shedding and the yields were less. The infestation of ectoparasites was also more in their animals. Considering this, they were shorn twice-a-year at an interval of 6 months, firstly in the month of April and secondly in October. There was no problem in following these shearing seasons in case of Angora. However, in crossbreds shedding occurred predominantly from August to October. To overcome this problem, the shearing seasons were changed to the months of January and July.

The shedding tendency in crossbreds was mostly associated with the proper type of crossbreds lacking uniformity in their fibres. The stress factors such as climate, nutrition and physiological conditions like pregnancy etc. contribute to enhance shedding in crossbreds. The shedding tendency goes on decreasing in higher grades. Also when strong fibre type bucks i.e. bucks having fibre diameter above 35 microns were used as sire, both in ¾ and $^{7}/_{8}$ Angora, the incidence of shedding was reduced to about half.

Lupton (1992) has given ranges for important characteristics of Cashmere goat and cashmere (Table 19.1). The characters related to mohair and cashmere production and quality are highly heritable and will respond to selection (Table 19.2).

Table 19.1 Ranges for Important Cashmere Goat and Cashmere Characteristics

Body weight (kg)	30 – 70
Fleece weight (hair-in, unscoured) (g)	80 – 1,000
Average fibre diameter (mm)	
Cashmere	11 – 20
Guard hair	50 – 100
Range of individual fibre diameters (mm)	
Cashmere	5 – 30
Guard hair	30 – 200
Scoured yield (%)	65 – 90
Grease content (%)	1 – 10
Suint (%)	1 – 10
Mechanical yield (% of hair-in, scoured)	
Fibre length (mm)	15 – 90
Cashmere	12 – 90
Guard hair	50 – 200
Medullation (%)	
Cashmere	0 – 3
Guard hair	90
Colour	White, tan, brown, gray, black
Style and character	Cashnere (Cashgora, Mohair)

Table 19.2 Ranges for Important Cashmere Goat and Cashmere Characteristics

Body weight (kg)	
Kids (12 months)	13 – 27
Young goats (18 months)	27 – 68
Mature does	29 – 57
Mature bucks	57 – 90
Greasy fleece weight (kg)	
Kids (12 months)	0.6 – 1.2
Young goats (18 months)	1.6 – 6.8
Mature does	1.8 – 3.2
Mature bucks	4.5 – 11.3
Yield %	
Kids (12 months)	70 – 76
Young goats (18 months)	72 – 78
Mature does	74 – 80
Mature bucks	
Average diameter (µm)	
Kids (6–12 months)	23 – 29
Young goats (18 months)	30 – 34
Mature does	32 – 38
Mature bucks	35 – 45
Lock length (mm)	100 – 150
Lock type	Ringlet, flat, intermediate, unclassified
Medullation (%)	0 – 2
Colour	White (brown, stained)
Specific gravity	1.32
Regain at 65% RH and 21°C (%)	14.50 – 17.5
Tenacity (cN/tex)	14.70
Elongation at break (%)	42.00
Work at rupture (cN/tex)	2.65
Initial modulus (cN/tex)	353.00

The AGMARK Grade Designation and Definition of Quality of various classes goat hair is given in Table 19.3–19.6.

Table 19.3 Grade Designation and Definition of Quality of Indian Clipped Goat Hair

Grade designation	Colour	Length	General characteristics
Extra long	White	Above 100 mm	Finres shall be clean, dry in feel and not moist and shall not contain any extraneous matter likedirt, dust, sticks etc., with a minimum yield of 85%
Long	White	75-100 mm	
Medium	White	35-75 mm	

Table 19.3 (Contd.)

Grade designation	Colour	Length	General characteristics
Short	White	Less than 35 mm	Extra long hair shall contain a mini mum of 60% of hair above 100 mm. Length long hair shall contain a minimum of 60% ofhair above 75 mm in length
Extra long	Black	Above 100 mm	
Long	Black	75-100 mm	
Medium	Black	35-75 mm	Medium hair shall contain a minimum of 60% of goat hair above 35 mm in length
Short	Black	35 mm less than	
Extra long	Grey	Above 100 mm	
Long	Grey	75-100 mm	
Medium	Grey	35-75 mm	
Short	Grey	Less than 35 mm	
Extra long	@ coloured	Above 100 mm	Any admixture of pulled and/or tannery hair shall not be permitted
Long	@ coloured	75-100 mm	
Medium	@ coloured	35-75 mm	
Short	@ coloured	less than 35 mm	

@ Coloured hair shall mean goat hair of all colours which do not fallunder the categories of white, black or grey

Note 1. The following colour tolerance limits shall be allowed:-

a) 10% of non-white in whites

b) 10% of non-black in blacks

c) White or black not falling under these categories will be treated as grey.

Note 2. Admixture with disease infested goat hair shall not be permitted

Table 19.4 Grade Designation and Definition of Quality of Indian mixed* Goat Hair Clipped-pulled)

Grade designation	Colour	Length	General characteristics
Extra long	White	Above 100 mm	Fibres shall be clean, dry in feel and not moist and shall not contain any extraneous matter like dirt, dust, sticks etc., with a minimum yield of 85%
Long	White	75–100 mm	
Medium	White	35–mm	
Short	White	Less than 35 mm	Extra long hair shall contain a mini mum of 60% of hair above 100 mm length. Long hair shall contain a minimum of 60% ofhair above 75 mm in length
Extra long	Black	Above 100 mm	
Long	Black	75–100 mm	
Medium	Black	35–75 mm	Medium hair shall contain a minimum of 60% of goat hair above 35 mm in length
Short	Black	Less than 35 mm	

Table 19.4 (contd.)

Grade designation	Colour	Length	General characteristics
Extra long	Grey	Above 100 mm	
Long	Grey	75-100 mm	
Medium	Grey	35-75 mm	
Short	Grey	Less than 35 mm	
Extra long	@ coloured	Above 100 mm	Any admixture of tannery hair shall not be permitted
Long	@ coloured	75-100 mm	
Medium	@ coloured	35-75 mm	
Short	@ coloured	Less than 35 mm	

* Mixed goat hair shall mean goat hair containing a mixture of clipped and pulled goat hair in any proportion.

@ Coloured hair shall mean goat hair of all colours which do not fall under the categories of white, black or grey

Note 1. The following colour tolerance limits shall be allowed:-

a) 10% of non-white in whites

b) 10% of non-black in blacks

c) White or black not falling under these categories will be treated as grey.

Note 2. Admixture with disease infested goat hair shall not be permitted

Table 19.5 Grade Designation and Definition of Quality of Indian Tannery Goat Hair

Grade designation	Colour	Length	General characteristics
Medium	White	Above 35 mm	The fibres shall be clean, dry in feel and not moist and shall not contain any extraneous matter like dirt, dust, sticks etc., with a minimum yield of 55%
Short	White	Less than 35 mm	
Medium	Black	Above 35 mm	
Short	Black	Less than 35 mm	
Medium	Grey	Above 35 mm	
Short	Grey	Less than 35 mm	Medium hair shall contain a minimum of 60% of minimum hair above 35 mm length
Medium	@ coloured	Above 35 mm	Tannery goat hair may contain other allied animal fibres to the extent of 50%
Short	@ coloured	Less than 35 mm	

@Coloured hair shall mean goat hair of all colours which do not fall under the categories of white, black or grey

Note 1. The following colour tolerance limits shall be allowed:–

a) 10% of non-white in whites

b) 10% of non-black in blacks

c) White or black not falling under these categories will be treated as grey.

Note 2. Admixture with disease infested goat hair shall not be permitted

Table 19.6 Grade Designation and Definition of Quality of Indian Pulled Goat Hair

Grade designation	Colour	Length	General characteristics
Extra long	White	Above 100 mm	Fibres shall be clean, dry in feel
Long	White	75–100 mm	and not moist and shall not contain
Medium	White	35–75 mm	any extraneous matter like dirt, etc.,
dust,sticks			with a minimum yield of 85%
Short	White	Less than 35 mm	Extra long hair shall contain a mini-
Extra long	Black	Above 100 mm	mum of 60% of hair above 100 mm
Long	Black	75–100 mm	in length. Long hair shall contain a
Medium	Black	35–75 mm	minimum of 60% of hair above
Short	Black	Less than 35 mm	75 mm in length
Extra long	Grey	Above 100 mm	Any admixture of tannery hair shall
Long	Grey	75–100 mm	not be permitted
Medium	Grey	35–75 mm	
Short	Grey	Less than 35 mm	
Extra long	@ coloured	Above 100 mm	
Long	@ coloured	75–100 mm	
Medium	@ coloured	35–75 mm	
Short	@ coloured	Less than 35 mm	

@Coloured hair shall mean goat hair of all colours which do not fall under the categories of white, black or grey

Note 1. The following colour tolerance limits shall be allowed:-

a) 10% of non-white in whites

b) 10% of non-black in blacks

c) White or black not falling under these categories will be treated as grey.

Note 2. Admixture with disease infested goat hair shall not be permitted

CHAPTER 20

BIOTECHNOLOGICAL APPROACHES FOR ENHANCING GOAT PRODUCTION

20.1 Introduction

Human population growth, increasing urbanization and rising incomes are fuelling a massive increase in demand for food of animal origin in developing countries. All over the world, livestock production is growing faster than any other sector, and by 2020 the livestock sector is predicted to become the most important agricultural sector. In our country, the demand for meat is being satisfied by small ruminant species, *viz.* sheep and goats. The vast biodiversity in sheep and goats in India is represented by 42 and 20 breeds, respectively. According to the estimated figures of FAO statistics (2000), the current populations in our country are about 57.9 million sheep and 123 million goats, producing meat production 0.70 MMT. Though the impressive figures, the per capita availability of meat in India is 9.72g against 25g per capita recommended by ICAR. The gap between requirement and availability is indicative of an urgent need for increasing meat production in the country to meet the ever-increasing internal demand of our huge population. The demand is again augmented by great scope for meat export and its potential to earn foreign exchange. The heavy demand can only be met by applying integrated approach for small ruminant production in conjunction with biotechnological methods.

Rapid advances in molecular biology and biotechnological tools and further developments in reproductive biology provide new powerful tools for enhancing the meat production form these animals. The likely areas of manipulation are as follows:

- Enhancing breeding practices
- Increasing animal growth and food production
- Embryo multiplication and transfer
- Transgenesis
- Increasing the levels and types of nutrients in animal feeds
- Improving animal health to produce healthier, more productive farm animals.

20.2 Breeding Practices and Biotechonology

The animal characteristics for meat production are determined by the combined interaction of many genes with the environment. In order to augment meat production from small ruminants, the genetic improvement of locally adapted breeds will be important to realizing sustainable production systems.

Genetic improvement of livestock depends on genetic variation and effective methods for exploiting this variation. Genetic diversity constitutes a buffer against changes in the environment and is a key in selection and breeding for adaptability and production on a range of environments. Traditional Breeding programmes based upon performance recording and cross breeding have led to substantial improvements in animal production during the past few decades.

20.3 Multiple Ovulation Embryo Transfer and Open Nucleus Breeding System

Breeding System

Multiple ovulation embryo transfer (MOET) is a composite technology, which includes superovulation, fertilization, embryo recovery, short-term *in vitro* culture of embryos, embryo freezing and embryo transfer. MOET offers large number of offsprings produced by valuable females, increasing the population base of rare or endangered breeds or species, *ex-situ* preservation of endangered populations, progeny testing of females and increasing rates of genetic improvement in breeding programmes. In any case, the generally small herds and uncontrolled breeding in communal grazing situations preclude implementation of progeny testing. Smith (1988) suggested the Open Nucleus Breeding System (ONBS), which can be especially valuable in developing countries where the use of AI has been a failure.

The ONBS concept comprises a nucleus herd established under controlled conditions to facilitate selection. The nucleus is established from the "best" animals screened from the base population for outstanding females. These are then recorded individually and the best individuals chosen to form the elite herd of the nucleus. If ET is possible, the elite female herd is used through MOET with superior sires to produce embryos, which are implanted to recipient females. The resulting offspring are reared, recorded and the males among them are evaluated using the performance of their sibs and paternal half sibs and their own performance. Such an elite group of males with high breeding values for the specific trait can be used in the base population for genetic improvement through natural service or AI. Although, MOET improves the rate of progress substantially, it is also possible to operate an ONBS without ET technology, especially in small ruminants, where reproductive rates are high. Such schemes are being tried for sheep in West Asia by FAO (Jasiorowski, 1990) and in Africa (Yapi *et al.,* 1994). However availability of AI and ET, in addition to increasing rates of genetic gain, enhances the flexibility of the system.

Modern biotechnology can help and develop hardier and more productive stock more quickly than possible with traditional methods. In classical breeding system, only animals that are closely related may be crossbred. However, new genetic engineering techniques allow individual genes to be swapped between unrelated species, so that animal breeders can incorporate new features that would normally not be available in the native animals. The result is "transgenic" animals-animals carrying genes transferred from another species.

Scientists have already produced transgenic pigs with superior growth rates and meat quality. Efforts are on to develop gene-transfer methods that could produce sheep that would grow more wool, or that would be better prime lamb producers.

Indicator traits

Indicator traits are characteristics that are genetically correlated to traits of economic importance and are easier to measure. Such traits are usually not the target of genetic improvements but provide an indirect means of improving a targeted trait. Blair *et al.* (1990) reviewed some physiological and metabolic characteristics, which might be considered as potential indicator traits. Testicular size in rams or bulls or FSH in ewe lambs (Bodin *et al.,* 1986) has potential as indirect predictors of fertility. Indicator traits can improve genetic response by increasing accuracy of selection and reducing generation interval. The value of an indicator trait will depend largely on the magnitude of co-heritability and the genetic correlation between the two traits (Woolliams and Smith, 1988). Woolliams and Smith (1988) concluded that with high co-heritability, selection for the indicator trait alone can result in greater rates of response than is possible with progeny testing, especially when breeding values are not accurately measured by progeny testing.

Genetic markers and marker-assisted selection

A genetic marker for a trait is a DNA segment, which is associated with the trait and segregates in a predictable pattern. Genetic markers facilitate the "tagging" of individual genes or small chromosome segments containing genes, which influence the trait of interest. Availability of large numbers of such markers has enhanced the likelihood of detection of major genes influencing quantitative traits. The method involves screening the genome for genes with a large effect on traits of economic importance through a procedure known as linkage analysis (Paterson *et al.,* 1988). The chances of major genes existing for most traits of interest, and of finding them are considered to be high (Mackinnon, 1992). The process of selection for a particular trait using genetic marker is called marker assisted selection (MAS). MAS can accelerate the rate of genetic progress by increasing accuracy of selection and by reducing the generation interval (Smith and Simpson, 1986). However the benefit of MAS is greatest for traits with low heritability and when the marker explains a larger proportion of the genetic variance than does the economic trait. MAS also facilitates increased rate of genetic gain by allowing measurement in early age thereby reducing generation interval.

Marker identification and use should enhance future prospects for breeding for such traits as meat tenderness, lean meat, meat genotype, immune response genes, tolerance or resistance to environmental stresses, including disease. Already, identification of carrier of genes of resistance and introduction of such genes into a population seems feasible for resistance against *Trichostrongylus colubiformis* and *Heamonchus contortus* (Gogolin Evens *et al*., 1990). It would also be possible to eliminate factors predisposing sheep to Listeriosis or Salmonellosis (Blancou, 1990). Research is currently underway at ILCA for resistance to endoparasites in Red Maasai sheep.

Transgenic animal

A transgenic animal is one, whose hereditary DNA has been augmented by addition of DNA from a source other than parental germplasm through recombinant DNA techniques. Transfer of genes or gene constructs allows the manipulation of individual genes rather than entire genomes. There have been dramatic advances in gene transfer technology in the past two decades since the first successful transfer was carried out in mice in 1980 (Palmiter *et al.,* 1982; Jaenisch, 1988). Successful production of transgenic livestock has been reported for pig, sheep, rabbits and cattle. Although, transgenic cattle and sheep have been successfully produced, the procedure is still inefficient in these species (Niemann *et al.,* 1994).

In livestock, the ability to insert new genes for such economically important characteristic such as fecundity, growth hormone, resistance for or tolerance of other environmental stresses would represent a major breakthrough in the breeding of commercially superior stock. To date, the majority of genes transferred into sheep have been growth hormone encoding gene constructs. Unfortunately, in most cases the elevated growth hormone level have resulted into a clinical diabetes situation leading to a early death of the trasgenic sheep (Rexroad *et al.*, 1990). Transgenic sheep have been generated which express the *visna* virus envelope gene (Clement *et al.*, 1994). Recent advances in *in vitro* fertilization and maturation technology will increase the number of zygotes available for gene transfer purposes. This, plus the utilization of embryonic stem cell (Stice *et al.*, 1994) and primordial germ cell (Stokes *et al.*, 1994) technologies should enhance the efficiency of gene transfer in cattle and sheep considerably.

Genetic characterization of animal genetic resources

Our country is endowed with vast majority of the domestic animal diversity including 42 sheep and 20 goat breeds over 18 micro-agro-climatic zones. Some livestock breeds in the country are in immediate danger of loss through indiscriminate crossbreeding with exotic breeds. The importance of indigenous livestock breeds lies in their adaptation to local biotic and abiotic stresses and to traditional husbandry systems. However, most of these animal genetic resources are still not characterized and boundaries between distinct populations are unclear. In such cases, breeds are defined on the basis of subjective data and information obtained from local communities. It is in this sphere that biotechnology has an important role in identification of different breeds.

The relative genetic distances of such populations from each other measure the genetic uniqueness of populations. Polymorphism in gene products such as enzymes, blood group systems and leukocyte antigens which have traditionally been used for measuring genetic distances, are being rapidly replaced by polymorphism at the level of nuclear (Jeffreys and Morton, 1987) and mitochondria (Loftus *et al.*, 1994) DNA as a source of information. The first DNA polymorphism to be used widely for genome characterization and analysis was the restriction fragment length polymorphism (RFLP) (Southern, 1975), which detects variations ranging from gross rearrangements to single bade changes. Minisatellites sequences of around 60 bases repeated many hundred or thousand times at one unique locus within the genome have been used to generate DNA fingerprints typical of individuals within species (Jeffreys and Morton, 1987). Microsatellites (Weber and May, 1989) repeats of simple sequences, the commonest being dinucleotide repeates are abundant in genomes of all higher organisms, including livestock. Ease of identification and of sequence determination (Moore *et al.*, 1992) and need for only small amounts of DNA, are some of the advantages of Microsatellites. Additionally, because microsatallite polymorphism can be described numerically, they lend themselves to computerized data handling and analyses (Teale *et al*., 1994). Microsatellites can be used in non-PCR systems in a way similar to minisatellite probes (Haberfeld *et al*., 1991)

Randomly amplified polymorphic DNA (RAPD) (Williams *et al*., 1990) has been extensively used for genetic characterization of a wide range of organisms. The technique uses short (up to 10 bases) primers to amplify nuclear DNA in the PCR. The procedure does not require knowledge of the sequence of and under study; primers are designed randomly. The basis of the polymorphism detected by this method is that products are either generated in PCR or not.

Complete sequencing of the genome is the ultimate form of genetic characterization. Sequencing has traditionally been expensive and laborious, but with the advent of automated sequencing, this is changing rapidly. However, sequencing is unlikely to be used as a technique of choice of genetic characterization.

20.4 Reproductive Biotechnologies

The main objective of biotechnologies in reproduction is to increase reproductive efficiency and increased output from the livestock sector. They also offer potential for greatly extending the multiplication and transport of genetic material and for conserving unique genetic resources in reasonably available forms for possible future use.

Artificial insemination (AI)

AI has already had a major impact on cattle, sheep, goat, pig, turkey and chicken improvement programmes of developed countries by accelerating breeding progress primarily through increase in intensity of selection of male and through diffusion of breeding progress, initially with fresh, and later with frozen semen, offering rapid world-wide transport of male genetic material. Globally, more than 3.3 million AI in sheep and 0.5 million in goats are performed annually, but still only in very few developing countries, AI is practiced to a level that impact substantially livestock production.

Embryo transfer (ET)

ET in the mammalian species is enhanced by multiple ovulation and oestrus synchronization (MOET), allows acceleration of genetic progress through increased selection intensity of females. MOET may also be used to produce crossbred replacement females whilst only maintaining a small number of the straight breeds. Worldwide, 40,000 ETs have been recorded in cattle, 17,000 in sheep, 1,200 in goat, and 2,500 in horses. Most of the bulls used in AI in the developed world are derived from ET. Despite the potential benefits of ET, its application is largely limited to developed countries. ET is also one of the basic technologies for the application of more advanced reproductive biotechnologies such as ovum pick-up (OPU) and *in vitro* maturation and fertilization (IVM/IVF), sexing of embryos, cloning, and of transgenics.

Cloning

IVM/IVF are a source of large numbers of low cost embryos required for biotechnologies such as cloning and transgenesis. Three different types of clones are distinguished, as a result of: (1) limited splitting of an embryo (clones are genetically identical): (2) introducing an embryonic cell into an enucleated Zona (clines may differ in their cytoplasmic inheritance); (3) introducing the nucleus of a somatic cell (milk, blood, dermal cells) after having reversed the DNA quiescence, into an enucleated Zona (clones may differ in their cytoplasmic inheritance and there is likely to already exist substantial knowledge of the phenotype of the parent providing the somatic cell). Cloning will be used to multiply transgenic founder animals. Cloning technologies offer potential as research tools and in areas of very high potential return. The sampling of somatic tissue may assist collection and transfer of breed samples from remote area for conservation purposes.

20.5 Improving Nutrition Through Biotechnology

Nutritional deficiencies represent one of the most serious limitations to production for small ruminants in developing countries. Feed resources are inadequate in both quality and quantity, particularly during the dry seasons. Biotechnological options are available for improving rumen fermentation and enhancing the nutritive value and utilization of agro-industrial by-products and other forages (Kundu and Kumar, 1987).

Fibrous feeds, including crop residues, of low digestibility constitute the major proportion of feeds available to most ruminants under smallholder situations in developing countries. The associated low productivity can be overcome to some extent by several means, among which are: balancing of nutrients for the growth of rumen microflora thereby facilitating efficient fermentative digestion and providing small quantities of by-pass nutrients to balance the products of fermentative digestion, enhancing digestibility of fibrous feeds through treatment with alkali or by manipulating the balance of organisms in the rumen and genetic manipulation of rumen micro-organisms. Rumen micro-organisms can also be manipulated by adding antibiotics as feed additives, fats to eliminate or reduce rumen ciliate protozoa (defaunation), protein degradation protectors, methane inhibitors, buffer substances, bacteria or rumen content and/or branched chain volatile fatty acids.

Increasing digestibility of low-quality forages

The characteristic feature of tropical forages is their low rate of microbial breakdown in the rumen. The slow rate of breakdown also results in reduced outflow rate of feed residues from the rumen, which consequently depresses feed intake. At present, the main treatment methods for forages such as cereal straws are either mechanical (*e.g.* grinding), physical (*e.g.* temperature and pressure treatment) or a range of chemical treatments of which sodium hydroxide or ammonia are among the successful (Greenhalgh, 1984). Lignin present in the cell wall prevents degradative action of cellulose or hemicellulose enzymes. Now it is possible to use lignase enzyme produced by the soft–rot fungus (*Phanerocheate chrysosporium*), which causes a high degree of polymerization of lignin (Tien and Kirk, 1983). The levels of the lignase enzyme produced by the *basidiomycete* fungi are insufficient for the treatment of straw of a commercial scale. However, recombinant DNA techniques will allow the modification of the lignase genes and associate proteins to increase their efficiency and stability. The lignin gene has been cloned and sequenced from *P. chrysosporium* (Zhang *et al.,* 1986; Tien and Tu, 1987).

Removing anti-nutritive factors from feeds

Anti-nutritive factors in plant tissues include protease inhibitors, tannins, phytoheamagglutinins and cyanogens in legumes, and glucosinolates, tannins and snapping in oilseed rape (*Brassica napus*) and other compounds in feeds belonging to the Brassica group. As with amino acid deficiencies, the adverse effects of these compounds are more marked in non-ruminants than in ruminants (Chubb, 1983).

Genetically manipulated rumen microbes could also play a role in the detoxification of plant poisons (Gregg, 1989) or inactivation of anti-nutritional factors. Successful introduction of a caprine rumen inoculum obtained in Hawaii into the bovine rumen in Australia to detoxify 3-hydroxy 4 (IH) pyridine (3, 4 DHP), a breakdown product of the non-protein amino acid mimosine found in Leucaena forage (Jones and Megarrity, 1986) demonstrates the possibilities for future.

Improving nutritive value of conserved feed

Silage making is traditionally being applied for conservation and fodder quality enhancement. The additives used in the process include mineral acids (hydrochloric acid, sulphuric acid), their salts and organic compounds like cresol, toluene and chloroform. These additives were used to restrict the activity of the micro-organisms, to stimulate the fermentation by the lactic acid bacteria or simply as nutrients by lowering the pH or inhibiting the growth of microbes. The hazardous nature of some of the chemical additives has necessitated a search for alternative compounds. These include sugar sources (*e.g.* molasses and whey), enzymes and inoculums of lactic acid bacteria. Enzymes are essential for the breakdown of cell-wall carbohydrates. Addition of enzymes derived from certain bacteria, *e.g. Aspergillus niger* or *Trichoderma viridi* (Henderson and McDonald, 1977) increases other amount of available sugars. Commercial hemicellulase and cellulase enzyme cocktails are now available to improve the fermentation process considerably (Hooper *et al.*, 1989). However, prices of these products preclude their viability for farm level application, especially in developing countries.

In order to improve the effectiveness of microbial inoculants in breaking down structural carbohydrates to glucose, detailed knowledge of the *lactobacilli* bacteria is essential. Work already undertaken on the molecular biology of *Lactobacillus plantarum* and other species (Armstrong and Gilbert, 1991) suggest that the rapid progress in this area will make it possible to construct novel genes encoding highly active fiber-degrading enzymes. Such genes could then be inserted into strains of *L. plantarum*.

Improving rumen function

Rumen biotechnology has the potential to improve the nutritive value of ruminant feedstuffs that are fibrous, low in nitrogen and of limited nutritional value for other animal species. Biotechnology can alter the amount and availability of carbohydrate and protein in plants as well as the rate and extent of fermentation and metabolism of these nutrients in the rumen. Current limitations to rumen manipulation includes; isolation and taxonomic identification of strains for inoculation and DNA recombination; isolation and characterization of candidate enzymes; level of production, localization and efficiency of secretion of the recombinant enzyme; stability of the introduced gene; fitness, survival and functional contribution of introduced new strains. Methods for improving rumen digestion in ruminants include the use of probiotics, supplementation with chelated minerals, and the transfer of rumen microorganisms from other species.

The major areas of rumen function which might benefit from transgenic technology are development of transgenic bacteria with enhanced cellulolytic activity, capability to cleave lignohemicellulose complexes, reduced methane production capability, decreased proteolytic and/or deaminase activities, increased capability for nitrogen "fixation" and increased ability for microbial production of specific amino acids. The first successful transfer of foreign gene into rumen bacteria (*Bacteriodes ruminicola*) was reported by Thomson and Flint (1989). However, we are still a long way from commercial production of genetically engineered rumen bacteria.

It can be concluded that there are several potential opportunities for utilizing a wider range of feeds than is currently possible. Modification of rumen microbial population (Hespell, 1987) is one such opportunity. However, technical difficulties associated with making genetic modification to individual species of rumen bacteria (Armstrong and Gilbert, 1991) hinder progress in this area.

Applications are being developed for improving the performance of animal through better nutrition. Enzymes can improve the nutrient availability from feedstuffs, lower feed costs and reduce output of waste into the environment. Prebiotics and probiotics or immune supplement can inhibit pathogenic gut microorganisms or make the animal more resistant to them. Administration of recombinant somatotropin (ST) results in accelerated growth and lean carcasses in meat animals. Immunomodulation can be used for enhancing the activity of endogenous anabolic hormones.

Plant biotechnology may produce forages with improved nutritional value or incorporate vaccines or antibodies into feeds that may protect the animal against disease.

Growth manipulation

Porcine somatotropin (PST) and recombinant growth hormone stimulatory peptides (*e.g.* growth hormone releasing factor, GRF) along with Bovine somatotropin have been shown to increase growth rates by 8–38% in cattle, sheep and pigs. In almost all cases, administration of exogenous growth hormones have been associated with increased carcass protein and reduced carcass fat (Hart and Johnson, 1986). Other growth promoting agents (*e.g.* anabolic steroids and beta agonists) have been shown to have deteriorating effect and even being banned in most developing countries. The use of anabolic implants is however, permitted in some countries such as the UAS.

20.6 Health Management and Biotechnology

Animal diseases are a major and increasingly important factor of livestock productivity in developing countries. Use of DNA biotechnology in animal health may contribute significantly to improved animal disease control, thereby stimulating both food production and livestock trade. Biotechnology is helping to control a range of economically important animal infectious diseases that can be disastrous to the farming industry. Only the immunogens are used in the vaccine, reducing the change of adverse side effects. Various molecular biotechnology applications are available in animal production and health, involving both on farm production and of farm production processing applications.

Drug treatment

Biotechnology can also help to produce better drugs to fight animal disease. Parasitic infections, which can include both internal parasites such as worms and external parasites such as tick and sheep blowfly are a major source of ill-health in farm animals. Biotechnology was used to develop anti-worming drugs called Avermectins from products secreted by the soil mould *Streptomyces avermitilis*. However, these drugs turned out also to remove some parasites from the external coat of treated cattle. This unexpected bonus led to a new class of pest controlling agents (endectocides) able to control both external and internal animal parasites.

20.7 Detection of Nucleic Acid

Diagnosis by restriction fragment length polymorphisms

The commonly employed serological tests are sometimes not sufficiently discriminatory to identify closely related micro-organisms. Restriction fragment length polymorphism analysis can detect differences in the genomes of closely related microbial species. The RFLP analysis may be used in epidemiological studies. The involvement of a particular isolate in a disease outbreak can be defined and other epidemiological tracing of isolates within a country or between countries should be possible. Except for use in epidemiological studies, the potential diagnostic value of RFLPs is not clear. The diagnostic use or other interpretation of RFLP data involves the construction of databases where RFLP profiles are linked to factors such as virulence, host range and clinical significance.

Polymerase chain reaction (PCR)

The Polymerase chain reaction (PCR) procedure simulates natural DNA replication *in vitro*, producing large quantities of a desired sequence of DNA from a complex mixture of heterogeneous sequences (Erlich *et al.*, 1988, Saiki *et al.*, 1988). With RNA viruses, a complementary DNA (cDNA) copy of the RNA must first be made using reverse transcriptase (RT), then the

PCR can be used for amplification. This technique, referred to as RT-PCR, is used to detect RNA viruses. The PCR product had a characteristic size and its identity is generally confirmed using DNA probes or restriction (enzyme or endonuclease) digests. The advent of automated cycle sequencing techniques have been used in the virulence typing of *Avian influenza* A virus (Horimoto and Kawaoka, 1995). The sensitivity of the PCR product of the first reaction is commonly referred as 'nested PCR'.

20.8 Vaccines

Traditional vaccines consist of killed microorganisms, inactivated toxins or lie organisms that have been deliberately weakened to reduce their ability to cause disease. Only a few molecules (known as immunogens) on the surface of the infecting organism are needed to trigger the immune response. Genetically engineered vaccines, on the other hand, can be produced by "tricking" microbes into making large quantity of only the specific immunogens needed to generate immunity to the disease. Molecular biology procedure has and will continue to have major impact on the preparation of vaccines. Products of recombinant technology may be easier to monitor because they may have specific markers introduced into or deleted from the vaccine. Even though genetic information is often manipulated in live recombinant vaccines, there are significant safety concerns surrounding their usage.

Gene-deleted vaccines

Virus genes coding for virulence can be deleted, leaving a virion that can be used in live virus vaccines with reduced virulence. These deletion mutants are frequently unable to replicate without the addition of specifically required nutrients. The vaccine agent undergoes restricted replication in the vaccinated animals but is unable to continue replication cycles. This results in a vaccine that is capable of inducing an immune response in a similar way to other live vaccines, but which has greater safety because it will not persist in the animal or in the environment. Because of the gene deletion, they are easy to differentiate from the field strains of their counterparts.

Marker vaccines

During the development of both conventional vaccines and those produced using recombinant technology, consideration must be given to the development of appropriate tests to differentiate vaccinated from infected animals. This has lead to the development of marker vaccines. A marker vaccine uses a specific deletion or an added microbial protein code into the genome in order to distinguish between vaccinated and infected individuals (Van Oirschot, 1999).

Virus-vectored vaccines

The most discussed vectored vaccines are the vaccinia-vectored viruses of rinderpest vaccine and the oral rabies vaccines. A plasmid is constructed containing the DNA of the donor rinderpest virus. Then the rinderpest cDNA segments in the plasmid are inserted into the TK locus of the vaccinia virus genome (Yilma *et al.*, 1988). The haemagglutinin (H) gene of rinderpest can be inserted into the heamagglutining region of the vaccinia vector, thus inactivating the corresponding H gene of the vaccinia. The recombinant *vaccina* virus grown in cell culture produce the rinderpest virus haemagglutinin protein that serves as protective antigen (Yamanouchi *et al.*, 1993). A single capripox recombinant vaccine was effective in protecting cattle against rinderpest and lumpy skin disease (Romero *et al.*, 1993).

DNA vaccines

DNA vaccination is the direct introduction into host cells of a bacterial plasmid DNA that is able to cause expression of an antigenic protein. Several DNA vaccines for veterinary use are currently being developed (Yamanouchi *et al.*, 1998). Such genetic immunization is a very promising new approach of vaccination (Robinson, 1997). The immune response can be further improved by simultaneous inoculation of immunostimulators such as CpG motif sequences and plasmids expressing cytokines (Kim *et al.*, 1998; Xiang and Eitl, 1995).

New developments in vaccine technology

Subunit vaccines contain purified protein or glycoprotein components of a pathogen that have been identified as carrying critical epitopes involved in inducing a protective immune response (Babiuk, 1997). These vaccines have distinct advantages in the realm of safety, and recent improvement in their production using recombinant DNA technology may facilitate more widespread usage (Desmettre, 1999). Synthetic peptide vaccines have been engineered with distinct advantages in handling properties and safety, and they are relatively inexpensive to produce (Meloen, 1997). A synthetic peptide vaccine for foot and mouth disease virus has been developed but only limited protection has been observed in host animals (Sobrino *et al.*, 1999).

Monoclonal antibodies

Antibodies are special proteins used by the body to fight disease. They are made by white blood cells in response to any substance, including disease-causing microorganisms, that the body recognizes as foreign. Antibodies work by binding to these foreign substances as they are circulated in the blood and preventing them form harming the body.

Because antibodies bind specifically to the substance that caused them to be produced, the past history of infection by disease can be determined by examining antibodies in an animal's blood. Antibodies obtained from the blood of immunized animals are used widely for both diagnostic and research purposes. However, obtaining commercial quantities of pure antibodies by traditional methods has often been difficult. The body makes antibodies in relativity small quantities. Further, because the immune system responds to a range of characteristics of foreign substances, it often produces a range of antibody to a single trigger substance.

CHAPTER 21

CONSERVATION OF GOAT GENETIC RESOURCES

21.1 Introduction

Goat as an animal is a renewable resource of great value for our country and has diverse ecological adaptability over a wide range of agro-climatic zones. It is a preferred animal over other livestock particularly in drought prone and tribal areas due to its small size, easy handling, high reproductive efficiency and proportionately high returns. They play an important role in most disadvantageous area by contributing significantly to the economy of resource poor people besides meeting their nutritional requirement. The management and conservation of goat genetic resources in this country addresses maintaining variability in breed, and characteristics for which it is traditionally being valued.

The chapter discusses current trends in the management of goat genetic resources for the purpose of enhancing productivity, management, conservation and research needs and utilization. It has been also discussed that it is necessary to plan and formulate breeding strategies in order to take greater advantage of the available genetic diversity for improving the productivity of goats.

21.2 Goat Genetic Resources

The goat breeds with varying capacities to produce meat, milk and fibre have been developed in India primarily through natural selection and adapted to diversified agro-climatic conditions based on their utility and production function. The goats of Temperate Himalayan Region (where rainfall is low) grow on fibres of good quality and possesses the finest quality undercoat called cashmere or pashmina. The goat breeds found in North and North-Western region are large in size, and primarily of milch type. In the Southern and Peninsular part of the country, goats with dual production of meat and milk are found. The highly prolific meat breeds are found in the Eastern region of the country. There are about 20 well defined goat breeds in the country which constitute about 20–25% of the total goat population. The remaining are non-descript having mixed features. Distribution of goat breeds-region wise is given in the following Table 21.1.

Table 21.1 Goat Breed of India

Temperate Himalayan	North-Western	Southern	Eastern
Gaddi	Jamunapari	Sangamneri	Ganjam
Changthangi	Marwari	Osmanabadi	Bengal
Chegu	Zalawadi	Kanni Adu	
	Beetal	Malabari	
	Kutchi		
	Sirohi		
	Barbari		
	Mehsana		
	Surti		
	Jakhrana		
	Gohilwadi		

Detailed description of these breeds have been provided by a number of authors

21.3 Production Systems

Management practices and system of goat production in the country are unique and diversified based on local traditional patterns depending on the geo-ecological conditions and agricultural cropping patterns. The flocks in Himalayan region migrate to mountains in the summer months for Alpine grazing and returns to foothill and lowland valleys during winter. The migration starts in April when they move to Alpine pastures at high altitude and start migrating down in the month of October. During the period of April to October, they graze on mountains and high lands and during winter are fed with stored dry fodder, tree leaves and concentrates to a limited scale. They are penned in closed house for protection from severe cold. Occasionally, goats are taken out when the weather is good.

Hot arid and semi-arid regions of Rajasthan has a very high temperature during the summer period and goats are raised on permanent migration. Flocks follow well established migratory routes year after year in search of food and water. Flocks from Western Rajsthan migrate for 6–9 months and are brought back to home base for rest of the period. During the period of migration, the flocks are grazed on natural vegetation on the road and canal verges, crop residues in harvested fields and in lean period on tree loppings enroute.

Goat in humid areas are raised by the farmers under traditional system in small numbers (2–5) and maintained under semi-intensive conditions. They are commonly grazed on tethers and housed at night.

Devendra (1986) have discussed small ruminant production systems in Asian countries in three categories as extensive, semi-intensive and intensive systems combining with cropping and system integrated with three cropping. The extensive grazing on natural range lands, community grazing lands, crop stubbles are most common both in nomadic and sedentary systems of management which is most unorganized leading to low production efficiency and economic losses. Integration of small ruminants with agricultural crops specially with tree cropping appears to be the most useful in the South Asian countries, which are confronted with the ecological and land degradation problems. The alternative way to avoid such problems by emphasizing to develop the lands including waste land by using grasses of high nutritional value on the ground, shrubs/bushes in the middle and trees on the top to create three tier canopy for maximizing the biomass availability to goats besides the timber and fuel wood for domestic use. The economic benefits of such an integrated system have been discussed by Devendra (1991).

Goat raising is primarily with the people who lack economic and technical resources and follow age-old traditional systems. There is ample need for making available improved technologies to the traditional farmers, which should be economical and acceptable. This needs to intensify the efforts to demonstrate the workable know-how in the rural areas and intensification of such efforts would be a positive step if the production efficiency is to be increased.

21.4 Evaluation, Conservation and Management

In the recent past, considerable emphasis was given on crossbreeding of indigenous goat breeds with exotic as well as with indigenous improver breeds in order to enhance their productivity. The resultant outcome was very inconsistent due to many shortfalls including inadequate infrastructure. However, there is now growing realization for the importance of indigenous breeds because of their suitability and adaptation to diversified agro-climatic and socio-economic situations. With this in view, the breeding strategies and improvement programmes were, therefore devised considering the efficiency of production in relation to physical environment, feed and fodder resource availability, management and health aspects. The information on goat genetic resources is not adequate and whatever information is available on institutional sector is not representing their production status and performance under their natural habitat. Efforts made in the direction to identify and evaluate the breeds and to bring them on record are also restricted, making it difficult to decide as what to conserve and what to leave. It is therefore necessary to view this process in an integrated manner comprising or establish to identify the breeds at the first instance, their evaluation and documentation of the data so that a format for conservation and management could be developed based on informed decision making process.

21.4.1 Evaluation

It is necessary to evaluate the breeds in their home tract under native management system through extensive surveys following appropriate sampling methods. The description and evaluation of breeds should involve studying the population structure including the population size and trends, flock size and structure, physical environment, climatic profile, soil and land use pattern, availability of feed resources, physical conformation and size parameters and economic traits related with body weights, growth pattern, production, reproduction and survival rate. This will provide the baseline information on breeds which may be helpful in determining the present status of goat breeds. National Bureau of Animal Genetic Resources (NBAGR) and AICRP on Goat Improvement after in length deliberation at their meetings have devised the format on breed descriptors which need to be adopted to generate the information for better understanding of the goat genetic resources of the country. Breed description and evaluation should be supported by the information on gene markers, karyotypes, blood groups and other variables of protein types inorder to utilise such information for identification of breeds/breed types.

In-situ conservation

The process of *in-situ* conservation of germ plasm refers to maintenance of live animals in their original and natural condition without altering the genetic potential as far as possible. This can be accomplished at institutional farms or in the places of their natural habitats, although the methods of conserving rare or endangered breeds in improvement programmes are very diverse and differ. Inbreeding poses a definite problem in case of smaller flocks and should be looked into very carefully. Efforts should be devised to conserve the genetic variability to avoid undesirable effects on the populations.

Ex-situ Conservation

Cryopreservation of gametes and embryos is an alternative method which gives an opportunity to preserve the genetic material and maintain the biodiversity for posterity. *Ex-situ* conservation of germ plasm for longer periods and its use at the time of need is necessary in order to prevent any breed from getting lost. The methodology of cryopreservation of semen and embryos has been developed for many years and application of such technologies is very common in many countries. Storage and preservation of semen and embryos can be used for *ex-situ* conservation of breeds that need to prevent the loss of breeds which are in very small numbers or on the verge of extinction.

Manipulation and splitting of embryos at an early developmental stage paved the way to produce identical twins, chimeras and animals in the cloned form. Gene transfer technologies in livestock production have also been developed in recent years and storage of genes in the form of genome DNA has become possible. Stored genomic DNA therefore, can be utilized by the application of DNA recombinant technology which make the production of transgenic animals possible. This approach of animal production to meet the specific objectives is very advance and most sophisticated in nature having great potential to create revolution in producing the animals of desired type.

Conservation in research organization

Mostly selective breeding has been practiced in Research Institutes and Universities for increasing the efficiency in goat productivity. In India, cross breeding in goat has been practiced for more than a decade. This resulted into very inconsistent results due to many shortfalls and inadequate infrastructure. Now a days, there is growing realization for the importance of indigenous genetic resources because they are more relevant to meet out the nutritional breeds under their natural habitat. Goats are also important species of livestock which contribute significantly to poverty alleviation and rural development.

The Central Institute for Research on Goats at Makhdoom has established a breed improvement and conservation programme of Jamunapari goat in Chakarnagar area of Uttar Pradesh, the hometract of this breed. This is a isolated area which bordered by a number of rivers and is a very arid and sandy area. This breed is a large dairy breed and able to survive in this ravenous topography. Farmers are keen to this process of conservation and improvement programme. The flocks are registered in the villages with appropriate identification is being recorded and selection of superior animals is made under the programme. Survey of goat breeds on the similar lines has also been initiated at different locations under Network programme of NBAGR and AICRP on Goat improvement.

Conservation Networks

An information network for dissemination technology/knowledge has been established by Indian Council of Agricultural Research (ICAR) and database development is in progress. The networking will offer a number of advantage such as promoting linkages, addressing diverse and complex problems, provide opportunity for better focus in terms of goals and priorities and promotes common approach to problem related with the management.

It is necessary to improve the prevailing management efforts and technologies for appropriate management of goat genetic resources in India. It is the need to develop, improve and make sustainable use of goat genetic resources using multidisciplinary approach with a farming system perceptives. It is essential to analyse constraints and opportunities for productivity and utilization of goats.

Survey of the breeds and on farm breed characterization have been initiated by different Institutes/Universities under goat improvement programmes and in collaboration with National Bureau of Animal Genetic Resources in India. Database of animal genetic resources is in progress in different countries.

21.5 Management

Proper and adequate utilization of goat genetic resources is much more important and needs greater emphasis on the productivity characteristics, while the surveys are being made for collection of information of breeds. There is a wide gap between production potential and present productivity in goats of the county and existing genetic diversity provide great scope for their development. Goats with better performance above the flock average should be identified and utilized for improvement programmes by organizing open nucleus breeding schemes. Programme should be devised in such a way so that positive participation of the farmers may be ensured. This necessarily will need certain incentives like subsidies on health care etc. Breeding bucks should be produced from the outstanding parents in the open nucleus flock and after initial performance testing of the individual for body weight growth should widely be utilized for the improvement of flocks.

21.6 Impact Assessment and Technology Transfer

The generation of new technologies by research Institute and Universities has been made in animal genetics, nutrition, health and resource management to improve small ruminants productivity and development in

India. The role of goats on ecological and deforestation has been well studied in the country. It is necessary to evaluate the breeds in different agro-ecological situations and testing of improved technologies in different situations.

21.7 Future Thrusts

India is very rich in goat genetic resources and number of breeds with great diversity is available having good potential for production of meat, milk and fibre. Very little work has been done for breed characterization and their proper evaluation. It is necessary and recommended that concerted efforts should be made to characterize and evaluate the indigenous breeds including genetic characterization. Further, it is suggested that regional programme for sustainable development and purposeful use of goat genetic resources need to be developed and put to function. The programme should be designed in such a manner so as to contribute both immediate production needs and long term improvement programme in order to take up breed characterization, evaluation and utilization. It is necessary that each State Government should establish an appropriate infrastructure at provincial level with adequate allocation of funds. National Bureau of Animal Genetic Resources as nodal agency on Animal Genetic Resources from time to time encourages Research Institutes/State Agricultural Universities, State Governments to give adequate priority to management and development of goat genetic resources for food security, poverty alleviation and rural development. The experts and technicians may be trained in the area of management of animal genetic resources conservation. Adequate support including funding should be provided by the Government to carry out the National Goat Genetic Resources Conservation work plan.

Following areas may need attention:

i) Breed characterization and evaluation.

ii) Recording system and data management.

iii) Development of suitable breeding strategies including ONBS.

iv) Gene markers and genetic identification.

v) Management and utilization of Goat Genetic Resources.

vi) Monitoring the population of those that are identified as at risk.

A National information network should be established to provide an opportunity for the exchange of information on Animal Genetic Resource including goats within country as well as to other countries.

REFERENCES

Acharya, R.M. (1992). Recent Advances in Goat Production. Goat Production Proceedingof the V international Conference on Goats. pp. 49-91.

Acharya, R.M. (1982). Sheep and goat breeds of India. F.A.O. Animal Production and Health Paper No. 30, F.A.O., Rome.

Acharya, R.M. (1979). Selection for improving milk production in Beetal goats. Paper presented at Goat Development in Asian countries, March 5–10, NDRI, Karnal.

Acharya, R.M. and Bhat, P.N. (1983). Genetic Researches in Livestock. XV International Congress of Genetics, New Delhi. December 12–21, 1983.

Acharya, R.M. and Patnayák, B.C. (1974). Role of sheep in desert ecosystem and drought proofing through improved sheep production with special reference to Rajasthan. Monograph, CSWRI, Avikanagar.

Acharya, R.M. and Sharma, V.D. (1980). A note on pashmina production and quality of Changthangi pashmina goat. *Indian J. Anim. Sci.* **50**: 586–587

Acharya, R.M., Gupta, U.D., Sehgal, M.J.P. and Singh, M. (1987). Coat characteristics of goat in relation to heat tolerance in hot tropics. Proc. of the IV in. conf. on Goats, Brazil, Vol. II. pp. 1373.

Acharya, R.M., Gupta, U.D. and Singh, M. (1987). Influence of coat characteristics on heat tolerance in water deprived Indian Sirohi goats. Proc. of the IV Int. Conf. on goats, Brazil, Vol. II, pp. 1372.

Acharya, R.M., Swain, N., Jain, P.M., Mishra, R.K., Mathur, P.B., Sharma, M.M. and Mudgal, R.D. (1980). Relative productivity of sheep and goats on free range grazing/browsing management on semi-arid rangelands of Rajasthan. Final Report, CSWRI, Avikanagar, India.

Achuthankutty, A. and Raja, C.K.S.C.C. (1971). Studies on the incidence of early embryonic loss in goats based on abattoir specimens. *Kerala Journal of Vet. Sci.* **2**: 13–16.

Agrawal, K.P. and Bhattacharyya, N.K. (1984). Barbari goat: temporal effect on the cells stage and location of embryo. *International Goat and Sheep Research* **3**: 227–230.

Agrawal, K.P. and Bhattacharyya, N.K. (1983). *In vitro* preservation of goat embryo at refrigerator temperature (5°C). *Indian. J. Anim. Sci.* **53**: 441–442.

Agrawal, K.P., Mongha, I.V. and Bhattacharyya, N.K. (1982). Collection and transfer of embryos in goats-surgical methods. *Indian Vet. J.* **59**: 298–303.

Agrawal, K.P., Mongha, I.V. and Bhattacharyya, N.K. (1979). Success in embryo transfer in indigenous goats in India. *Current Science* **48 (17)**: 792–794.

Agrawal, K.P., Mongha, I.V. and Bhattacharyya, N.K. (1983). Relative efficacy of some media in short time preservation for goat embryos at room temperature. *Indian J. Anim. Sci.* **53**: 850–851.

Ahuja and Rathore, M.S. (1987). Goats and Goat Keepers Institute Development Studies. Printwell Publisher, Jaipur.

Agrawal, K.P. and Bhattacharyya, N.K. (1978). Yield and composition of milk in Black Bengal, Barbari and Black Bengal × Barbari nannies. *Indian. J. Anim. Sci.* **48**: 882–886.

Aitenkirch, W. (1957). Comparative studies of growth and production of different breeds of goat. Arch. Gelflugelz. Kleinterk. **6**: 295–333 (fide *Anim. Breed. Abstr*. **27**: 1911).

Ali, S.Z., Hoque, M.M. and Hasnath, M.A. (1973). A study on the growth and reproductive performance of Black Bengal goats under farm condition. *Indian Vet. J.* **50**: 438–440.

Alkov, G.V. (1982). A new breed of down producing goat. *Ovtsevododstvo* **8**: 29–32.

Alkov, G.V. and Kraskova, Z.K., 1981. Altai Mountain cashmere goats. *Ovtsevododstvo* **3**: 32. (fide *Anim. Breeding Abstr.* **49**: 3976).

Alkov, G.V. and Tadkin, N.C. (1974). Shearing the kid fleece is advantageous. *Ovtsevododstvo* **6**: 28.

Amble, V.N., Khandekar, N.C. and Garg, J.N. (1964). Statistical studies on breeding data of Beetal goats. Indian Council of Agricultural Research, Research Series No. 38, pp. 30. ICAR, New Delhi.

Anderson, G.B. (1987). Use of chimeras to study development. *J. Reprod. Fert. Suppl.* **34**: 252–259.

Annual Reports 1985–1990–91. Central Institute for Research on Goats, Makhdoom, Mathura, U.P., India.

Anonymous (1987). Report of the Committee on Fodder and Grasses. National Wastelands Development Board, New Delhi.

Armstrong, D.G. and Gilbert, H.J. (1991). The application of biotechnology for future livestock production. In: Proceedings of the 7th International Symposium on Ruminant Physiology, Sendai, Japan. Academic Press, San Diego, USA. pp. 737–761.

Armstrong, D.T. and Evans, G. (1983). Factors influencing success of embryo transfer in sheep and goats. *Theriogenology* **19**: 31–42.

Armstrong, D.T., Pfitzner, A.P. and Seamark, R.F. (1982). Ovarian responses and embryo survival in goats following superovulation and embryo transfer. *Theriogenology* **17**: 76.

Arora B.G. and Iyer, P.K.R. (1966). Observations on the pathology of caprine and bovine liver infested with *Schistosoma indicum*. *Indian J. Animal Health* **7**: 69–72.

Arora, C.L. and Swain, N. (1989). Relative performance of sheep and goats under range and management system. Rangelands resource and management IGFRI Jhansi, pp 436–442. Recent advance in goat production (1992). Proceedings International conference on goat, March 1992, New Delhi.

Asdell, S.A. and Smith, A.D. (1928). *J. Hered.* **19**: 425–430.

Babiuk, L.A. (1997). Subunit Vaccines. *In*: Veterinary Vaccinology, Pastoret, P.P., Blancou, J., Vannier, P. and Verschueren, C., *eds*. Elsevier Science BV, Amsterdam, The Netherlands, 262–264.

Badger, S. and Lee, R. (1987). Embryo transfer experience. NZ farmer, May 28th. pp. 3839.

Banerjee, G.C. (1978). Animal Nutrition. pp. 415 Oxford and IBH Co., Calcutta.

Barhat, N.K. and Chowdhary, M.S. (1978). Factors affecting some production traits in Rajasthani goats. *Indian J. of Dairy Sci.* **31**: 185–188.

Baril, B., Casamitjana, P., Perrin, J. and Vallet, J.C. (1989). Embryo production, freezing and transfer in Angora, Alpine and Saanan goats. *Zuchtyg*. **24**:101–115.

Beheshi, R.D. and Saleh, B.A. (1979). Performance of dairy goats in Iran. In International Dairy Congress. Volume E, Paris, France. *Dairy Science Abstr*. **40**: 4357.

Bell, H.M. (1978). Rangeland management for livestock production, 2nd edn. University Oklahoma Press, Norman.

Berardino, D.D., Ronne, M., Burguere, I., Lioi, M.B., Taibi, L. and Metassiono, D. (1987). R-banding pattern of the prometaphase chromosomes of the goats. *Journal of Heredity* **78(4)**: 225–230.

Betteridge, K.J. (1981). A historical look at embryo transfer. *J. Reprod. Fert*. **62**: 1–13.

Bhat, P.N. (1988). Animal health production in the tropics. *Indian Science Congress Association Proceedings* **75**: 1–127.

Bhat, P.P. and Khan, B.U. (1984). Reproduction and production traits in Indian goat breeds and their crosses. Annual Report, CIRG. pp. 6–7.

Bhat, P.N., Bhat Pran, P., Khan, B.U., Goswami, O.B. and Singh, B. (1981). Animal Genetic resources in India. Publication No. 192, NDRI, Karnal.

Bhat, P.N., Bhat, P.P., Khan, B.U., Goswami, O.B. and Singh, B. (1980). Animal Genetic Resources in India. Publication No. 192. NDRI, Karnal, India.

Bhatia, B.B. and Pande, B.P. (1969). Emerian infections in goats. *Indian Vet. J.* **46**: 399–403.

Bhatia, S. and Shanker, V. (1991). Cytogenetic analysis of Gaddi goats. *Indian J. Anim. Sci.* **61(6)**: 646–648.

Bhatia, S. and Shanker, V. (1990). A case report on XO/XX/XXX mixoploidy in goats. *Veterinary Record* **126**: 312–313.

Bhatia, S. and Shanker, V. (1989). Chromosome analysis of peripheral leukocytes of Bengal goat. *Indian J. Anim. Sci.* **59(10)**: 1340–1342.

Bhatia, D.R., Mohan, M., Patnayak, B.C. and Sheikh, Q.D. (1973). Nutritive value of the available pasture in the semi-arid areas of Rajasthan. *Indian J. Anim. Sci.* **43: 838.**

Bhatnagar, D.S., Chawla, D.S. and Sharma, R.C. (1980). Crossbreeding goats for milk production . Annual Report, NDRI, Karnal.

Bhattacharyya, N.K. (1989). An Overview – Goats. In Animal Productivity. Oxford and IBH Publishing Co. Pvt. Ltd. New Delhi. pp. 451–470.

Bhattacharyya, N.K. (1984). Progress in embryo technology and its application in animal science. National Animal Physiology Research Workers Conference at NDRI, Karnal, pp. 6.

Bhattacharyya, N.K. and Khan, B.U. (1988). Goat meat production in India. Proc. Workshop on "Goat meat production in Asia". Tando Jam, Pakistan, 13–18 March.

Bhattacharyya, N.K. and Moullick, D.N. (1995). Comparative study of mechanical factors in ruminant digestion: normal mortality of rumen in the Indian farm ruminants. *Indian J. Physiol. and Allied Sci.* **29**: 127–134.

Bhattacharyya, N.K. and Prasad, S.P. (1974). Ovulation characteristics in nulliparous Barbari nanies. *In* Proceedings of the 26th International Conference of Physiological Science, New Delhi.11: 392.

Bhattacharyya, N.K. and Sharma, K. (1993). Silvi-pastoral system in relation to goat farming. Keynote address to the IDRC Agroforestry Project Seminar on Agroforestry in 2000 A.D. for the semi-arid and arid tropics. Held at NRC for agroforestry, Jhansi, March 4–5.

Bhattacharyya, N.K. and Sharma, K. (1992). Forages for Goat Production. Strat. Forage Production and Imp. pp. 237–253.

Bhattacharyya, B.K., Mazumadar, N.K. and Luktuke, S.N. (1983). Certain aspects of oestrous observations on parturition process in Pashmina goats. *Indian Vet. J.* **60**: 287–90.

Bhattacharyya, N.K. (1982). Physiology of reproduction in Indian small and dwarf goats. Presidential address: Section of physiology, 69th Session of the Indian Science Congress, Mysore, India.

Bhattacharyya, B.K., Mazumadar, N.K. and Luktuke, S.N. (1981). Studies on certain aspects of oestrous behaviour in Pashmina goats. *Indian J. Anim. Sci.* **51**: 67–69.

Billingham, R.E. and Silvers, W.K. (1960). Studies on tolerance of the Y-chromosome antigen in mice. *J. Immunol.* **85**: 14–26.

Bilton, R.J. and Moore, N.W. (1976). *In vitro* culture, storage and transfer of goat embryo. *Australian J. Bio. Sci.* **29**: 125–129.

Blair, H.T., McCutcheon, S.N. and Mackenzie, D.D.S. (1990). Concept drawing, components of the somatotropic axis as predictors of genetic merit for growth. *In*: Proceedings of the 4th world congress on genetics and Applied Livestock Production, Edinburgh, Scotland. Vol XVI. International Committee for world Congress on Genetics applied to Livestock Productivity, Edinburgh, UK. pp. 246, 255.

Blancou, J. (1990). Utilization and control of biotechnical procedures in veterinary science. Revue Scientific et Technique de l'Office International des Epizooties **9**: 641–659.

Blanken, G. Kuehner, H. and Sebastian, F. (1970). Arbeitsverfahren in der schafhaltung. KTEL – Flugshrift Nr.20.

Bligh, J. (1967). A Thesis concerning the Process of Secretion and Discharge of Sweat. Environmental Research 1: 28–45.

Blood, D.C. and Henderson, J.A. (1974). Veterinary Medicine, 4th Edn. Bailliere, Tindall.

Bodin, L., Bibe, B., Blance, M.R. and Ricordean, G. (1986). Parameters genetiques de la concentration plasmatique en FSH des agrelles de race Lacaune viade. *Genetics, Selection, Evolution* **18**: 55–62.

Bodo, L. (1961). The milk production of goats in Hungary. Karanah hozel (Godollo). pp. 317–324. (fide *Anim. Breed. Abstr*. SO: 1854).

Bondioli, K.R., Ellis, S.B., Pryou, J.H., Williams, M.W. and Harpold, M.M. (1989). The use of male specific chromosomal DNA fragments to determine the sex of bovine preimplantation embryos. *Theriogenology* **31**: 95–104.

Bonner, J.C. (1947). The Angora goat: A foot note in Southern Agriculture history. *Agric. Hist.* **21**:42–46 (*Anim. Breed. Abst.* **16**: No. 575).

Booman, P. (1986). Control of sex ratio by sexing sperm and embryos. Exploiting New Technologies in Animal Breeding. pp. 13–22, (Des) Smith. King and Mckay. Oxford University Press.

Borgohain, A.C., Deka, B.C. and Rajkonwar, C.K. (1985). Preservation of buck semen. *Indian Vet. J.* **62**: 81–82.

Bose, A.B. (1963). Place of goat in desert economy. Agric. Res., New Delhi **3**: 56.

Bouillon, J. and Ricordeau, G. (1975). Genetic parameters of growth and milk production in the goat at a testing station estimates of direct and indirect response to selection. I eres journees de la recherche ovine et caprine. pp. 124–132 (*Anim. Breed. Abstr*. **44**: 3247).

Bouillon, J. and Ricordeau, G. (1975). Testing Alpine, Sannen, and Chamois bucks. III. Genetic correlation between milk yield and rate of flow. I eres journees de la recherche ovine et Caprine, 2–3 et 4 decembre, 1975. pp. 98–104.

Brendan, C.F. and Thomson, L.U. (1991). Lectin-tannin interaction and their influence on pancreatic amylase activity and starch digestibility. *J. Agric. Food. Chem*. **39**: 727–31.

Bretzlaff, K.N. and Madrid, N. (1989). Clinical use of Norgestomet ear implants or intaravaginal pessaries for syncronization of oestrus in anoestrous dairy goats. *Theriogenology* **31**: 419–23.

British Goat Society (1972). Brief Report on goat breeding in Israel. *J. Br. Goat Soc.* **65**: 57–58.

Brout, A., Dmiel, R. and Shkolnik, A. (1979). Heat balance of resting and walking goats: Comparison of climatic chamber and exposure in the desert. *Physiol. Zool.* **32**: 105–112.

Bunch, T.D., Foote, W.C. and Spillett, J.J. (1976). Translocations of acrocentric and their implications in the evolution of sheep (*Ovis*). *Cytogenet. Cell Genetics* **17**: 122–136.

Calmohoy, I. L., Arganosa, V.G. and Poland, O.H. (1959). Exploratory goat diary with naove and gradxe goats. *Phillpp. Agric.* **43**: 243–249 (fide *Anim. Breed. Abstr*. **29**: 901).

Charan, K., Singh, S.D. and Singh, R.P. (1973). Tuberculosis in goats. *Indian Vet. J.* **50**: 486–487.

Chauhan, M.S. and Anand, S.R. (1990). Effect of egg-yolk lipids on the freezing of goat semen. *Theriogenology* **34**: 1003–1013.

Chawla, D.S., Bhatnagar, D.S. and Sundaresan, D. (1981). Dairy Goats at Karnal. (Dd.) Bhatia, K.K Pub. No. 195. NDRI, Karnal.

Chemineau, P., Procureur, R., Cognie, Y., Fefevre, P.C., Locatelli, A. and Chupin, K. (1986). Production, freezing and transfer of embryos from a blue tongue infected goat herd without blue tongue transmission. *Theriogenology* **26**: 279–290.

Choueiri, E. (1973). The Damascus goat. The future milking animal of the Mediterranean countries. World Conference on Animal Production. Vol. I. Short contribution, Melbourne, Australia 1(b): I .

Chubb, L.G. (1983). Anti-nutritive factors in animal feedstuffs. *In*: Haresing W. (ID), Recent Advances in Animal Nutrition Butterworths, London, UK. pp. 21–37.

Cianci, D., Bufano, G. and Bellitti, E. (1967). Milk production of goat population in Southern Italy. Anmati. Fac. Agric. Univ. Iari pp. 211–244 (fide *Anim. Breed Abstr.* **33**: 538).

Ciolca, N., Stefanescu, C., Stoiano Vivi V., Raich, L. and Schneider, A. (1959). Morpho-productive character of the Banat white goat tuce. *Sti. Inst. Cerelootch.* *(Bucuresti). **17**: 407.

Clement, J.E., Eall, R.J., Narayan, O., Haver, D., Sheffer, D., Powell, A.M., Zink, M.C. and Rexroad, C.E. (1994). Transgenic Sheep that express the visna virus envelope gene. *Thriogenology* **41**: 180 (abstr.).

Corteel, J.M. (1977). Production, storage and insemination of goat semen. Proc. Symp. On Management of Reprod. in sheep and goat. Madison (Wisc. USA) July 24–25 pp. 41–57.

Crampton, E.W. (1961). Applied Animal Nutrition. W.H. Freeman and Co., San-Fransisco.

Crampton, E.W. and Harris, L.S. (1969). Applied Animal Nutrition 2nd edn, pp. 174–177. W.H. Freeman and Co., San Fransisco, CA.

Creas Britain (1961). Herd book with foundation book, supplementary registration etc. No. 86. British Goat Society Year book (fide *Anim. Breed. Abstr.* **29**: 1527).

Cui Lizhong (1990). Contrast analysis of cashmere yield performance of white cashmere goat for different types. Sheep husbandry in China No. 1.

Das Anubrata, Kakatim, B.N., Rajkonwar, C.K. and Borgohain, N.N. (1985). Survival of buck spermatozoa in egg-yolk-citrate-milk and tris diluents at 4°C, *Indian J. Anim. Sci.* **55**: 670–672.

Deka, B.C. and Rao, A.R. (1989). Effect of extenders on freezability of buck semen. *Indian J. Anim. Sci.* **59**: 1038–1040.

Deka, B.C. and Rao, A.R. (1989). Effecting of washing on the fertility of frozen goat semen. *Indian Vet. J.* **66**: 268–269.

Deoghare, P.R., S.B. and Ganga Ram (1990). Economics of goat rearing under different farming conditions. Annual Report 1989–90. Central Institute for Research on Goats, Makhdoom, Mathura, U.P., India.

Deshpande, S.B. and Mehta, V.M. (1991). Effect of dilutors used for buck semen freezing on conception rate in goat. First National Seminar on Small Ruminant Reproduction, Central Sheep and Wool Research Institute, Avikanagar, Rajasthan (India) 10–12 January 1991. Abstr. pp. 12.

Desmetttre, P. (1999). Diagnosis and prevention of equine infectious diseases: Present status, potential, and challenges for the future. *Adv. Vet. Med.* **41**: 359–377.

Devendra, C. (1992). Goats and Rural Prosperity. Planary Paper presented at the Fifth International Conference on Goats, New Delhi. 2–8 March, 1992.

Devendra, C. (1991). Goat production: An International perspectives. Proc. of the International Goat Production Symposium. Tallahassee, Florida.

Devendra, C. (1986). Small Ruminant Production Systems. Proc. IDRC/SR – CRSP workshopon small ruminant production systems in south and south-east Asia, Bogor, Indonesia.

Devendra, C. (1979b). Goat production in the Asian region: Current status available on genetic resources and potential prospects. *Indian Dairy Management* **31(8)**: 513–533.

Devendra, C. (1979a). The protein requirements for maintenance of indigenous Kambing Kaqang goats in Malaysia, MARDI. Research Bull.

Devendra, C. (1970). "Ist Nat. Sem. Goat Prod." Barquisemeto, Venezuela, 11–15th Nov, 1970, 11 pp.

Devendra, C. (1962). Upgrading of local goats by the Anglo Nubian at the Federal Experiment Station, Serdang, *Malaya. Agric. J.* **43**: 265–280 (fide *Anim, Breed. Abstr.* **32**: 313).

Devendra, C. (1966). The importance of goats in Malaya (fide *Anim. Breed. Abstr.* **35**: 1556).

Devendra, C. and Burns, M. (1970). Goat production in the tropics Commonwealth Bureau of Animal Breeding and Genetics, Technical Communication No. 19, Farnham Royal, Bucks, England.

Devendra, C. and Nozawa, K. (1976). Goats in south-east Asia – their status and production. *Zeitschrift fur Tierzuchtung und Zuchtungsbiologie* **93** (**2**): 101–120.

Devendra, C. and Rankine (1971). The present status and future for goat production in Trinidad and Tosego. *J. Agric. Soc. Trin.* **17**: 455–463. (fide *Anim. Breed. Abstr.* **40**: 4743).

Devendra, C., Narine Singh, H.P. and Iton, L. (1969). Unpublished data reported in Devendra and Burns (1970). Goat production in the Tropics.

Devendra, M.R., Sharma, G.L. and Bhalla, N.P. (1959). Contagious agalactia in goats and sheep. *Indian J. Vet. Sci.* **29**: 62–68.

Dharia Mohan (1987). Afforestation in India. Vanarai Trust, Pune.

Dickinson, F.N. and King, G. (1977). Ann. Meet Am. Dairy. Sci. Ass. IOWA.

Disset, R and Ricordeau, G.J. (1977). Phenotypic parameters of dairy goat lactation records. *J. Dairy Sci.* **60**(supplement No. I): 104.

Dmiel, R., Prevulotzky, A. and Shkolink, A. (1980). Is a Black coat in the desert a means of saving metabolic energy? *Nature*, London **283**: 761–762.

Dorn, G.G., Wolfe, B.A., Bessoudo, E. and Kraemer, D.C. (1989). Follicular detection in goats by ultrasonography. *Theriogenology* **31**: 185.

Drobnis, E.Z., Nelson, E.A. and Lin, T.Y. (1982). The effect of freezing rate on motility and Got release from frozen goat spermatozoa. IInd Introduction Conf. on Goat Prod. A diseases. Arizona, USA. pp. 232

Duil, M. (1968). A preliminary not on some of the economic characters of Barbari goats. *Indian J. Anim. Hlth.* **7**: 15–18.

Durrant, B.S., Oosterhuis, J.E. and Hoge, M.L. (1986). The application of artificial reproduction technologies to the propagation of selected endangered species. *Theriogenology* **25**: 25–32.

Dutt, M. (1968). A preliminary note on some of the economic characteristics of Barbari goats. *Indian J. of Animal Health* **7**: 15–18.

Dyrinr, Y. (1976). Genetic and phenotypic parameters of production traits in goats. Melding rfra Norges, I andru As. Hogohole **56**(**4**): 19 (fide *Anim. Breed. Abstr.* **4**: 57551).

East, N.E. and Rowe, J.D. (1989). Subcutaneous progestin implants versus intravaginal sponges for dairy goat oestrus synchronization during the transitional period. *Theriogenology* **32**: 921–927.

Eker, M., Tuncel, M., Askin, Y. and Yener, S. (1976). Some milk production characteristics of Saanen × Kilis crossbred dairy goats raised at the Faculty of Agriculture University of Ankara. Ankara University. *Zirdat Fak Fak. Yillisi* **26**: 451–462.

Eldridge, F. and Blazak (1977). Comparison between the Y-chromosomes of Chianina and Brahman crossbred steers. *Cytogenetics and Cell Genetics* **18**: 57–60.

Ellerman, J.R. and Morrison Scott, T.C.S. (1951). Checklist of Prearctic and Indian Mammals 1758 to 1946. British Museum, London.

Epstein, H. (1953). *East Afric. Agric. J.* **18**: 123–132.

Epstein, H. (1946). *J. Hired.* **11**: 345–352.

Erlich, H.A., Gelfand, D.H. and Saiki, R.S. (1988). Specific DNA amplification. *Nature* **331**: 461–462.

Evans, H.J., Buckland, R.A. and Summer, A.T. (1973). Chromosome homology and hererochromatic in goat, sheep and ox studied by banding techniques. *Chromosomes* **42**: 383–402.

Fan Jingzhao *et al.* (1988). Relation of white cashmere goats to climate condition in inner Mongolia. Sheep Husbandry in China No. 1.

FAO (1974). Production Year Book, Vol. **28**. FAO, Rome.

FAO (1976). Production Year Book, Vol. **29**. FAO, Rome.

FAO (1977). Production Year Book, Vol. **31**. FAO, Rome.

FAO (1992). Production Year Book, Vol. **46**: FAO, Rome.

Finch, V.A., Dmiel, R., Boxman, R.R., Shkolnik, A., Andiavior, C.R. (1980). Why black goats in hot desert. Effect of coat colour on heat exchange of wild and domestic goats. *Physiol. Zool.* **53**: 19–25.

First, N.L. and Ruteledge, J.J. (1990). The use of new biotechnologies in livestock improvement. Proc. 5th AAAP, *Anim. Sci. Cong., Taipei.* **2**: 247–267.

Ford, C.E., Pollock, D.L. and Gustavsson, I. (1980). Proceedings of the first International Conference for the Standardization of Banded Karyotypes of Domestic Animals. *Herediats* **92**: 145–162.

French, M.H. (1970). Observations on the goats. FAO publication No. 80, Rome Italy

Gains, W.L. and Davidson, F.A. (1923). Relation between percentage fat content and yield of milk. Correction milk yield for fat content. Illinois Agriculture Experiment Station Bulletin. No. **245**: 577–621.

Gall, C. (1975). Milk Production from sheep and goats. *World Animal Review, FAO* **13**:1–8.

Gall, C. (1973). Relationship between body conformation and milk yield in German improved Fawn goats, L' Institue Technique de I' Elevege ovine el Caprine. PO. 287–294.

Garcia, B.O. (1971). Heritabilities of milk and buffaloes production in goats. Revn. Lat. Am. Prod, Anim Boaota, **13** (fide *Anim. Breed. Abstr.* **30**: 2008).

Garcia, O. Castillo, J. and Gado, C. (1972). *Agr. Trop.* **22**: 239–250.

Garcia, B.O., Berde, S.O., Castillo, J. and Paraza, F. (1970). Production of four imported goat breeds in the topics. *Anim. Breed Abstr.* **46**: 1361.

Garg, L.K., Rana, P.S. and Lal Chand (1980). Estimation of Genetic Trends in Beetal Goats. Indian Agricultural Statistics Research Institute (ICAR), New Delhi.

Garren, W.N., Mayer, J.H. and Foftreen, G.P. (1959). The comparative energy requirements of sheep and camel for maintenance and gain. *J. Anim. Sci.* **18**: 228–247.

Gearheart, W.W., Simth, C. and Teepker, G. (1989). Multiple ovulation and embryo manipulation in the improvement of beef cattle: Relative theoretical rates of genetic change. *J. Anim. Sci.* **67**: 2863–2871.

Ghosh, P.K. and Khan, M.S. (1980). The goat in desert environment Research Bulletin No. 12, CAZRI, Jodhpur.

Gibert, D.E., Coonrod, S.A., Whiting, C.J. and Pashen, R.L. (1990). Comparison of progesterone intravaginal device (CIDR) with flunixine meglumine (Gindyane) for reducing the effects of corpora lutea regression in the goat. *Theriogenology* **33**: 230.

Gill, G.S. and Dev, D.S (1972). Performance of two exotic breeds of goats under Indian conditions, *J. Anim. Prod.* **3(4)**: 173–178.

Gill, J.S., Singh, N. and Kwatra, M.S. (1983). An outbreak of fascioliasis in goats in the Punjab state and its treatment. *Indian J. Parasitol.* **7**: 215–216.

Goat Research Project (1979–80). Marathwada Agriculture University. Parabhani, (M.S.).

Goatcher, W.D. and Church, D.C. 1970. Taste responses of ruminants. *J. Anim. Sci.* **31**: 373.

Goel, A.K. and Agrawal, K.P. (1988a). Seasonality in breeding behaviour in goats under tropical cins. Proceedings of VII Annual convention of Indian Society for the Study of Animal Reproduction, Mannuthy, Trichur (Kerala), India. August 22–24.

Goel, A.K. and Agrawal, K.P. (1988b). Parturitionin goats. Bulletin No.13 CIRG, Makhdoom, Farah, Mathura, Uttar Pradesh.

Goel, A.K. and Agrawal, K.P. (1989a). Postpartum oestrous interval in Jakhrana goats. *Indian J. Anim. Reprod.* **10**: 175.

Goel, A.K. and Agrawal, K.P. (1989b). A study on the pubertal oestrus in Jakhrana goats. National Symposium on Applied Reproduction in Farm Animals and VIII Annual Convention of Indian Society for the Study of Animal Reproduction, Nov. 10–12, Anand (Gujarat). India. pp. 30.

Goel, A.K. and Agrawal, K.P. (1990). Superovulation and embryo collection Jamunapari goats. *Theriogenology* **33**: 232.

Goel, A.K. and Agrawal, K.P. (1991). Ovulation characteristics in Jakhrana goats. First National Seminar on Small Ruminants Reproduction. January 10–12, CSWRI, Avikanagar, Jaipur (Rajasthan), India. pp. 31.

Gogolin-Ewens, K.J., Meeusen, E.N.T., Scott, P.C., Adams, T.E. and Brandon, M.R. (1990). Genetic selection for disease resistance and traits of economic importance in animal production. Revue Scientifique et Technique de l'Office international des Epizooties **9**: 865–896.

Goyal, S.P. (1983). Coat colour and body temperature regulation in Indian desert goats. *Animals of the Arid Zone* **22**: 59–63.

Gray, A.P. (1972). "Mammalian Hybrids: A check list with bibliography". Commonwealth Agricultural Bureaux, England.

Greenhalgh, J.F.D. (1984). Upgrading crop and agricultural by products for animal production. *In*: Gilchrist, F.M.C. and Mackie, R.I. (*eds*), Herbivore Nutrition in the Subtropic and Tropics. The Science Press, Pretoria, South Africa. pp. 167–184.

Gregg, K. (1989). Enhancement of microbial detoxification by gene transfer. *In*: Proceedings of the 7th international Symposium on Ruminant Physiology, Sendai, Japan, Academic Press, San Diego, USA. p. 58.

Gribovskii, V.A. (1956). Breeding highly productive wool goats. *Ovtsevododstvo* **6**:27–29 (*Anim. Breed. Abstr*. **25**: 236).

Guha, H., Gupta, S., Mukherjee, A.K., Moulick, S.K. and Bhattacharya, S. (1968). Some causes of variation in the growth rates of Black Bengal goats. *Indian J. Vet. Sci. and Animal Husb.* **38**: 269–278.

Guha, S., Kohli, M.L. and Bhattacharya, P. (1951). Artificial insemination in sheep and goat at Izatnagar. *Indian. J. Vet. Sci.* **21**: 171–176.

Gupta, V.K., Mehrotra, S.N. and Batra, Y.K. (1985). Building Research Note No. **37**. C.B.R.I., Roorke.

Haberfeld, A., Cahaner, A., Yoffe, O., Plotsky, Y. and Hillel, J. (1991). DNA fingerprints of farm animals generated by microsatellite and minisatellite DNA probes. DNA probes. *Animal Genetics* **22**: 299–305.

Hahn, R. (1972). A contribution to deep freezing preservation of goat buck and ram semen. *World Rev. Anim. Prod.* **8**: 80.

Hamano, K., Sasada, H., Sugawara, S. and Masaki, J. (1985). *In vitro* fertilization in goats: *In vitro* fertilization of goat oocytes. *Jap. J. Anim. Reprod.* **31**: 32–39.

Handa, A. and Pao, S. (1984). *In vitro* fertilization and two-cell embryo transfer in the goat. *Jap. Soc. Zootech. Sci. Ann. Meet. Abstr.* p. 123.

Handa, B.N. (1946). The Punjab. *Indian Fmg.* **7**: 32–33.

Hang, J.C., Wen, S.H., Shieh, R.C. and Cheng, W.T.K. (1989). *In vitro* fertilization and early development of caprine oocytes and embryos. *Taiwan Livestock Research*. **22**: 39–45.

Hansen, K.M. (1973). Q-band karyotype of the goat (*Capra hircus*) and the relation between goat and bovine Q-bands. *Hereditas* **75**: 119–130.

Harak, G. (1969). Assessment of Milk Yield of gats in individual lactation. *Mas zin. Chor. Zvi*., pp. **88**–95 (fide *Anim. Breed. Abstr.* **40**: 3304).

Harris, D.R. (1962). The distribution and ancestry of the domestic goat. *Proc. Linn. Soc. London* **173**: 79–91.

Harsh, L.N. and Shankarnarayan, K.A. (1982). Performance of sheep and goat under different intensities of grazing on *Cenchrus pasture* dominated by *Zizphus nummularia*. Seminar on problems and prospects on Livestock Development. CAZRI, Jodhpur.

Hart, I.C. and Johnson, I.D. (1986). Growth hormone and growth in meat producing animals. *In*: Buttery, P.J., Haynes, N.B. and Lindsay, D.B. (*eds*), Control and Manipulation of Animal Growth. Butterworths, London, UK. pp. 135–159.

Helmke, E. (1952). Goat keeping in Germany. *Br. Goat. Soc. Mon. J.* **45**: 55–58.

Henderson, A.R. and McDonald, P. (1977). The effect of cellulose preparations on the chemical changes during the ensilage of grass in laboratory silos. *Journal of the Science of Food and Agriculture* **28**: 486–490.

Herr, C.M., Halt, N., Matthaei, K.I., Bradely, M.P. and Reed, K.C. (1990a). Sex of progeny from bovine embryos sexed with a rapid Y-chromosome-detection assay. *Theriogenology* **33**: 247.

Herr, C.M., Matthaei, K.I. and Reed, K.C. (1990b). Rapid, accurate sexing of livestock embryos. Proceedings of 4th World Congress on Genetics applied to Livestock Production, Edinburgh **16**: 334–343.

Hespell, R.B. (1987). Biotechnology and modifications of the rumen microbial system. *Proceedings of the Nutrition Society* **46**: 407–413.

Hofmayer, M. (1972). The goats of S. Africa. *Proc. S. Africa. Soc. Prod.* **S**: 117.

Holm, P., Petersen, B.A., Hepburn, J., Krogh, K., Dagnaes-Hansen, F. and Callesen, H. (1990). Transfer of Angora goat embryos imported into Denmark from New Zealand under quarantine conditions. *Theriogenology* **33**: 251.

Holmberg, S.D., Osterholm, M.T. and Senger, K.A. (1984). Drug-resistant salmonella from animals fed antimocrobials. *N. Engl. J. Med.* **311**: 617–622.

Honker (1952). *Berl. Munch. Tierarztl Wschr* **65**: 60 (fide *Anim. Breed. Abstr.* **21**: 278).

Hooper, P.G., Sharp, R., Rowlinson, P. and Armstrong, D.G. (1989). An Assessment of the Potential of a Lactobacilli Inoculant as a Silage Additive. British Society of Animal Production, Winter Meeting, 13–15 March. Paper 63. Scarborough, UK.

Horak, F. and Pindak, A. (1969). Persistence of lactation in goats Acta Univ. Agric. Facv. Agron. Brno. IZ: 299–306 (fide *Anim. Breed. Abstr.* **30**: 3875).

Horimoto, T. and Kawaoka, Y. (1995). Direct reverse Transcriptase PCR to determine virulence potential of influenza A virus in birds. *J. Clin. Microbiol.* **33**: 748–751.

Hsu, T.C. and Benirschke, K. (1969). "An Atlas of Mammalian Chromosomes". Vol. Lil Folio 141.

Hutchinson, I.C.D. and Morris, S. (1936). The digestibility of dietary protein in the ruminant. I. endogenous nitrogen excretion on low-N diet and in starvation. *Biochem. J.* **30(2)**: 168–247.

Illmensee, K. and Hoppe, P.C. (1981). Nuclear transplantation in *Mus musculus*: Development potential of nuclei from preimplantation embyos. *Cell* **23**: 918.

Indian livestock Census (1951, 1961, 1966, 1972, 1977, 1982, 1987, 1997, 2003). Directorate of Economics and Statistics, Ministry of Agriculture, New Delhi.

Ishwar, A.K. and Pandey, J.N. (1990). Oestrus synchronization and fertility behavior in Black Bengal goats following either progesterone or prostaglandin in treatment. *Theriogenology* **34**: 1015–1024.

Jaenisch, R. (1988). Transgenic Animals. *Science* **240**: 1468–1473.

Jain, S.K. (1977). Pathology of uterine chlamydial infection and perinatat mortality in goats. Agra University, Agra, India. Ph.D. thesis.

Jainkiraman, K. and Metha, D. (1990). Final Tech. Report of PL-480 Goat Project, Gujarat Agril. University, Anand, Gujarat.

Jasiorowski, H.A. (1990). Open nucleus breeding schemes – New challenge for the developing countries. *In*: Zurkowski, M. (*ed*), Proceedings of a Conference on Open Nucleus Breeding Systems, Biatobrzegi, Poland, 1–19 January 1989. FAO., Rome, Italy. pp. 7–12.

Jeffreys, A.J. and Morton, D.B (1987). DNA fingerprints of dogs and cats. *Animal Genetics* **18**: 1–15.

Jones, R.J. and Megarrity, R.G. (1986). Successful transfer of DHP degrading bacteria from Hawaiian goats to Australian ruminants to overcome the toxicity of *Leucaena*. *Australian Veterinary Journal* **63**: 259–262.

Jenkinson, D.Mc.E. and Robertshaw, D. (1971). Studies on the nature of sweat gland fatigue in the goat. *J. Physiol.* **212**: 445–665.

Johri, C.B. and Talpatra, S.K.(1971). Growth studies with Jamnapari goats. 1: Early growth of Jamnapari goats. *Indian Vet. J.* 48: 389–393

Jollans, J.L. (1959). Meat preferences of people in the central region of Ghana. *J. W. African Sci. Assoc.* **5**: 64.

Jolles, P. and Jolles, J. (1984). What is new in Lysozyme research? Always a model system today and yesterday. *Mol. Cell. Biochem.* **63**: 165–189

Jufen, Q., Zhiming, H., Yong, Z. and Jianchen, W. (1991). *In vitro* capacitation of ejaculated spermatozoa and *in vitro* fertilization in dairy goats. *Theriogenology* **35**: 219.

Jurik, J. (1971). Goat husbandry improvement by breeding. *Chov. Hospod. Zviedr* **31**: 32–35.

Kaditski, E. (1952). Study of indigenous goats in North-West. Bulgaria. *Dairy Sci. Abstr.* 1954, **16**: 797 (fide *Anim. Breed. Abstr.* **2S**: 240).

Kaditski, E. (1957). Studies of the results of crossing local goats with Saanen goats in North-West Bulgaria Nauch. Trud Seiskostop, Inst. "G" Dimitror. *Dairy Sci. Abstr.* **20**: 2243.

Kaura, R.L. (1943). Some common breeds of goats in India. *Indian Ins.* **4**: 549–552.

Kaura, R.L. (1952). Indian Breeds of Livestock Including Pakistan Breeds. pp. 19. Prem publishers, Lucknow (Uttar Pradesh), India.

Kelly, C.F., Bond, T.E. and Ittner, N.R. (1950). *Agric. Engrg.* **31**: 601.

Khan, B.U. (1980). Production and reproduction in Jamunapari goats. A review. *Agric. Rev.* **1**: 65–70.

Khan, B.U. and Deoghare, P.R. (1998). Village milk production from goats. Proc. Second Pan Common Wealth Veterinary Conference. Bangalore, 22–27 February.

Khan, B.U. and Garg, R.C. (1983). Post-weaning body weights in Jamunapari goats. *Indian J. Anim. Sci.* **17**: 72–76.

Khan, B.U. and Rai, B. (2000). Goat Breeds of India. Central Sheep and Wool Research Institute, Avikanagar.

Khan, B.U. and Raut, P.K. (1998). Small ruminant genetic Resources Conservation and Management in SAARC region. Proc. Fourth Global Conference on Domestic Animal Genetic Resources. Kathmandu, Nepal, 17–21 August.

Khan, B.U. and Sahni, K.L. (1982). Note on evaluation of certain non-genetic factors affecting body weight gains in Jamunapari goats under semi-arid conditions. *Indian J. Anim. Sci.* **52**: 1256–1257.

Khan, B.U. and Sahni, K.L. (1979). A study on carcass characteristics in Jamunapari kids. *Indian Vet. Med. J.* **3**: 77–80.

Khan, B.U. and Saini, A.L. (1989). Prospects of economic goat rearing in India. Proc. National Seminar on "Importance of goat in rural economy". Central Institute for Research on Goats, Makhdoom, 15 April.

Khan, B.U. and Singh, S.K. (1995). Genetics of feed conversion efficiency in Barbari goats. *Small Ruminant Research* **15**: 283–285.

Khan, B.U. and Singhal, R.A. (1980). Factors affecting body weights and measurements at birth in Jamunapari kids. Proc. 67th Session, *Indian Science Congress*, Calcutta, 1–5 Feb.

Khan, B.U. and Taneja, V.K. (1996). Current Development in Management of Small Ruminant Genetic Resources: South and South-East Asia. Lead paper. Proc. of IGA/Round Table on the Global Management for small ruminant Genetic Resources. Beijing, China, 6 May.

Khan, B.U., Rai, B., Raut, P.K. and Taneja, V.K. (1998). Small Ruminant Genetic Resources in Himalyan Hindu Kush region. Proc. Fourth Global Conference on Domestic Animal Genetic Resources. Kathmandu, Nepal, 17–21 August.

Khan, B.U., Roy, R., Prakash, B. and Singh, S.K. (1987). Annual Report, Central Institute for Research on Goats, Mathura, U.P. pp. 15–32.

Khan, B.U., Sinha N.K. and Sahni, K.L. (1982). Note on gestation length in Jamunapari goats. *Indian J. Anim. Sci.* **52**: 1258–1259.

Khan, B.U., Sinha, N.K., Wani, G.M. and Sahani, K.L. (1981). A note on breeding performance of Jamunapari goat. *Indian Vet. J.* **58**: 251.

Khan, B.U., Sinha, N.K., Wani, G.M. and Sahni, K.L. (1980). A note on breeding performance in Jamunapari goats. *Indian Vet. J.* **58**: 251–252.

Khan, B.U., Sinha, N.K., Wani, G.M. and Sahni, K.L. (1980). Lactation performance in Jamunapari goats under farm conditions. *Indian Vet. Med. J.* **4**: 129–130.

Khan, B.U., Sinha, N.K. and Sahni, K.L. (1979). Evaluation of certain live and carcass traits in Jamunapari kids slaughtered at two ages. *J. Rem. Vet. Corps*. **18**: 201–206.

Khan, B.U., Sinha, N.K., Singhal, R.A. and Sahni, K.L. (1979). Variability in body weight and size in Jamunapari kids at birth. *Indian J. Anim. Res*. **13**: 19–22.

Khan, B.U., Vihan, V.S. and Sahni, K.L. (1978). A note on mortality in Jamunapari kids. *Indian Vet. Med. J*. **2**: 225.

Khan, B.U., Vihan, V.S. and Sahni, K.L. (1978). A note on mortality in Jamunapari kids. *Indian Vet. Med. J*. **2**: 225.

Khattar, S.S. (1975). Studies on performance of Alpine, Beetal and Alpine × Beetal crossbred goats. M.Sc. dissertation, Kurukshetra University, Kurukshetra, India.

Khavary, H. (1973). A proposed method for classifying chromosomes of *Capra hircus*. *Journal of Veterinary Faculty*, Unversity of Tehran **29(1)**: 69–71.

Khera, R.C. (1965). Sheep and goat breeding and wool technology coordination of animal husbandry research in India. I.C.A.R., Vol. II, No. 3.

Kiessling, A.A., Hughes, W.H and Blankevoort, M.R. (1986). Superovulation and embryo transfer in the dairy goat. *J.A.V.M.A*. **188**: 829–832.

Kijatkin, P.E.(1949). Segeration through crossing in goats with black down. *Angrobiologija* **2**: 112–115 (fide *Anim. Breed. Abstr*. 17, No. 1383).

Kim, J.J., Tivedi, N.N., Nottingham, L.K., Morrison, L., Tsai, A., Hu, Y., Mahalingam, S., Dang, K.M., Ahn, L., Doyle, N.K., Wilson, D.M., Chatergoon, M.A., Chalian, A.A., Byer, J.D., Agadjayan, M.G. and Weiner, D.B. (1998). Modulation of amplitude and direction of *in vivo* immune response by co-administration of cytokine gene expression cassettes with DNA immunogens. *Eur. J. Immunol*. **28**: 1089–1103.

King, W.A. (1984). Sexing embryos by cytological methods. *Theriogenology* **21**: 7–17.

Krishna, L. and Rajya, B.S. (1985). Incidence of perinatal chlamydiosis. *Indian J. Pathol*. **9**: 62–65.

Kulshrestra, R.C., Kalra., D.S. and Kapur, M.P. (1978). Seroepidemiological studies on Brucellosis in mammals and management in Haryana State. *Indian Vet. J*. **55**: 1–3.

Kumar, S. and Narang, M.P. (1991). Effects of different planes of nutrition on the performance of gaddi male kids using concentrate rations and kikuyu grass hay. *Indian J. Anim. sci*. **61**: 869-872

Kumar, Puneet, Singh, D. and Singh, Khub (1990). Heat tolerance in three breeds of Indian goats. Proc. of the Congr. Asian and Oceanian Physiol. Soc. AIIMS, New Delhi.

Kumar, S., Singh, G. and Kumar, A.N. (1962). Indian J. Dairy Science. 15, 101-104.

Kundu, S.S. and Kumar, V. (1987). Biotechnological processes to convert crop residues to quality feeds for livestock. *Indian Dairyman* **39**: 113–114.

Kurar, C.K. and Mudgal, V.D. (1979). Effect of plane of nutrition on the utilization dietary nitrogen in Beetal goats. Paper presented in Dairy Industry Conference, Hyderabad, Feb. 24–27.

Kurar, C.K. and Mudgal, V.D. (1990). Effect of plane of nutrition on the utilization of dietary energy in Beetal goats. *Indian J. Anim. Sci.*

Kuriakose, K.K. (1981). Observation on gestation and parturition in goat (*Capra hircus*). M.V.Sc. Thesis, Kerala Agricultural University, Mannuthy, Trichur, India.

Lal Krishna, Paliwal, O.P. and Kulshrestha, S.B. (1978). Joint ill (Arthritis) in lambs and kids. *Indian Vet. Med. J*. **2**: 101.

Lall, H.K. (1947). Some common breeds of goats in India. III. *Indian Fmg*. **8**: 322–327.

Lall, H.K. (1968). Crossbreeding for Indian type Angora goat. *Indian Fmg*. **17**: 45–47.

Laltim, H. and Singh, J. (1949). Some observations on mortality in goats. *Indian J. Vet. Sci*. **19**: 261–273.

Lee, K.W., Choi, R.S., Tak, T.S. and Sul, D.S. (1974). Improvement of Korean native goats by grading up with Saanen. II. Change in milk performance of crossbreeds between Korean natives goats and Saanen. Research Reports of the ou ce Rural development, Suwon, *Livestock* **6(8)**: 7–14.

Lie, O. (1980). Genetic variation in serum lysozyme activity in cattle. *Acta Veterinaria Scandinavica* **21**: 448–450.

Lie,O. and Solube, H. (1983). Evidence for a major gene regulating serum lysozyme activity in cattle. *Zietschrift fur Tierzuchtung und Zuchtbiologie* **100**: 134–138

Lie, O. (1980). Genetic variation in serum lysozyme activity in cattle. *Acta Veterinaria Scandinavica* **21**: 448–450.

Lemaitre, Y. (1969). The Angora goat and mohair production in tuleae province. *Terra malquache* **5**: 185–219 (fide *Anim. Breed. Abstr*. **38**, No. 1514).

Levi, M.F. (1946). Crossing woolled goats of the Caucasus and Central Asia with Angora goats. Dokl. Nauc. Konf. Timivajazev, *Selijskohoz Akad.* **4**: 202–207 (fide. *Anim. Breed. Abstr.* **17**: p. 42, No. 132).

Loftus, R.T., MacHugh, D.E., Negere, L.O., Balain, D.S., Badi, A.M., Bradley, D.G. and Cummingham, E.P. (1994). Mitochondrial genetic variation in European, African, and Indian cattle populations. *Animal Genetics* **25**: 265–271.

Lohan, I.S., Singhal, S.P., Kker, M.L. and Deshpande, L. (1989). Superovulatory response in anoestrous Black Bengal goats using Synchromate B and Rolligon. *Indian J. Dairy Sci.* **42**: 358–363.

Louca, A., Mavzogenis, A. and Lawlor, M. (1975). The effect of early weaning on the lactation performance of Damascus goats and the growth rate of the kids. *Anim. Prod.* **20(2)**: 213–214.

Lu, C.D. (1988). Grazing behaviour and diet selection of goats. *Small Ruminant Research.* **1**: 205–216.

Lupton, C.J. (1992). Characterization and end uses of Mohair and Cashmere. Proceedings of Fifth International Conference on Goats, March 2–8, 1992, New Delhi, India.

Lyon, M.F., Whittingam, D.G. and Glenister, P. (1977). Long-term storage of frozen mouse embryos under increased background irradiation. The freezing of Mammalian embryos. pp. 273–282. (*Eds*) Elliot, K. and Whelan, J. Ciba Fondation Symposium No. 52 (new series). Elsevier, Amsterdam.

Mackenzie, D. (1967). "Goat Husbandry", Faber and Faber, London.

Mackenzie, D. (1970). Goat Husbandry. 3rd edn. Faber and Faber Ltd., London, England.

Mackinnon, M.J. (1992). Use of genetic markers in genetic improvement programmes. *Proceedings of the Australian association of Animal Breeding and Genetics* **10**: 245–251.

Magdub, A.B., Baccari (Jr.), Concalves, H.C. and Polastere, R. (1987). Effects of heat stress on the growth performance of kid. Proc. of the IV introduction Conf. on Goats, Brazil, Vol. II. pp. 1371.

Mahmud, A.B. and Devendra, C. (1966). Repeatability of milk yield and birth weights of goats in Malaya. I. Milk yield and length of lactation. Expt. Auric. 29.

Majumdar, B.N. (1960a). Studies on goat nutrition. 1. Minimum protein requirements of goats for maintenance-endogenous urinary nitrogen and metabloic faecal nitrogen excretion studies. *J. Agric. Sci.*, Camb. **54**: 329–334.

Majumdar, B.N. (1960b). Studies on goat nutrition. 2. Digestible crude protein requirements for maintenance from balance studies. *J. Agric. Sci.*, Camb. **54**: 335–340.

Mali, P.C., Patnayak, B.C. and Bedekar, A.R. (1983). Plane of nutrition of Marwari ewes in different seasons grazing on natural range pastures of area region. *Ann. Arid Zone* **22**: 223–226.

Maltz, D., Silanikove, N. and Shlokink, A. (1982). Energy cost and water requirement of Black Bedouin goats at different levels of production. *J. Agril. Sci.*, Camb. **98**: 499–504.

Marincowitz, G. (1958). The mohair industry. I. Mass slaughter saved South-African goats population. Fmg. *S. Afr.* **34**: 24–25 (fide *Anim. Breed. Abstr*. **27**: 319).

Markendya, N.M. and Pargaonkar, D.R. (1990). Studies on biochemistry of Osmanabadi buck semen. *Indian J. Anim. Reprod.* **11**: 36–38.

Mason, I.L. (1951). A world dictionary of Breeds, Types and Varieties of Livestock. Technical Communication No. 8. Farnham Royal, Bucks, Commonwealth Bureau of Animal Breeding and Genetics.

Mason, I.L. (1988). A world dictionary on livestock breeds, types and varieties. CABI, Farnham Royal, England.

Mason, I.L. and Maule, I.P. (1960). The Indigenous Livestock of Eastern and Southern Africa. Farnham Royal Buck. U.K. Commonwealth Agricultural Bureau, pp: 151.

Mathai, E. and Nair, B.R.K. (1981). Effect of age semen on conception rate in goat. Kerala *J. Vet. Sci.* **12**: 220–223.

Maule, J.P. (1966). A note on dairy goats in the tropics. *Anim. Breed. Abstr.* **38**(2): 153–158.

Maynard, L.A. (1951). Animal Nutrition. McGraw-Hill Book Co., New York, USA.

McDowell, R.E. (1974). Animal Production of in World Food Supplies. Cronell International Aeric. Mimeo. No. 45. Cornell Univ., Ithaca N.Y.

McDowell, R.E. (1972). Improvement of livestock production in warm climates. W.H. Freman and Co., San Francisco.

McDowell, R.E., Cestero, H., Rivera-Anaya, J.E., Roman, F., Garcia Arroyo-Agrilu, J.A., Berrocal, C.M., Soldevile, I.U., Lopez-Abberty, J.C. and Metz, S.W. (1975). Tropical grass Pastures with and without supplement for lactating cow's in Puerto-Rico. University. Puerto Rci, Agol. Experiment Sta, Bull No. 23s.

McGarth, J. and Solter, D. (1983). Nuclear transplantation in the mouse embryo by microsurgery and cell fusion. *Science* **200**: 1300–1302.

Mckelvey, W.A.C. and Bhattacharyya, N.K. (1992). Embryo biotechnology in goats. Planary papers and invited lectures. V introduction. Conf. on Goats, New Delhi, 2–8 March. pp. 51–70.

Mckelvey, W.A.C., Robinson, J.J. and Aitken, R.P. (1958b). A simplified technique for the transfer of ovine embryos by laparoscopy. *Vet. Record.* **117**: 492–494.

Mckelvey, W.A.C., Robinson, J.J., Aitken, R.P. and Robertson, I.S. (1986). Repeated recoveries of embryos from ewes by laparoscopy. *Theriogenology* **25**: 855–865.

Mckelvey, W.A.C. (1992). Unpublished data. Ref. Mckelvey and Bhattacharyya.

McLaren, A. (1976). Mammalian Chimears. Cambridge University Press.

McNatty, K.P., Hudson, N.L., Ball, K., Mason, A. and Simmons, M.H. (1989). Superovulation and embryo recovery in goats treated with Ovagen and follitropin. *New Zealand Vet. J.* **37**: 27–29.

McVicar, J.W. and Sutmoller, P. (1972). Experimental foot and mouth disease in sheep and goats. An epizootiological model. *Archive fuer die Gesamte Virusforschung.* **38**: 85–96.

Meinecke, B. and Meinecke-Tiilman, S. (1986). Comparison of laparoscopy for controlled transmural ultra-uterine insemination in superovulated sheep and goats. *Tieratztliche Praxis* **14**: 35–41.

Meloen, R.H. (1997). Synthetic peptide Vaccines. *In*: Veterinary Vaccinology, Pasturet p.p. Blancou, J., Vannier, P. and Vershueren, C., *eds*. Elsevier Science VB, Amsterdam, The Netherlands, pp. 272–276.

Meyer, E.H. (1984). Classification of *bos indicus* cattle breeds in Southern and Central Africa and Sanga or Zeby type by means of Y-chromosome morphology. VI European Colloquium on Cytogenetics of Domestic Animals. Zurich, Switzerland. pp. 96–103.

Misarev, S.S., Zaporozhtnev, E. and Orlyansku, V.A. (1974). The importance of selection Don goats for milk production. Trudy, Vsesoyuznyl Nauchoro-Issledovatel Sku. Institute. *Ovtsevododstvo Kozovodstva* **33** (**1**): 43–47.

Mishra, R.K. (1985). All India Coordinated Research Project. Terminal Report. Central Institute for Research on Goats, Makhdoom, Mathura, India.

Mishra, R.R.(1978). Management practices for Goats. NDRI, Karnal, Haryana. Pub. No. **147**.

Mishra, R.R. (1976). Management practices for goats. NDRI. (ICAR), Karnal.

Mishra, H.R. and Biswas, S.C. (1966). A study on the distribution of oestrus in desi goats. *Indian J. Dairy Sci.* **19**: 132–134.

Mishra, R.K., Nivsarkar, A.E. and Arora, C.L. (1979). A note on the analysis of gestation length in Sirohi goats. *Indian J. Anim. Sci.* **49**: 967–968.

Mishra, R.R., Bhatnagar, D.S. and Sundaresan, D. (1976). Heterosis of various economic traits in Alpine × Beetal crossbred goats. *Indian J. Dairy Sci.* **76**(29): 235.

Misra, R.K. and Mukundan, G. (1990). Milk production potential in small ruminant production in India by the year 2000. Proceedings of the National Seminar held in Tirupati, 13–17 Nov.

Mitchell, H.H. (1929). The minimum protein requirement of cattle. NRC bulletin No. **67**.

Mittal, J.P. (1991). Performance of Kutchi goat in arid western Rajasthan. *Indian J. Anim. Sci.* **61**: 904–995.

Mittal, J.P.(1988). Breed Characterization of Marwari goat of arid western Rajasthan. *Indian Journal of Animal Sciences* **58**: 357-361

Mittal, J.P. (1987). Male reproductive characteristics of indigenous and crossbred goats under Indian arid zone. *Indian J. Anim. Sci.* **57**: 158–161.

Mittal, J.P. (1985). Libido and semen quality of Jamunapari bucks under arid conditions. *Indian Vet. J.* **62**: 179.

Mittal, J.P. (1976). A study on mortality in kids. *Indian Vet. J.* **53**: 681.

Mittal, J.P. and Pandey, M.D. (1978). A Study on Growth Rate in Barbari Goats. Indian Veterinary Journal 55: 470-474.

Mittal, J.P. and Pandey, M.D. (1972). Evaluation of Semen quality of Barbari and Jamunapari bucks. *Indian J. Anim. Prod.* **2**: 14–19.

Mohmood, S., Kaul, G.L. and Biswas, J.C. (1989). Superovulatory response in Pashmina goats. *Vet. Rec.* **125**: 117.

Moore, E.L. (1966). Practice stall feeding of dairy goats. Paper presented at the dairy goat seminar on May 18, at the international house, Lodi Estate Road, New Delhi.

Moore, S.S., Barendse, W., Berger, K.T.M, Armitage, S.M. and Hetzel, D.J.S. (1992). Bovine and ovine DNA microsatellites form the EMBL and GENBANK databases. *Animal Genetics* **23**: 463–467.

Morand-Fehr, P. and Sauvant, D. (1978). Nutrition and optimum performance of dairy goats.

Moulick, S.K. and Systrad, O. (1970). Genetic and environmental causes of variation in birth weight of Black Bengal goats. *J. Agric. Sci.*, Cambridge 74: 409–414.

Livestock Prod. Sci. **5**: 203–213.

Moulik, S.K., Guha, S.H., Gupta, S., Mitra, D.K and Bhattacahraya, S. (1966). Factors affecting multiple birth in Black Bengal goats. *Indian J. Vet. Sci. and Anim. Husb.* **36**: 154–163.

Mukherjee, R.D. (1963). On two new species of amphistomes from Indian sheep and goats. *Indian J. Helminth.* **15**: 70–76.

Mukundan, G.(1976). "Proc. 2nd Workshop on All India Coordinated Research Project on Goat Breeding." NDRI, Karnal, 22-23.

Mukundan, G., Bhat, P.N. and Khan, B.U. (1983). Reproductive traits in Malabari goats and their halfbreeds with Saanen. *Indian J. Anim. Sci.* **52**: 90–93.

Na, J.S., Kim, Y.S. and Kang, B.K. (1987). Embryo transfer in Korean native goat. Korean *J. Anim. Sci.* **29**: 295–302.

Nair Krishana, B.R. (1978). Goats for milk production performance of crossbreds. Presented at Seminar on Progress of Research in animal genetics and breeding, held at NDRI, Karnal, 12–15 June.

Nair, K.P., Mathaim, E. and Nand Kumaran, M. (1982). Note on sodium and potasium in semen of bucks. *Indian J. Anim. Sci.* **52**: 608.

Nanda, P.N. 1942. The Punjab. *Indian Fmg.* **3**: 41–42.

Nandy, D.K., Agrawal, K.P. and Bhattacharyya, N.K. (1989). Totipotency of 2–4 and 8 cell caprine embryo. Abstr., 5th API Conference held at Matupatti, Kerala, 27–30 Dec. pp 1–21.

Nandy, D.K., Agrawal, K.P. and Bhattacharyya, N.K. (1989). Viability of single blastomere of 2 cell embryo (ovine). Proc. 76th Session, Indian Sci. Congr., Section physiology, Madurai. pp. 1–21.

Nandy, D.K., Agrawal, K.P. and Bhattacharyya, N.K. (1990). Ovarian response to superovulation in non-descript does. *Indian J. Anim. Sci.* **60**: 972–973.

Naryanan, L.R.A. and Ayyangar, R.S. (1987). Remote sensing of rangeland development and management. Souvenir, third International Rangeland Congress, Range management Society of India, IGFRI, Jhansi, India. pp. 1–5.

National Dairy Research Institute (1978). Annual Report, NDRI, Karnal (Haryana).

National Dairy Research Institute (1980). Annual Report, NDRI, Karnal (Haryana).

Niekerk von, C.H., Barry, D.M., Rust, J.M., Van der walt, T. and Langenhoven, J. (1990). Cervical softening with prostaglandin E2 and oestradiol cypionate for embryo collection in goats. *Theriogenology* **33**: 348.

Niemann, H., Halter, R. and Paul, D. (1994). Gene transfer in cattle and sheep: A summary perspective. *In*: Proceedings of the 5th World Congress on genetics applied to Livestock Production, Guelph, Canada, 712 August 1994. Vol. 21. International Committee for World Congress on Genetics Applied to Livestock Production, Guelph, Ontario, Canada. pp. 339–346.

NRC (1981). Nutrient requirement of goats. No.15, National Academy of Sciences, Washington D.C.

NRC (1981). Nutrient requirements of goats: Angora, dairy and meat goats in temperate and tropical countries. National Academy Press Washington, D.C.

NRC (1971). Nutrient requirement of Dairy Cattle. 4th edn. National Academy of Sciences, Washington, D.C.

Ocampo, M.B., Uenishi, R.S., Valdez, C.A Muralla, Z., Hishimuma, M. and Kanagawa, H. (1988). Superovulation response of upgraded indigenous Philippine goat (*Capra hircus*). *Japanese J. Vet. Research*. **36**: 249–255.

Opstvedt, J. (1967). Studies on the effect of energy level of feed combination on feed intake, milk yield composition and orgnaoleptic properties, bull. No. 134. Inst. of Ani. Nutr., the Afric. Colleage of Norway. p. 114. *Ovtsevododstvo* **3**: 32 (fide *Animal Breeding Abstracts* **49**: 3976).

Paliwal, O.P., Lal Krishna and Kulsherstha, S.B. (1978). Studies on the mortality in lambs and kids. An incidence of mortality. *Indian Vet. Med. J.* **2**: 191.

Palmiter, R.D., Brinster, R.L., Hammer, R.E., Trumbauer, M.E., Rosenfeld, M.G., Birnberg, N.C. and Evans, R.M. (1982). Dramatic growth of mice that develop from eggs microinjected with metallothionine growth hormone fusion genes. *Nature* **300**: 611–615.

Pant, K.P. (1969). Medullated Mohair fibres of Angora, Gaddi and their crossbred goats. *Indian Vet. J.* **46**: 125–132.

Pant, K.P. (1968). Studies on birth weight Mohair yield and Mohair fibre length of Angora and Angora X Gaddi goats. *Indian Vet. J.* **45**: 929–939.

Pant, K.P. and Kapri, B.D. (1968). Fineness of Mohair fibres of Angora Gaddi and their crossbred goats. *Indian J. Vet. Sci.* **38**: 112–116.

Parendekar, D.R., Pargaonkar, D.R., Bakshi, S.A. and Markendya, N.M. (1991). Cytometry of buck spermatozoa. *Indian J. Anim. Reprod.* **12**: 95–96.

Paterson, A.H., Lander, E.S., Hewitt, J.D., Peterson, S., Lincoln, S.E. and Tanksley, S.D. (1988). Resolution of quantitative traits into Mendelian factors. *Nature* **355**: 721–726.

Pathak, R.C. (1968). The incidence of Brucellosis in domestic animals in Uttar Pradesh with a note on a preliminary survey of human Brucellosis in Mathura district. *Bulletin of the Indian Society for Malaria and Other Communicable Disease* **5**: 214–216.

Pathak, R.C. and Singh P. (1984). The genital mycoplasm infection in domestic animals with particular reference to prevalence and pathogenicity. Final DST project report on Mycoplasma. College of Veterinary Science, Mathura, India. pp. 111.

Pathak, S. and Kiefer, N.M. (1979). Sterility in hybrid cattle. 1. Distribution of constitutive heterochromatin and nucleolar organizer ferions in somatic and meiotic chromosomes. *Cytogenetics and Cell Genetics* **24**: 42–52.

Patil, R.V. and Raja, C.K.S.V. (1973b). The fructose, citric acid and ascorbic acid content in the semen of Malabari bucks. *Kerala J. Vet. Sci.* **4**: 94–95.

Patil, R.V. and Raja, V.K.S.V. (1973a). Biometrics of sperms of Malabari bucks. *Kerala J. Vet. Sci.* **4**: 162–164.

Patil, R.V. and Raja, C.K.S.V. (1978). Effect of the season on semen characteristics of Malabari bucks. *Indian Vet. J.* **55**: 761–766.

Patil, V.K. (1978). A Review on Goats for Fibre (Mohair, Pashmina). A paper presented at the Seminar on progress of research in animal genetics and breeding, NDRI, Karnal, India.

Patil, V.K. (1992). Genetic potential of goat for wool production research in goats: Indian Experience. (*Eds*) Bhattacharyya, N.K., Singh, Lokeshwar, R.R. CIRG, Makhdoom, Mathura, India.

Patre (1966). Milk performance of recorded goats. No. **139**: 23 (Fr). *Anim. Breed. Abstr.* **S5**: 1561.

Pattabhiraman, D. (1955). Milch goats of Malabar and their development. *Indian Vet. J.* **31**: 398–402.

Pegg, S.E. (1957). Production recording of goats. *Agric. J.* **83**: 659–660. (fide *Anim. Breed. Abstr.* **Y6**: 878).

Pinheiro, L.E.L., Moraes, J.C.F., Mattevi, M.S., Erdtnmann, B., Sakzabd, F.M. and Filho, A.M. (1980). Two types of Y-chromosomes in a Brazillian cattle breed. *Caryologia* **33**: 25–32.

Pizada, W.H., Anwar, M. and Mehmood, A. (1988). Possible role embryo transfer technology can play in improvement of sheep and goat breeding in Pakistan. *J. Anim. Health Prod.* **8**: 33–39.

Pradhan, S.L. (1979). Goat development in Nepal. Presented at the workshop "Goat development in Asia": March 5–10, NDRI, Karnal.

Prajapati, M.C., Tieari, A.K., Nambiar, K.T.N., Singh, J.P., Malthora, B.M. and Sharda, V.N. (1988). Effect of different intensities of grazing by goats on vegetation cover, run off and soil loss in a forest water shed in Yamuna ravines *vis-a-vis* animal production. Abstract Vol. II. 3rd International Range land Congress. RMSI, Jhansi pp. 419–422.

Prakash, B. (1977). Choromosome study in goats. M.Sc. disseration, Kurukshetra University, Kurukshetra, India.

Prakash, B. and Khan, B.U. (1988). Postpartum oestrus in Barbari goats. National Symposium and IV Annual Conference of SAPI, CIRG, Makhdoom, Farah, Mathura, Uttar Pradesh, India. Sept. 24–26, (Abstr.) 7.19, pp. 65.

Prakash, C. and Khanna, R.S. (1972). Effect of order of lactation and lactation length on milk production in close herd of Beetal goats. *Indian J. Anim. Prod.* **2**(4): 10–13.

Prakash, C., Acharya, R.M. and Dhillon, G.S. (1971). Sources of variation in milk production in Beetal goats. *Indian J. Anim. Sci.* **41**: 356–360.

Prasad, S.P. (1979). A note on the characteristics of postmartum oestrus in Barbari nannies. *Indian J. Anim. Sci.* **49**: 583–585.

Prasad, S.P. (1974). Studies on some aspects of reproduction in Barbari nannies. Ph.D. thesis. Agra University, Agra, Uttar Pradesh, India.

Prasad, V.S.S.A. and Agnihotri, M.K. (1991). Annual Report, 1990–91, CIRG, Makhdoom, U.P., India.

Prasad, S.P. and Bhattacharyya, N.K. (1979a). Oestrus cycle and behaviour in different seasons in Barbari nannies. *Indian Vet. J.* **58**: 380–387.

Prasad, S.P. and Bhattacharyya, N.K. (1979b). Incidence of onset and cessation of oestrus at different day hours in nullipara Barbari nannies. *Indian. J. Anim. Sci.* **49**: 1055–1057.

Prasad, S.P. and Bhattacharyya, N.K. (1977). A note on the characteristics of pubertal oestrus and anestrous cycle in Barbari nannies. *Indian J. Anim. Sci.* **49**: 969–970.

Prasad, S.P. and Pandey, M.D. (1981). Observation on sexual activity and foetal membranes in Barbari nannies. *Indian Vet. J.* **58**: 380–387.

Prasad, S.P., Das, S.C. and Bhattacharyya, N.K. (1980). A study on the characteristics of Graafian follicles and ovulaion sequence in multipara Barbari nannies. *Indian J. Anim. Sci.* **50**: 53–57.

Prasad, S.P. Roy, A. and Pandey, M.D. (1971). Influence of age, body weight, para and season on the reproductive performance of Barbari goats. *Agra University J. Research*

(Sci.). **20**: 31–39.

Prasad, S.P., Roy, A. and Pandey, M.D. (1970). Effect of age on semen quality and development of sex libido in Barbari males. *Agra University J. Research. (Sci.)* **19**: 23–30.

Prasad, V.S.S., Sinha, N.K., Singh, S.K. and Khan, B.U. (1992). Influence of age-by carcass weight classification on carcass traits of Barbari and Jakhrana male goats. *Indian. J. Anim. Sci.* **62**: 374–377.

Prather, R.S., Barnes, F.L., Sims, M.L., Robl, J.M., Eyestone, W.H. and First, N.L. (1987). Nuclear transfer in the bovine embryo: Assessment of donor nuclei and recipient oocyte. *Biol. Reprod.* **37**: 859–866.

Prather, R.S., Sims, M.L. and First, N.L. (1989). Nuclear transplantation in early pig embryos. *Biol. Reprod.* **41**: 414–481.

Project Coordiantor's Report. (1989). All India Coordinated Research Project on Goats. Central Institute for Research on Goats, Mathura, U.P., India.

Puranik, S.V. (1991). Andrological investigation and cytometry of spermatozoa. Personal communication, M.A.U., Parvani.

Rahimov, A.A. (1957). Breeding wool goats at the collection farm Komsomol in summerkanal Province *Ovtsevododstvo* **3**: 35–38.

Rahman, A., Hossain, A., Ahmed, M.U. and Sen, M.M. (1977). Studies on some reproductive performances and biomerty of female genital tract of Black Bengal goat. *Indian J. Anim. Sci.* **47**: 724–725.

Rai, A.K., Mehtha, B.S., Gour, D. and Singh, M. (1979). Sweating in sheep and goats. *Indian J. Anim. Sci.* **49**: 546–548.

Rai, G.S. and Chorey, P.A. (1965). Lactation performance of Jamunapari and Barbari goats. *Indian Vet. J.* **42**: 958–961.

Raicu, P., Tumensan, H. and Lungeanu, A. (1977). Comparative study of the G- and C-model bands on 3 bovine species. Introduction. *Congr. Genetic (Moscow)* **29**: 2677.

Rajpoot, R.L., Sengar, O.P.S. and Singh, S.N. (1979). Data generated in pursuance of Ph.D. programme energy and protein in goat nutrient W of the Senior author (unpublished). Department of Animal Husbandry and Dairying, R.B.S. College, Bichpuri, Agra, India.

Rako, A. (1950). Performance of crosses between indigenous and Saanen goats. *Schweiz. Arch. Tierheilh* **92**: 305–320.

Rana, C.L. (1966). Souvenir of first dairy industry seminar held on May 18 at the Internationl house, Lodi Estate Road, New Delhi, under the auspices of Indian Dairy Goat Association.

Randhawa, M.S. (1983). A history of agriculture in India. Vol. III, ICAR, New Delhi, pp. iii-v.

Rao, B.V. (1971). Biology and control some tape worm infection domestic animals. Ph.D. thesis, Madras University, Madras, India.

Rao, V.H. (1984). *In vitro* fertilizer of ova and preservation of zygote embryo of goats. Ph.D. thesis, Madras University, Madras, India.

Rao, V.H. and Bhattacharyya, N.K. (1980). Ovulation in Bengal nanny goats. *J. Reprod. Fertil., Cambridge* **48**: 67–69.

Rao, V.H., Agarwal, K.P., Mongha, I.V. and Bhattacharyya, N.K. (1982). Note on the effect of certain stimuli on the ovulation in Black Bengal nanny goats. *Indian J. Anim. Sci.* **52**: 117–118.

Rao, V.H., Sarmah, B.C. and Bhattcharyya, N.K. (1984). Xenogenous fertilization of goat ova in the rabbit oviduct. *J. Reprod. Fertil., Cambridge* **71**: 377–379.

Rao, V.H., Sarmah, B.C., Ansari, M.R. and Bhattcharyya, N.K (1983). Survival of goat embryo after rapid freezing and thawing. Proc. V World Conf. Anim. Prod. Japan. Abstr. No. II-64.

Randhawa, M.S. (1982). A History of Agriculture in India; Vol. II. ICAR., New Delhi, pp. x-xi.

Rathore, A.K. (1970). Goats-milk production and the effect of age on milk quality. *Aust. Goat. Wld.* **22** (1): 1–2.

Raut, K.C. and Nadkarni, U.G. (1974). Cost of rearing sheep and goats under migratory and stationary conditions. *Indian J. Anim. Sci.* **33**: 459–463.

Read, W.S. (1940a). The Angora and the mohair industry, II. *Indian Fmg.* **1**: 328–333.

Read, W.S. (1940b). The angora goat in the mohair industry. *Indian Fmg.* **1**: 162–165.

Rebsburg, S.J. (1970). Reproductive physiology and endocrinology of normal and habitually aborting Angora goats. A thesis for Doctor of Veterinary Science, Dept. of Physiology, Faculty of Veterinary Science, University of Protoria, South Africa.

Rekib, A. and Patil, B.D. (1981). Feed resources in the plain and coastal areas for sheep and goats. Paper presented in the First National Seminar on sheep and goat production and utilization, Jaipur. 11–14 April.

Rexroad Jr., C.E., Hammer, R.E., Behringer, R.R., Palmiter, R.D. and Brinster, R.L. (1990). *J. Reproduction and Fertility (supplement)* **41**: 199–124.

Richards, K.L. (1959). Annual Report for 1955 of the agriculture office. Annexure-II, Department of Agriculture, Cyprus.

Ricordeau, G. (1963). Possibilities of selection in the goat. *Bull. Tech. Ing. Ser. Agric.* No. **179**: 16.

Ricordeau, G. (1964). Improvement of goat breeding in France and problems connected. *Tierziichter* 133–134 (German) (fide *Anim. Breed. Abstr.* **32**. 215).

Robinson, H.L. (1997). Nucleic acid Vaccines: an overview. *Vaccine* **15**: 785–787.

Rodinson, C. and Vida, C. Anim. Breed. Abstr. **37**: 2664.

Rodriguez, M. and Schudel, A.A. (1993). Nucleic acid Hybridization and Polymerase chain reaction in the diagnosis of infectious animal disease. *Rev. Sci. Tech. Off. Int. Epiz.* **12**: 405–423.

Romero, C.H., Barrett, T.M., Evans, S.A., Kitching, R.P., Gershon, P.D., Bostock, C. and Black, D.N. (1993). Single capripoxvirus recombinant vaccine for the protection of cattle against rinderpest and limpy skin disease. *Vaccine* **11**: 737–742.

Rong, R., Guangya, W., Jufen, Q. and Jiancher, W. (1989). Simplified quick freezing of goat-embryos. *Theriogenology* **3**: 252.

Ronningen, K. (1965). Causes of variation in milk yield in 90 goats. Meld. Norg., Lndbar Hogok **43**(18): 20.

Ronningen, K. (1967). Study of genetic parameters for milk characteristics in goats. *The Agricultural College of Norway, Institute of Animal Genetics and Breeding.* Report No. **232**: 46–16.

Roy, A. (1957). Egg coagulating enzymes in the semen and cowper's gland of the goat. *Nature* **179**: 318.

Roy, A. and Gupta, D.N. (1959). Studies on reproduction physiology of goat. *Indian J. Vet. Sci.* **29**: 55.

Roy, A., Dutta, I.C., Sahni, K.L., Singh, B. and Sengupta, B.P. (1962). Studies on certain aspects of sheep and goat husbandry, artificial breeding telescoping the breeding season and certain reproductive phenomenon in sheep and goats. *Indian J. Vet. Sci.* **32**: 269–275.

Roy, R. (1991). Chambal queen – Jamunapari goat. *Dairy guide* **13**: 48–49.

Sabnis, S.R. (1991). Results of A.I. in rural goats using freshly diluted semen on same day of storage (12 hours). Personal communication, Rural Agriculture Institute, Naraingoan, (M.S.).

Saha, R.C. and Gupta, B.N. (1987). Tree leaves as feed for dairy cattle in India. *Indian Dairyman* **39**: 489–492.

Sahni, K.L. (1967). Studies on quality, preservation and fertility of semen on sheep and goats. Ph.D. Thesis, Agra University, Agra, India.

Sahni, K.L. (1987). Practical aspect of artificial insemination of goats in India. Proceedings, IV International Conference on goats, Vol. II Brasilia, Brazil, March 8–13, pp. 549.

Sahni, K.L. (1979). Reproductive physiology and constraints to A.I. in goats. *Assam Livestock* **4**: 5–6.

Sahni, K.L. and Roy, A. (1972). A study on the effect of deep freezing (-79°C) on post thaw revival of sheep and goat spermatozoa. *Indian J. Anim. Sci.* **42**: 214–216.

Sahni, K.L. and Roy, A. (1967). A study on the sexual activity of the Barbari goat (*Capra hircus*) and conception rate through A.I. *Indian J. Vet. Sci. and Anim. Husb.* **37**: 269–276.

Sahni, K.L. and Tiwari, S.B. (1992). Caprine semen evaluation, processing and artificial insemination in India. Research on Goats – Indian Experience: (*Eds*) Bhattacharyya, N.K., Singh, K. and Lokeswar, R.R., CIRG, Mkhdoom, Mathura pp. 94–107.

Saiki, R., Gefland, D., Stofferl, S., Schanf, S., Higuchi, R., Horn, G., Mullis, K., and Erlich, H. (1988). Primer – directed enzymatic amplification of DNA with thermostable DNA polymerase. *Science* **293**: 487–491.

Saini, A.L., Khan B.U. and Singh, Khub (1988). Growth performance of goats under three system of feeding management. *Indian J. Anim. Sci.* **58**: 604–609.

Saini, A.L., Singh, K. and Bhattacharya, W.K. (1986). Economics and prospects of goat rearing under farm and field condition in semi-arid tract of India. Extension bulletin No. 8, CIRG, Makhdoom.

Sands, M. and McDowell, R.E. (1978). The Potential of the Goat for Milk Production in the Tropics. Mimeograph. Cornell University, Itahara, New York, USA.

Sanfiorenzo, J.H. (1957). A study of milk production by native, Barbodos and crossbred goats in Puerto Rico. Bull. Univ. P.R. Agric. Experiment Sta. No. 139 (pp. 37)

Sanfiorenzo. J.H. (1962). Evaluation of sire breeding lines of milk goats. *J. Agric. Univ. Puerto Rico*. **46**: 205.

Saraja Ortiz (1955). "In Enciclopedia de la Leche" Ageup Cecila C. Espagne, Calpe, S.A.

Sengar, O.P.S. (1976). Sheep and goat production system in India. Paper presented at the International Workshop on the role of Sheep and Goats in Agriculture development at the Winrok International Livestock Research and Training Centre, Morrilton (Arkansas), USA. Nov. 15–17.

Sengar, O.P.S. (1979). Goat breeding and management for meat production. International Workshop on "Development of Goats in Asia" held at NDRI, Karnal (Haryana), India, March 5–11.

Sengar, O.P.S. (1980). Indian research on protein and energy requirements of goats. Paper presented at the International Symposium on Dairy goats. Sponsored by the agency for International Development and American Dairy Science Association, held at Logal (Utah.) USA on June 26–27, 1979. *J. Dairy. Sci.* **63**(10): 1655–1670.

Seth, O.N., Saraswat, K.C., Chorey, P.A., Naithani, S.P. and Roy, A. (1968). The effect of alterations in management practices on kidding percentage and lactational performance of Barbari goats as well as the survival of new-born kids. *Indian J. of Vet. Sci.* 38: 93–100.

Shakaranarayanan, K.A., Bohra, H.C. and Gosh, P.K. (1985). The goat and appropriate animal for arid and semi-arid regions. *Economics and Political Weekly* **20**: 1965–1972.

Sharifuddin, W., Rosnina, Y. and Jainudeen, M.R. (1988). Superovulation and embryo transfer in goats. The Malaysian experience. Proceedings of regional workshop on biotechnology in animal production and health, Bangkok. pp. 158–162.

Sharma, K. (1987). Goat rearing. (*Ed*) Bhattacharyya, N.K. CIRG, Makhdoom.

Sharma, K. and Ogra, J.L. (1987). Reaction of component plant species of synthesized pasture under three tier system to high intensity of grazing by goats and sheep in semi-arid zone. Proceedings of IV International conference of goats, Brazil.

Sharma, K. Singh, B. and Sahni, K.L. (1982). Goat farming an alternative to economic upliftment of real poor farmer **17**: 27–28.

Sharma, K., Singh, N. and Ogra, J.L. (1989). Potentialities of silvipasture development for goats in sandy soils of Yamuna ravines. Abstract Vol. II. International Symposium on managing sandy soils. CAZRI. Jodhpur. India. Feb. 6–10. pp. 607–612.

Sharma, K.M. and Dutt, N.S. (1968). Studies on foot and mouth disease virus in newly born kids. *Indian J. Anim. Health*. **7**: 173–178.

Sharma, N.C. (1985). Reproductive performance of Jamunapari does under tropical conditions. *Indian Vet. J.* **9**: 177.

Shastry, K.N.V. and Ahluwalia, S.S. (1972). Helminth infections in goats at Mathura. *Indian Vet. J.* **49**: 217–218.

Sheltion, M. (1961). Factors affecting kid production of Anogra does. Misc. Punl. Tex. Agric. Experiment. Sta. No. MP-496.

Sheltion, M. and Grorf. J.L. (1974). Reproductive efficiency in Angora goats. Tex. Agri. Experiment. Sta. B-1136.

Shirbhate, R.N. and Honmode, J. (1982). Observations on certain factors affecting semen quality during deep freezing of buck semen. Lilrd Introduction. Conf. on Goat production. Disease, Arizona, USA, pp. 294.

Shkolnik, A., Bnorut, A. and Choshniak, J. (1972). Water economy of the Bedouin goat. *Symp. Zool. Soc. London* **31**: 229–242.

Shrivastava, V.K. (1979). Uterine involution in postmortem goat. Proceedings of summer institute on goat production publication 147, pp. 154. (*Eds*) Sahni, K.L. and Singh, R. CIRG, Makhdoom, Mathura, Uttar Pradesh, India.

Shukla, K.K. and Bhattacharya, P. (1952). Seasonal variation in reaction time and semen quality in goat. *Indian J. Vet. Sci.* **22**: 179–190.

Simons, J.P. and Land, R.B. (1987). Transgenic livestock. *J. Reprod. Fert. Suppl.* **34**: 237–250.

Singh, S.N. and Sengar, O.P.S. (1990). Final technology report, pl 480 research. Project on goats, R.B.S College, Bichpuri, Agra, India.

Singh, B.B. and Singh, B.P. (1974). Performance of Jamunapari goats. *Indian Vet. J.* **51**: 326–311.

Singh, B.B. and Singh, B.P., (1974). Performance of Jamunapari goats. *Indian Vet. J.* **51**: 326–332.

Singh, C.M. (1988). Retroviruses as biological threat to man and animals. *Indian Sci. Congr. Assoc. Proc.* **75**: 1–44.

Singh, C.S.P., Shrivastav, S.N. and Singh, S.K. (1978). A note on the incidence of oestrus in goats. *Indian J. Dairy Sci.* **19**: 132–134.

Singh, D.K. (1992). Goat breeding and meat production. Summer Instt. on goat breeding and management, 1–20 June. Animal Genetics and Breeding, Birsa Agri. Univ., Ranchi, Bihar, India, pp. 35–50.

Singh, D.K., Singh, R.A., Sinha, M.P. and Singh, C.S.P. (1986). Postpartum oestrus in Black Bengal goats. *Indian J. Anim. Sci.* **56**: 232–234.

Singh, K. and Ram, K. (1987). Economic analysis of goat keeping in goat breeding areas of Punjab, *Indian J. Anim. Sci.* **87**: 317–323

Singh, K., Misra, R.P. and Bhattacharyya, N.K. (1992). Housing requirements of Indian goats. Research on goats. Indian experience (*Eds*) Bhattacharyya, N.K., Singh, K. and Lokeshwar, R.R. CIRG Makhdoom, Mathura, India.

Singh, Khub and Sexena, S.K. (1991). Heat tolerance in three breeds of Indian goats. Proc. Physiology Sec. 78th Session of Indian Sci. Cong. Indore, January 3–8.

Singh, Khub, Kumar, Punnet; Saini, A.L. and Bhattacharyya, N.K. (1987). Study of animal housing in semi-arid climate. Annual report. Central Institute for Research on Goats, Makhdoom, Mathura, Uttar Pradesh, India.

Singh, Khub, Kumar, Punnet, Saini, A.L. and Bhattacharyya, N.K. (1988a). Micro-environment with in convertible shed with different roofing materials during hot-day macro-environment. IV Annual Conference S.A.P.I., Central Institute for Research on Goats, Makhdoom, Mathura, Uttar Pradesh, India.

Singh, Khub, Kumar, Punnet, Saini, A.L. and Bhattacharyya, N.K. (1988b). Study of animal housing in semi-arid climate. Annual report. Central Institute for Research on Goats, Makhdoom, Mathura, Uttar Pradesh, India.

Singh, Khub, Kumar, Punnet, Saini, A.L., Bhattacharyya, N.K., Mishra, R.P. and Singh, D. (1991). Animal housing in semi arid climate. Annual Scientific Report. Central Institute for Research on Goats, Makhdoom, Mathura, Uttar Pradesh, India.

Singh, L.B. (1986). Development of goat in desert of Rajasthan for Mohair. *Indian J. Anim. Health* **25**: 73–77.

Singh, N. (1961). M.V.Sc. thesis, Agra University, Agra, Uttar Pradesh, India.

Singh, N. (1973). Disorders of the female genital system in goats: Pathoanatomical and histochemical studies. Ph.D. Thesis. Agra University, Agra, India.

Singh, N. (1985). Some useful information on goats: Vaccination schedule for goats. In new Dimension in goat production. (*Eds*) Misra, R.K. and Khan, B.U. Centarl Institute for Research on Goats, Makhdoom, Mathura, Uttar Pradesh, India. pp. 51.

Singh, N. and Mudgal, V.D. (1978). Studies on endogenous urinary nitrogen, metabolic faecal nitrogen and maintenance requirement for protein in dairy goats. 20th International Dairy Congress, Paris. pp. 1111.

Singh, N., Rajya, B.S. and Mohanty, G.C. (1974). Granular vulvovaginitis (GVV) in goats associated with Mycoplasma agalactiae. *Cornell Veterinarian.* **64**: 435–442.

Singh, N., Rajya, B.S. and Mohanty, G.C. (1975). Pathology of Mycoplasma agalactiae induced granular vulvovaginitis in goats. *Cornell Veterinarian* **65**: 363–373.

Singh, N., Singh, S.V. and Vihan, V.S. (1986). Prevalence of the clinical form of Johnes disease in goats. Paper presented at the National Symposium on Immunodeficiencies and Emerging disease of Animals and Birds, Bombay Veterinary College, Bombay, India. 5–7 March.

Singh, N., Singh, S.V. and Vihan, V.S. (1987). Annual Report. Central Institute for Research on Goats, Makhdoom, Farah, Mathura, Uttar Pradesh, India.

Singh, N., Singh, V.S. and Vihan, S.V. (1987). The effect of management practices on the prevalence of Johne;s disease in goat. Paper presented at the 8th Annual conference of the Indian Assoc. Vet. Microbiol and Immunol. (IAVMI), Izatnagar, India. 21–23 December. pp. 78.

Singh, N.P. (1987). Utilization of top feeds for sheep and goats. Proc. of the first National Seminar on Sheep and Goat production and Utilization, Jaipur.

Singh, Panjab. (1988). Forage research-present status and future strategy. Pasture and Forage crops research a stated of knowledge report rangeman society of India. IGFRI, Jhansi, India. pp. 1–16.

Singh, R.N. Acharya, R.M. and Biswas, D.K. (1970b). Selection index for goat expected direct and corelated responses. *Acta Agric. Scand.* **20**: 61–64.

Singh,L.B., Singh, O.N. andDolling, C.H.S.(1970a)Inheritance and occurence of wattles in malpura and crossbred sheep. Ind. Vet. J., 47: 959-964

Singh, R.N. and Acharya, R.M. (1982). Genetic and environmental trend of milk production in a closed flock of Beetal goats. *J. Dairy. Sci.* **65**: 2015–2017.

Singh, S.N. (1976). Nitrogen balance their significance and limitations. An extra lecture presented at the U.P. Institute on Agriculture Sciences, Kanpur, on Nov. 6, 1971. BVJ *Agric. and Sci. Research* **4(2)**: 106–116.

Singh, S.N. and Sengar, O.P.S. (1970). Investigation on milk and meat potentialities of Indian Goats. Final Technical PL-480 *Research Project N.* A7-AH-18.

Singh, S.N. and Sengar. O.P.S. (1970). Investigation on milk and meat potentialities of Indian goats. R.B.S. College, Bichpuri, Agra, India. Dept. Anim. Husbandry Dairying Tech. Report.

Singh, S.N. and Sengar, O.P.S. (1979). Final Technical Report of PL-480 Research Project on Goats. Raja Balwant Singh College, Bichpuri, Agra, India.

Singh, S.N. and Sengar, O.P.S. (1978). Milk and meat potentialities of Indian goats. R.B.S. College, Bichpuri, Agra, India. Dept. Anim. Husbandry Dairying Tech. Report.

Sinha, N.K., Wani, G.M. and Sahni, K.L. (1981). Effect of season and age on seminal attributes of Jamunapari bucks. *Indian Vet. J.* **58**: 963-965.

Sinha, S.N., Singh, B.K. and Sinha, A.K. (1987). Post-thaw motility and fertility of frozen sperms of bucks of different breeds. *Indian J. Anim. Repro.* **8**: 28–31.

Siskin. Ju.N. (1950). Breeding Numabgan fleece goats. Sociallist Zivotn. 1, **49**: 52 (fide *Anim. Breed. Abstr.* **18**: 1018).

Slater, A.E. and Bhatia, S.S. (1935). Milk goats: Indian breeds and varieties. *Allahabad Farming* **9**: 156–159.

Smith, A.V. and Ploge, C. (1950). Recovery of motile goat spermatozoa after freezing to –70°C. *Vet. Rec.* **62**: 9.

Smith, C. (1988). Genetic improvement of livestock in developing countries using nucleus breeding units. *World Animal Review* **6**: 2–10.

Smith, C. and Simpson, S.P. (1986). The use of genetic polymorphisms in livestock improvement. *J. Anim. Breeding and Genetics* **103**: 205–217.

Smith, J.A.B. and Associates. (1965). The Nutrient Requirements of Farm Livestock. No. 2 Ruminants. Technical Reviews and Summaries. Publications of the Agriculture Research Council, U.K.

Smuts, D.B. (1935). General Nutrition 9: 403. Fide Majumdar, 1960a. *J. Nutr*. **9**: 400.

Smuts, K.B. and Marais, J.S.C. (1939). The endogenous nitrogen metabolism of young sheep with reference to the estimation the maintenance requirement of sheep. Onderstepoort. *J. Vet. Sci. Anim. Ind.* **13**: 219–235.

Sobrino, F., Blanco, E., Garcia-Briones, M. and Ley, V. (1999). Synthetic peptide vaccines: Foot and mouth disease virus`as a model. *Dev. Biol. Stand.* **101**: 39–43.

Song. H.B. and Iritani, A. (1985). *In vitro* fertilization of goat follicular oocytes with epididymal spermatozoa capacitated in a chemically defined medium. Proc. *3rd AAAP Anim. Sci. Congr*. **1**: 463–465.

Song, H.B. and Iritani, A. (1988). *In vitro* fertilization of goat follicular oocytes with epididymal spermatozoa. *Korean J. Anim. Sci.* **30**: 636–642.

Southern, E.M. (1975). Detection of specific sequences among DNA fragments separated by gel electrophoresis. *Journal of Molecular Biology* **98**: 503–517.

Siramamurthy, V. (1977). Anna data Magazine. December 1977, pp 6-14.

Srivastava, V.K., Raizada, B.C. and Kulkarni, V. (1968) Carcass quality of Barbari and Jamnpari goats. *Indian Vet. J.* 45: 219–225.

Stice, S., Strelchenko, N., Betthauser, J., Scott, B., Jurgella, G., Jackson, J., David, V. and Mathews, L. (1994). *Theriogenology* **41**: 301 (abstr.).

Stokes, T.M., Cherny, R.A. and Brandon, M.R. (1994). Production of chimeric bovine embryos. *Theriogenology* **41**: 303 (abstr.).

Sundarsanan, V. and Rajan, C.K.S.V. (1973). Observation on conception, gestation period, multiple birth and infertility in Malabari goats. *Kerala J. Vet. Sci.* **4**: 96–98.

Swain, N., Jain, P.M. and Acharya, R.M. (1983). Relative economics of sheep and goats. Third international conference on goat production and disease, January 11–15, 1982. Tucson Arizona, USA.

Swain, N., Maitra, D.N., Singh, M. and Acharya, R.M. (1986). Relative performance and economics of sheep and goats under production remuneration management system. *Indian J. Anim. Sci.* **56**: 597–605.

Taneja, M., Pawshe, C.H., Guron, C.S., Singh, G., Totey, S.M. and Talwar, G.P. (1991). Superovulation on Barbari goats with follitrophin. The effect of dose. *Theriogenology* **35**: 280.

Teale, A.J., Tan, S.G. and Tan, J.H. (1994). Applications of molecular genetic and reproductive technologies in the conservation of domestic animals diversity. *In*: Proceedings of the 5th World Congress on Genetics Applied to Livestock Production, Guelph, Canada, 7–12 August 1994. Vol. 21. International Committee for World Congress on Genetics Applied to livestock Production, Guelph, Ontario, Canada. pp. 493–500.

Terminal Report (1985). All India Coordinated Research Project on Goats. Central Institute for Research on Goats, Mathura, U.P., India.

Tervit, H.R., Goold, P.G. and McKenzie, R.D. (1986). Development of an effective goat embryo transfer regime. *Proc. New Zealand Soc. Anim. Prod.* **46**: 233–236.

Tervit, H.R., Goold, P.G., McKenzie, R.D., Clarkson, H.J. and Drummond, J. (1985). Embryo transfers in Angora and Saanen goats. *New Zealand Vet. J.* **33**: 78–80.

Tervit, H.R.(1987). Factors affecting the success of goat embryo transfer. Proc. 4th AAAP Anim. Sci. Congr. pp. 262–266.

Thomson, A.M. and Flint, H.J. (1989). Electroporation induced transformation of *Bacteroides ruminicola* and *Bacteroides uniformis* by plasmid DNA. *FFMS Microbiology Letters* **61**: 101–104.

Tien, M. and Kirk, K.T. (1983). Lignin degrading enzyme from the Hymenomycete *Phanerocheate chrysosporium* birds. *Science* **221**: 661–663.

Tien, M. and Tu, C.P.D. (1987). Cloning and sequencing of a cNDA for a ligninase from *Phanerochaete chrysosporium. Nature (London)* **326**: 520–523.

Tiwana, J.S., Singh, N. and Kwatra, M.S. (1984). Isolation of Mycoplasma and Acholeplasma from Vulvovaginitis in goats. *Indian J. Comp. Microbial. and Immunol. Infect. Diseases* **5**: 17–19.

Tiwari, S.B. (1967). Studies on certain aspects of semen production and artificial insemination goat, M.V.Sc. thesis, Agra University, Agra.

Tiwari, S.B. (1988). Cryopreservation of buck semen: problems and prospects. Lead paper. National symposium and IV Annual Conference Soc. Anim. Physio. of India (SAPI). Central Institute for Research on Goats, Makhdoom. Mathura. India. pp. 61–65.

Tiwari, S.B. (1990). Libido and semen ejaculation in goats (unpublished). Central Institute for Research on Goats, Makhdoom, Mathura, India.

Tiwari, S.B. (1991). Cryopreservation of buck semen: problems and prospects. Lead paper. First National symposium on small ruminant reproduction, Central Sheep and Wool Research Institute, Avikanagar, 10–12 January.

Tiwari, S.B. and Bhattacharyya, N.K. (1987a). Deep freezing of buck semen using egg yolk-citrate-gulcose and TRIS diluents Proc. *IV. Introduction. Goat Conf. Brazil*, Vol. II. pp. 1523.

Tiwari, S.B. and Bhattacharyya, N.K. (1987b). Effect of breeds and extenders on freezability of buck semen. Proc. IIIrd National Symp. on Animal Physiology, Ranchi, Abstr. 6.6.

Tiwari, S.B. and Bhattacharyya, N.K. (1988). Post-thaw motility of frozen semen and conception rate in Jamunapari and Barbari goat. VII Annual Convention of Indian Soc. Study on Anim. Reprod., Mannuthy, Kerala, 22–24 August.

Tiwari, S.B., Sharma, R.P. and Roy, A. (1968). Effect of frequency of semen collection on seminal attributes and fertility in goats. *Indian J. Vet. Sci.* **38**: 607–615.

Tiwari, S.B., Sharma, R.P. and Roy, A. (1969). Process of parturition in sheep and goats and morphological characteristics of their placenta. *Indian Vet. J.* **46**: 576–595.

Tiwari, S.B., Srivastava, R.K., Bhattacharyya, N.K. and Sinha, N.K. (1986). Deep freezing of buck semen using egg-yolk-citrate glucose diluent. VI National Cong. *Anim. Reprod. Guwahati*. Abstr. pp. 103.

Trodahl , S. (1975). Goat recording in 1974. *Sau. Og. Geit* **20** (**2**): 55–57.

Tsulloda, Y. and Sugie, T. (1989). Superovualation in non-seasonal Japanese goats with special reference to the development progression of embryo in goats. *Theriogenology* **31**: 991.

Tsunoda, Y., Tokunaga, T., Sugie, T. and Katsumata, M. (1985). Production of monozyogotic twins following the transfer of bisected embryos in goats. *Theriogenology* **24**: 337– 343.

Tyagi, S., Goel, R.K., Sood, S.B.and Agrawal. K.P.(1992). Factors affecting pre-weaning and post weaning growth in Jhakhrana goats. Indian J. of Animal Sciences 62(10): 999-1001

Udy, G.B. (1987). Commercial splitting of goat embryos. *Theriogenology* **28**: 837–847.

Ukeckermann, L., Joubert, D. and Seyn, G. (1974). The milking capacity of Boer goat does. *World Rev. Anim. Prod.* **10**(4): 73–83.

Upadhye, A.S., Krishanppa, G., Ahmed, S.N. and Keshavamurthy, B.S. (1980). Leptospiral antibodies in sheep and goats in Karnataka State. An epidemiological survey. *Indian Vet. J.* **57**: 968–970.

Van Oirschot, J.T. (1999). Bovine Viral Vaccines, Diagnostics, and Eradication. *In*: Veterinary Vaccines and Diagnostics, Schultz R.O., (*ed.*) Academic Press, London, UK, 197–213.

Verma, A.R. and Parasar, G.C. (1991). Gaot keeping for profit. *Indian Farming*. **41**: 31–34.

Vihan, V.S. and Singh, N. (1988). Epidemiological and clinicopathological studies on colibacillosis in kids. *Indian J. Anim. Sci.* **58**: 233–236.

Vihan, V.S., Saini, A.L. and Khan, B.U. (1986). Note on preliminary investigation of mortality in Jamnapari kids. *Indian Vet. Med. J.* **10**: 166–169.

Vihan, V.S., Singh, N. and Singh, S.V. (1987). Annual Report, Central Institute for Research on Goats, Makhdoom, Farah, Mathura, Uttar Pradesh, India.

Villegas, V. (1932). Goat raising. Phillz. *Agric.* **22**: 36–52.

Vohra, B.B. (1986). Land and water-toward a policy for life support system of India. The INTACH environmental series. The Indian National Trust for Art and Cultural Heritage, New Delhi.

Wahid, A. (1965). Goat raising in Pakistan – Pakistan Goats. Mimeograph No.8, University of Karachi, Pakistan.

Wakler, S.K., Warnes, G.M., Quinn, P., Seamark, R.F. and Smith, D.H. (1985). Laparoscopic technology for the transfer of embryos in sheep. *Aust. Vet. J.* **62**: 105–106.

Wani, G.M. Sinha, N.K. and Shani, K.L. (1980b). Note on seasonality in breeding behaviour in Jamunapari does under tropical conditions, *Indian J. Anim. Sci.* **50**:1153–1155.

Wani, G.M., Singha, N.K. and Sahni, K.L.(1978). Investigation on dilution, preservation and storage of buck semen. Annual Report, NRC on Goats. Makhdoom, Mathura, India.

Wani, G.M., Sinha, N.K. and Sahni, K.L. (1980a). Note on postpartum oestrus in Jamunapari does. *Indian J. Anim. Sci.* **50**: 663–665.

Weber, J.L. and May, P.E. (1989). Abundant class of NDA polymorphisms which can be typed using polymerase chain reaction. *American J. Human Genetics* **44**: 388–396.

Willadsen, S.M. (1986). Nuclear transplantation in sheep embryos. *Nature* **320**: 36–65.

Williams, J.G., Kubelik, A.R., Livak, K.J.C., Rafalski, J.A. and Tingey, M.S.V. (1990). DNA polymorphisms amplified by arbitrary primers are useful as genetic makers. *Nucleic Acids Research* **18**(22): 6531–6535.

Williamson, G. and Payne, W.J.A. (*eds*) (1978). An introduction to animal husbandry in the tropics. 3rd ed. London, Longman, pp. 755.

Wilmut, I., Schnieke, A.E., Mcwhir, J., Kind, A.J. and Campbell, K.H.S. (1997). Viable offspring derived from faetal and adult mammalian cells. *Nature* **385**: 81–113.

Wilson, R.T. (1976). *Trop. Anim. Hlth Prod.* **8**: 221–223.

Wilson, P.N. (1960). J. Agric. Sci. 54: 105-130

Wilson, P.N. (1958a). J. Agric. Sci. 50: 198-210

Wilson, P.N. (1958b). The Effect of Plane of Nutrition on the Growth and Development of the East African Dwarf Goat. Part II. Age changes in the Carcass Composition of female kids. Journal of Agricultural Science, Cambridge 51: 4-21.

Woolliams, J.A. and Smith, C. (1988). The value of indicator traits in the genetic improvement of dairy cattle. *Anim. Prodn.* **46**: 333–345.

Xiang, Z. and Ertl, H.C. (1995). Manipulation of the immune response to a plasmid-encoded viral antigen by coinoculation with plasmids expressing cytokines. *Immunity* **2**: 129–135.

Yamanouchi, K., Battett, T. and Kai, C. (1998). New approaches to the development of virus vaccines for veterinary use. *Rev. Sci. Tech. Off. Int. Epiz.* **17**: 641–653.

Yamanouchi, K., Inui, K., Sugimoto, M., Asano, K., Nishimaki, F., Kitching, R.P., Takamastu, H. and Barrett, T. (1993). Immunization of cattle with a recombinant vaccinia vector expressing the haemaglutin in gene of rinderpest virus. *Vet. Rec.* **132**: 152–156.

Yapi, C.V., Oya, A. and Rege, J.E.O. (1994). Evaluation of an open nucleus breeding programme for growth of the Djallonke sheep in Cote d'lvoire. *In*: Proceedings of the 5th World congress on Genetics Applied to livestock Production, Guelph, Canada, 7–12 August. Vol. 20. International Committee for World Congress on Genetics Applied to Livestock Production, Guelph, Ontario, Canada. pp. 421–424.

Yarkini, I. and Sonmez, R. (1962). Studies on the breed characteristics, conformation and growth of Killis milch goats. A. V Zir, Fak: 1961-YilliGi, Fasik-I-18-lurkey (fide *Anim. Breed. Abstr.* **31**: 1319).

Yarkini, L. and Eker, M. (1961). A native dairy goat in Trukey. Vlil. Inter Congress on Animal production (Hamburg, 1961), General Report. pp. 187–189 (fide *Anim. Breed. Abstr.* **29**: 220).

Yeo, J.S. (1984). Choromosomes analysis of the Korean native goat. *Korean J. of Anim. Sci.* **26**(3): 231–235.

Yilma, T., Hsu, D., Jones, L., Owens, S., Grubman, M., Mebus, C., Yamanaka, M. and Dale, B. (1988). Protection of cattle against rinderpest with vaccinia virus recombinant expressing the HA or of gene. *Science* **242**: 1054–1061.

Yin Changan (1974). Investigation report on white chasmere goats in Alashanzuoqi. Ningxia Agric. News, No.1.

Yin Changan (1978). Research on development of white cashmere goats of goats of coarse wool type. Ningiza Agric. Sci and Tech. No. 5.

Yin Changan (1988). The goat of China. pp. 15–61.

Yin Changan (1990). Fleece character, its development and utilization of zhongwei goat. Thesises of international conference on goat. FL, USA.

Yin Changan (1991). Correlation analysis of wool quality and character between cashmere goats and indigenous ones in Liaoning province. Sheep husbandry in China, No.1.

Yong, Z. and Jianchen, W. (1990). Production of monozyotic goat twins by bisection of frozen-thawed hatched blastocysts. *Theriogenology* **33**: 361.

Yong, Z., Jiachen, W., Jufen, Q. and Zhiming, H. (1991). Nuclear transplantation in goats. *Theriogenology* **35**: 299.

Yong, Z., Jianchen, W. and Xu (1990). Production of monozygotic goat twins by bisection of frozen-thawed hatched blastocysts. *Theriogenology* **33**: 360..

Yuswiati, E. and Holtz. W. (1990). Successful transfer of vitrified goat embryos. *Theriogenology* **34**: 629–632.

Zamen, M.S. (1979). Country report on goat. Presented at FAO/APHCA Seminar on "Goat development in Asia" NDRI, Karnal.

Zanderkin, A.I. and Pasekȧn, I.K. (1941). Angora goats and their crosses in Kazakhstan. Alma-Ata, pp. 42 (fide *Anim. Breed. Abstr.* **12**: 31).

Zanderkin, A.I. (1952). Breeding goats for Mohair in Kazatkhstan Socialist. *Zivotn.* **14**. 72–75 (fide *Anim. Breed. Abstr.* **21**: 812).

Zelenskii, G.C. (1950). Improving the fleece qualities in goats. Socialist Zivotn. **1**: 42–48 (fide *Anim. Breed. Abstr.* **19**: 227).

Zeuner, F.E. (1963). A History of Domestic Animals. Hutchinson, London.

Zhang, Y.Z., Zylstra, G.J., Olsen, R.H. and Reddy, C.A. (1986). Identification of cDNA clones for ligninase from Phanerocheate chrysosporium using synthetic oligonucleotide probes. *Biochemical and Biophysical Research Communications* **137**: 649–656.

Zhiming, H., Jianchen, W. and Jufen, Q. (1990). *In vitro* maturation of follicular oocytes from the mouse and dairy goat. *Theriogenology* **33**: 364.

INDEX

A

Abortions 195, 196
Age at first kidding 19, 20, 22, 23, 25, 26, 27, 28, 29, 30, 31, 32, 44, 45, 60, 79, 95
Artificial Insemination (AI) 81, 141, 250
AICRP 47, 169, 170, 238, 262, 263
Angora 10, 14, 42, 46, 47, 48, 53, 65, 78, 165, 166, 238, 239, 240
Arthropods 206

B

Back crossing 70, 75, 77, 78
Barbari 5, 9, 11, 27, 36, 40, 43, 44, 49, 51, 52, 54, 58, 60, 61, 62, 63, 64, 65, 67, 68, 79, 81, 82, 83, 84, 85, 87, 92, 93, 94, 96, 109, 110, 112, 113, 115, 116, 160, 161, 162, 166, 167, 168, 170, 171, 173, 174, 175, 198, 254
Beetal 5, 9, 21, 34, 39, 40, 41, 42, 43, 44, 45, 46, 50, 51, 52, 54, 58, 60, 61, 78, 79, 83, 84, 85, 86, 94, 110, 112, 113, 115, 116, 162, 167, 168, 169, 170, 173, 176, 178, 179, 181, 198, 260
Biochemical polymorphism 60
Biotechnological approaches 245
Birth weight 31, 39, 83, 114, 192, 197
Black Bengal 9, 10, 30, 31, 38, 39, 40, 42, 49, 50, 51, 52, 54, 60, 63, 64, 65, 79, 82, 83, 93, 94, 110, 112, 113, 115, 116, 162, 167, 170, 198
Blood smear 210, 211
Breeding bucks 11, 21, 42, 81, 138, 140, 264
Breeding season 92, 138, 140, 142, 151, 194, 219, 228

C

Changthangi 5, 9, 10, 14, 19, 33, 42, 50, 53, 260
Chegu 5, 9, 10, 14,19 33, 42, 50, 53, 260
Chevon 144, 212, 237
Chromosomal profile 59
Cloning 251
Coccidiosis 154, 195, 196, 199, 204, 205, 206
Colibacillosis 154, 195, 196, 202
Common salt 114,115,
Conservation 2, 9, 20, 80, 156, 251, 252, 259, 261, 262, 263, 264, 265
Contagious ecthyma 197, 200
Crossbreeding 6, 11, 42, 43, 46, 47, 48, 68, 75, 76, 79, 170, 171, 181, 249, 261
Cultivated fodders 101, 104, 105, 130

D

Degradation of land 155
Desertification 2, 134, 156
Diarrhoea 136, 145, 197, 203, 204, 205, 206
Domestication 4, 6, 7
Dressing percentage 40, 69, 160, 162, 163, 164, 165, 166, 167, 190
Dry does 136, 139

E

Eastern region 6, 8, 9, 18, 30, 170, 260
Economics 16, 195, 214, 215, 216, 218, 219, 220, 221, 222, 223, 224, 225
Embryo transfer 246, 250

Embryonic loss 195, 196
Energy requirement 98, 101, 104, 113, 119, 227
Enterotoxaemia 137, 202, 213
Export 1, 146, 147, 157, 184, 185, 188, 230, 235, 236, 245
Extensive system 28, 40, 50, 99, 120, 133, 212

F

Faecal examination 208
Fattening stock 136
Feed resources 101, 251, 262
Feeders 137, 152
Feeding of young stock 114
Female reproduction 89
Fibre production 39, 42, 53, 191
Fleece 66, 143, 144, 191, 193, 231, 240, 241

G

Gaddi 5, 9, 10, 18, 19, 33, 42, 47, 48, 160, 167, 260
Ganjam 5, 11, 31, 32, 38, 260
Genetic characterization 249, 250, 265
Genetic markers 248
Geographical distribution 8
Gestation period 94, 95, 140
Goat genetic resources 259, 262, 264, 265
Goat pox 195, 202
Gohilwadi 5, 13, 25, 260
Grading 21, 44, 70, 75, 75, 76, 77, 78, 143, 144, 146, 161, 181, 230, 253
Grasses 2, 19, 99, 102, 104, 121, 134, 148, 156, 227, 261
Growth 7, 14, 17, 39, 40, 42, 67, 68, 69, 80, 88, 92, 97, 98, 104, 111, 113, 114, 117, 118, 119, 125, 129, 130, 131, 134, 137, 139, 157, 159, 160, 161, 163, 169, 170, 174, 175, 177, 178, 182, 188, 190, 192, 193, 197, 204, 206, 245, 246, 247, 249, 251, 253, 254, 262, 264

H

Hair 1, 9, 14, 18, 19, 21, 24, 25, 26, 27, 28, 29, 30, 31, 42, 48, 66, 97, 98, 140, 143, 144, 145, 172, 183, 188, 190, 191, 192, 193, 205, 231, 235, 238, 240, 241, 242, 243, 244
Hay racks 153, 235, 237
Health 84, 136, 150, 153, 192, 195, 197, 212, 227, 228, 234, 236, 246, 254, 255, 262, 264
Heat detection 81, 141, 142
Helminths 206, 208, 210
Heritability 39, 41, 42, 44, 45, 68, 93, 178, 190, 191, 248
Heterosis 45, 75, 76, 78, 79
Hides 43, 146, 184
Housing 21, 25, 26, 29, 120, 132, 147, 151, 226, 227

I

Immunodiagonostic tests 211
Inbreeding 70, 71, 72, 73, 74, 76, 262
Indo-Swiss 171
Intensive system 49, 120, 121, 134, 135, 147, 153, 218, 261
Inventory 191

J

Jakhrana 5, 9, 11, 23, 35, 41, 50, 52, 60, 61, 62, 64, 65, 82, 83, 166, 167, 168, 170, 260
Jamunapari 5, 9, 11, 20, 21, 22, 34, 39, 40, 41, 42, 49, 50, 51, 52, 60, 61, 62, 63, 64, 67, 79, 82, 83, 84, 85, 87, 92, 94, 96, 104, 108, 109, 110, 112, 113, 115, 116, 160, 161, 162, 164, 167, 168, 170, 173, 174, 175, 182, 182, 198, 260, 263

K

Kanni Adu 12, 30, 38
Kid mortality 151, 196, 197, 198
Kidding does 135
Kidding interval 19, 21, 22, 23, 24, 27, 28, 29, 30, 31, 32, 44, 45, 79, 95, 96, 174, 179, 219
Kidding pens 151, 229
Kids 7, 19, 20, 21, 22, 23, 24, 27, 28, 29, 30, 31, 32, 39, 40, 46, 67, 68, 69, 83, 94, 113, 114, 119, 121, 136, 137, 138, 139, 147, 150, 151, 153, 154, 159, 160, 161, 163, 164, 166, 167, 174, 175, 191, 193, 196, 197, 199, 200, 201, 202, 203, 204, 205, 206, 213, 217, 219, 220, 221, 223, 224, 228, 239, 241
Kutchi 5, 9, 13, 26, 27, 36, 41, 49, 51, 52, 60, 64, 83, 96, 165, 168, 170, 171, 260

L

Lactation length 29, 41, 42, 43, 44, 45, 52, 79, 95, 168, 169, 171, 176, 218
Lactation yield 21, 22, 23, 28, 41, 42, 44, 45, 46, 52, 79, 95, 168, 169, 171, 176
Legumes 100, 102, 104, 138, 252

M

Maintenance 45, 46, 97, 98, 101, 104, 107, 108, 109, 110, 111, 112, 113, 115, 116, 117, 118, 135, 180, 216, 219, 234, 262
Malabari 5, 12, 28, 37, 40, 41, 43, 44, 50, 52, 58, 84, 86, 169, 170, 181, 260
Male reproduction 80
Marwari 5, 9, 11, 14, 23, 24, 35, 41, 47, 49, 52, 60, 61, 62, 64, 69, 83, 84, 85, 94, 96, 168, 170, 171, 260
Meat 1, 2, 3, 5, 9, 14, 19, 20, 27, 29, 30, 31, 39, 40, 41, 43, 46, 51, 61, 65, 79, 92, 93, 104, 120, 121, 134, 138, 139, 144, 145, 157, 158, 159, 161, 162, 163, 164, 166, 170, 173, 174, 181, 182, 183, 188, 191, 197, 211, 216, 217, 229, 230, 231, 232, 233, 234, 235, 236, 237, 245, 246, 247, 248, 254, 259, 260, 265
Mehsana 5, 13, 25, 260
Milk 1, 2, 3, 5, 9, 14, 19, 21, 22, 23, 24, 25, 26, 27, 28, 29, 30, 31, 32, 39, 41, 42, 43, 44, 45, 46, 52, 55, 57, 58, 60, 61, 62, 65, 68, 78, 79, 81, 92, 97, 104, 107, 112, 113, 114, 115, 117, 118, 120, 135, 136, 137, 138, 145, 147, 151, 154, 159, 168, 169, 170, 171, 172, 173, 174, 175, 176, 177, 178, 179, 180, 181, 182, 183, 190, 191, 197, 200, 206, 215, 216, 218, 223, 251, 259, 260, 265
Mineral 98, 104, 112, 113, 114, 115, 119, 122, 131, 186, 252, 253
Mohair 9, 14, 42, 46, 47, 48, 53, 78, 166, 188, 238, 239, 240
Moulds 123, 126, 237
Mutton 78, 144, 158, 236

N

Natural breeding 80, 140
Naval ill 200, 201
New breeds 74, 76
Nutrient requirement 97, 104, 116, 130, 228

O

Oestrous cycle 81, 91, 93
Open Nucleus Breeding System (ONBS) 11, 246, 247, 265
Orientation of sheds 147
Osmanabadi 5, 9, 12, 13, 29, 37, 50, 64, 86, 160, 164, 260
Outbreeding 70, 74, 75
Outcrossing 70, 75, 76

P

Parturition process 94
Pashmina 9, 14, 19, 20, 42, 53, 63, 64, 66, 67, 188, 238, 260
Pneumonia 19, 28, 197, 199, 200, 203
Population distribution 14
Post-partum interval 95
Pregnant does 136, 150, 151, 153, 154
Prolificacy 18, 22, 25, 27, 30, 96, 190, 191, 192
Protein requirement 98, 104, 107, 108, 109, 110, 111, 113, 130
Protozoa 123, 126, 128, 129, 130, 206, 208, 252
Puberty in does 92

R

Ration formulation 119
Records 45, 67, 138, 173, 175, 176, 178, 190, 191, 192, 194, 229
Repeatability 41, 178, 179
Requirements for growth 104, 113
Requirements for pregnancy 113
Roofing 148, 149
Rotational grazing 138

S

Sangamneri 5, 12, 29, 37, 47, 50, 51, 53, 260
Segregation shed 152
Semen quality 83, 84, 87
Semi-extensive system 40
Shearing 143, 144, 192, 193, 218, 239, 240
Silvipasture grazing 139
Sirohi 5, 11, 22, 34, 41, 49, 51, 52, 60, 61, 62, 63, 64, 68, 79, 83, 96, 162, 168, 170, 171, 260
Skin 1, 3, 30, 31, 79, 98, 98, 137, 162, 163, 172, 193, 200, 202, 211, 233, 235, 239, 257
Southern region 12, 28
Space requirement 150, 219
Species Hybridization 70, 75, 77
Surti 5, 9, 13, 26, 28, 36, 41, 52, 94, 168, 260

T

Taxonomy 4
Teaser bucks 81, 93, 138, 141
Temperate himalayan region 18, 226, 259
Top crossing 70, 75, 76
Traditional grazing 99
Transgenic animal 248
Tree leaves 19, 21, 22, 30, 31, 101, 102, 227, 260

V

Vaccines 212, 254, 256, 257
Ventilation 148, 149, 227, 235
Vitamin requirement 98

W

Water requirement 118
Water troughs 152
Weaning weight 40, 192, 193, 229
Weight gain 40, 49, 50, 69, 159, 178, 191
Wool 5, 66, 137, 139, 143, 144, 157, 188, 190, 191, 192, 193, 218, 231, 235, 239, 247

Y

Yearling weight 40, 192
Yeasts 123, 126

Z

Zalawadi 5, 13, 24, 35, 260